STAR WARE

The Amateur Astronomer's
Ultimate Guide to
Choosing, Buying, and Using
Telescopes and Accessories

SECOND EDITION

Philip S. Harrington

John Wiley & Sons, Inc.

New York • Chichester • Weinheim • Brisbane • Singapore • Toronto

For my daughter Helen, the star of my life

This book is printed on acid-free paper. ♾

Copyright © 1998 by Philip S. Harrington. All rights reserved
Published by John Wiley & Sons, Inc.
Published simultaneously in Canada

Library of Congress Cataloging-in-Publication Data:

Harrington, Philip S.
    Star ware : the amateur astronomer's ultimate guide to choosing,
  buying, and using telescopes and accessories / Philip S. Harrington.
  —2nd ed.
        p.   cm.
    Includes index.
    ISBN 0-471-18311-3 (pbk. : alk. paper)
    1. Telescopes—Purchasing—Guidebooks.   2. Telescopes—Amateurs'
  manuals.   I. Title.
  QB88.H37   1998   97-46930
  522'.2—dc21

Printed in the United States of America

10 9 8 7 6 5 4 3 2 1

Contents

Preface to the Second Edition

If the pure and elevated pleasure to be derived from the possession and use of a good telescope ... were generally known, I am certain that no instrument of science would be more commonly found in the homes of intelligent people. There is only one way in which you can be sure of getting a good telescope. First, decide how large a glass you are to have, then go to a maker of established reputation, fix upon the price you are willing to pay—remembering that good work is never cheap—and finally see that the instrument furnished to you answers the proper tests for telescopes of its size. There are telescopes and there are telescopes ...

With these words of advice, Garrett Serviss opened his classic work, *Pleasures of the Telescope*. Upon its publication in 1901, this book inspired many an armchair astronomer to change from merely a spectator to a participant, actively observing the universe instead of just reading about it. In many ways, that book was an inspiration for the volume you hold before you.

The telescope market is radically different than it was in the days of Serviss. Back then, amateur astronomy was an activity of the wealthy. The selection of commercially made telescopes was restricted to only one type of instrument—the refractor—and sold for many times what their modern counterparts cost today (after correcting for inflation).

By contrast, we live in an age that thrives on choice. Budding amateur astronomers must now wade through an ocean of literature and propaganda before being able to select a telescope intelligently. For many, this chore appears overwhelming.

That is where this book comes in. You and I are going hunting for telescopes. After opening chapters that explain telescope jargon and history, today's astronomical marketplace is dissected and explored. Where is the best place to buy a telescope? Is there one telescope that does everything well? How should a telescope be cared for? What accessories are needed? The list of questions goes on and on.

Happily, so do the answers. Although there is no single set of answers that are right for everybody, all of the available options will be explored so that you can make an educated decision. All of the chapters that detail telescopes, binoculars, eyepieces, and accessories have been fully updated in this second edition to include dozens of new products. Reviews have also been expanded, based on comments that I have received from readers around the world.

Not all of the best astronomical equipment is available for sale, however; some of it has to be made at home. Thus, ten new homemade projects are outlined in the book. The book concludes with a discussion of how to care for and use a telescope and finally, some suggestions of what to look for in the night sky.

Yes, the telescope marketplace has certainly changed in the past century (even in the four years since the first edition of *Star Ware* was released), and so has the universe. The amateur astronomer has grown with these changes to explore the depths of space in ways that our ancestors could not have even imagined.

Acknowledgments

Putting together a book of this sort would not have been possible were it not for the support of many other players. I would be an irresponsible author if I relied solely on my own humble opinions about astronomical equipment. To compile the reviews of telescopes, eyepieces, and accessories, I solicited input from amateur astronomers around the world by placing announcements in astronomical periodicals, at various astronomical conventions, and on the Internet. The responses I received were very revealing and immensely helpful. Unfortunately, space does not permit me to list the names of the hundreds of hobbyists who contributed, but you all have my heartfelt thanks. I also wish to acknowledge the contributions of the companies and dealers that provided me with their latest information, references, and other vital data. Glenn Jacobs of Pocono Mountain Optics in Moscow, Pennsylvania, deserves special recognition for allowing me to borrow and test equipment.

As you will see, chapter 7 is a selection of build-at-home projects for amateur astronomers. Several were invented and constructed by people who were looking to enhance their enjoyment of the hobby; they were kind enough to supply me with information, drawings, and photographs so that I can pass their projects along to you. For their invaluable contributions, I wish to thank Terry Alford, Steve Bygren, Chuck Carlson, Charlie Dilks, Bill Elison, Don Fox, Dave Goldberg, Randall McClelland, Rob O'Toole, Ed Stewart, and Dan Ward.

All the marvelous celestial photographs found throughout this book were taken by amateur astronomers. Astrophotography is not easy, and so I must thank those accomplished photographers who graciously allowed me to use some of their work: Brian Kennedy, Richard Sanderson, Gregory Terrance, and George Viscome.

I wish to pass on my sincere appreciation to my proofreaders Dave Kratz, Sue French, Alan MacRobert, Matt Marulla, Jack Megas, Cal Powell, Richard Sanderson, and John Shibley. I am especially indebted to them for submitting constructive suggestions while massaging my sensitive ego. Many thanks also to Kate Bradford of John Wiley & Sons for her diligent guidance and help.

Finally, my deepest thanks, love, and appreciation go to my ever-patient family. My wife Wendy and daughter Helen have continually provided me with boundless love and encouragement over the years. Were it not for their understanding my need to go out at three in the morning or drive an hour or more from home just to look at the stars, this book could not exist. I love them both dearly for that.

You, dear reader, have a stake in all this, too. This book is not meant to be written, read, and forgotten about. It is meant to change, just as the hobby of astronomy changes. As you read through this occasionally opinionated book (did I say "occasionally?"), you may take exception to a passage or two. Or maybe you own a telescope or something else astronomical with which you are happy or unhappy. If so, great! This book is meant to kindle emotion. Drop me a line and tell me about it. I want to know. Please address all correspondence to me in care of John Wiley & Sons, Inc., 605 Third Avenue, New York, NY 10158. If you prefer, email me at starware@juno.com. And please check out additions and addenda on the Star Ware Home Page:

http://ourworld.compuserve.com/homepages/pharrington.

I shall try to answer all letters, but in case I miss yours, thank you in advance!

1

Parlez-Vous Telescope?

Before the telescope, ours was a mysterious universe. Events occurred nightly that struck both awe and dread into the hearts and minds of early stargazers. Was the firmament populated with powerful gods who looked down upon the pitiful Earth? Would the world be destroyed if one of these deities became displeased? Eons passed without an answer.

The invention of the telescope was the key that unlocked the vault of the cosmos. Though it is still rich with intrigue, the universe of today is no longer one to be feared. Instead, we sense that it is our destiny to study, explore, and embrace the heavens. From our backyards we are now able to spot incredibly distant phenomena that could not have been imagined just a generation ago. Such is the marvel of the modern telescope.

Today's amateur astronomers have a wide and varied selection of equipment from which to choose. To the novice stargazer, it all appears very enticing but very complicated. One of the most confusing aspects of amateur astronomy is telescope vernacular—terms whose meanings seem shrouded in mystery. "Do astronomers speak a language all their own?" is the cry frequently echoed by newcomers to the hobby. The answer is yes, but it is a language that, unlike some foreign tongues, is easy to learn. Here is your first lesson.

Many different kinds of telescopes have been developed over the years. Even though their variations in design are great, all fall into one of three broad categories according to how they gather and focus light. *Refractors,* shown in Figure 1.1a, have a large lens (the *objective*) mounted in the front of the tube to perform this task, whereas *reflectors,* shown in Figure 1.1b, use a large mirror (the *primary mirror*) at the tube's bottom. The third class of telescope, called *catadioptrics* (Figure 1.1c), places a lens (here called a *corrector plate*) in front

1

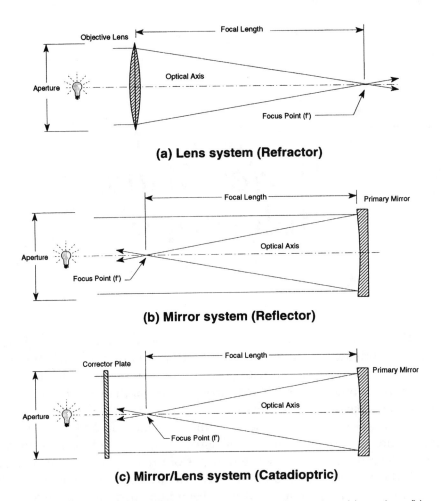

Figure 1.1 *The basic principles of the telescope. Using either a lens (a), a mirror (b), or a combination (c), a telescope bends parallel rays of light to a focus point, or prime focus.*

of the primary mirror. In each instance, the telescope's *prime optic* (objective lens or primary mirror) brings the incoming light to a *focus* and then directs that light through an *eyepiece* to the observer's waiting eye. Although chapter 2 addresses the history and development of these grand instruments, we will begin here by exploring the many facets and terms that all telescopes share. As you read through the following discussion, be sure to pause and refer to the diagrams found in chapter 2. This way, you can see how individual terms relate to the various types of telescopes.

Aperture

Let's begin with the basics. When we refer to the size of a telescope, we speak of its *aperture*. The aperture is simply the diameter (usually expressed in inches, centimeters, or millimeters) of the instrument's prime optic. In the case of a refractor, the diameter of the objective lens is cited, whereas in reflectors and catadioptric instruments, the diameters of their primary mirrors are specified. For instance, the objective lens in Galileo's first refractor was about 1.5 inches in diameter; it is therefore designated a 1.5-inch refractor. Sir Isaac Newton's first reflecting telescope employed a 1.33-inch mirror and would be referred to today as a 1.33-inch Newtonian reflector.

Many amateur astronomers consider aperture to be the most important criterion when selecting a telescope. In general (and there are exceptions to this, as pointed out in chapter 3), the larger a telescope's aperture, the brighter and clearer the image it will produce. And that is the name of the game: sharp, vivid views of the universe.

Focal Length

The *focal length* is the distance from the objective lens or primary mirror to the *focal point* or *prime focus,* which is where the light rays converge. In reflectors and catadioptrics, this distance depends on the curvature of the telescope's mirrors, with a deeper curve resulting in a shorter focal length. The focal length of a refractor is dictated by the curves of the objective lens as well as by the type of glass used to manufacture the lens.

As with aperture, focal length is commonly expressed in either inches, centimeters, or millimeters.

Focal Ratio

Looking through astronomical books and magazines, it's not unusual to see a telescope specified as, say, an 8-inch f/10 or a 14-inch f/4.5. This f-number is the instrument's *focal ratio,* which is simply the focal length divided by the aperture. Therefore, an 8-inch telescope with a focal length of 56 inches would have a focal ratio (*f-ratio*) of f/7, because 56 ÷ 8 = 7. Likewise, by turning the expression around, we know that a 6-inch f/8 telescope has a focal length of 48 inches, because 6 × 8 = 48.

Readers familiar with photography may already be used to referring to lenses by their focal ratios. In the case of cameras, a lens with a faster focal ratio (that is, a smaller f-number) will produce brighter images on film, thereby allowing shorter exposures when shooting dimly lit subjects. The same is true for telescopes. Instruments with faster focal ratios will produce brighter images on film, reducing the exposure times needed to record faint objects. However, a telescope with a fast focal ratio will *not* produce brighter images when used visually. The view of a particular object through, say, an 8-inch f/5 and an 8-inch f/10 will be identical when both are used at the same magnification. How bright an object appears to the eye depends only on telescope aperture and magnification.

Magnification

Many people, especially those new to telescopes, are under the false impression that the higher the magnification, the better the telescope. How wrong they are! It's true that as the power of a telescope increases, the apparent size of whatever is in view grows larger, but what most people fail to realize is that at the same time, the images become fainter and fuzzier. Finally, as the magnification climbs even higher, image quality becomes so poor that less detail will be seen than at lower powers.

It's easy to figure out the magnification of a telescope. If you look at the barrel of any eyepiece, you will notice a number followed by *mm*. It might be 26 mm, 12 mm, or 7 mm, among others; this is the focal length of that particular eyepiece expressed in millimeters. Magnification is calculated by dividing the telescope's focal length by the eyepiece's focal length. Remember to first convert the two focal lengths into the same units of measure—that is, both in inches or both in millimeters. A helpful hint: There are 25.4 millimeters in an inch.

For example, let's figure out the magnification of an 8-inch f/10 telescope with a 26-mm eyepiece. The telescope's 80-inch focal length equals 2,032 millimeters ($80 \times 25.4 = 2,032$). Dividing 2,032 by the eyepiece's 26-mm focal length tells us that this telescope/eyepiece combination yields a magnification of 78× (read *78 power*), because $2,032 \div 26 = 78$.

Most books and articles state that magnification should not exceed 60 × per inch of aperture. This is true only under *ideal* conditions, something most observers rarely enjoy. Due to atmospheric turbulence (what astronomers call *poor seeing*), interference from artificial lighting, and other sources, many experienced observers seldom exceed 40× per inch. Some add a caveat to this: Never exceed 300× even if the telescope's aperture permits it. Others insist there is nothing wrong with using more than 60× per inch as long as the sky conditions and optics are good enough. As you can see, the issue of magnification is always a hot topic of debate. My advice for the moment is to use the lowest magnification required to see what you want to see, but we are not done with the subject just yet. Magnification will be spoken of again in chapter 5.

Light-Gathering Ability

The human eye is a wondrous optical device, but its usefulness is severely limited in dim lighting conditions. When fully dilated under the darkest circumstances, the pupils of our eyes expand to about a quarter of an inch, or 7 mm, although this varies from person to person—the older you get, the less your pupils will dilate. In effect, we are born with a pair of quarter-inch refractors.

Telescopes effectively expand our pupils from fractions of an inch to many inches in diameter. The heavens now unfold before us with unexpected glory. A telescope's ability to reveal faint objects depends primarily on the diameter of either its objective lens or primary mirror (in other words, its aperture), not on magnification; quite simply, the larger the aperture, the more light gathered. Doubling a telescope's diameter increases light-gathering power by a factor of four, tripling its aperture expands it by nine times, and so on.

A telescope's *limiting magnitude* is a measure of how faint a star the instrument will show. Table 1.1 lists the faintest stars that can be seen through some popular telescope sizes. Trying to quantity limiting magnitude is anything but precise due to a large number of variables. Apart from aperture, other factors affecting this value include the quality of the telescope's optics, meteorological conditions, light pollution, excessive magnification, apparent size of the target, and the observer's vision and experience. These numbers are conservative estimates; experienced observers under dark, crystalline skies can better these by perhaps half a magnitude or more.

Resolving Power

A telescope's *resolving power* is its ability to reveal fine detail in whatever it is aimed at. Though resolving power plays a big part in everything we look at, it is especially important when viewing subtle planetary features, small surface markings on the Moon, or searching for close-set double stars.

A telescope's ability to resolve fine detail is always expressed in *arc-seconds*. You may remember this term from high-school geometry. Recall that in the sky there are 90° from horizon to the overhead point, or zenith, and 360° around the horizon. Each one of those degrees may be broken into 60 equal parts called *arc-minutes*. For example, the apparent diameter of the Moon in our sky may be referred to as either 0.5° or 30 arc-minutes, each one of which may be further broken down into 60 arc-seconds. Therefore, the Moon may also be sized as 1,800 arc-seconds.

Regardless of the size, quality, or location of a telescope, stars will never appear as perfectly sharp points. This is partially due to atmospheric interference and partially due to the fact that light consists of slightly fuzzy waves

Table 1.1 **Limiting Magnitudes**

Telescope In.	Aperture mm	Faintest Magnitude	Telescope In.	Aperture mm	Faintest Magnitude
2	51	10.3	14	356	14.5
3	76	11.2	16	406	14.8
4	102	11.8	18	457	15.1
6	152	12.7	20	508	15.3
8	203	13.3	24	610	15.7
10	254	13.8	30	762	16.2
12.5	318	14.3			

rather than mathematically straight lines. Even with perfect atmospheric conditions, what we see is a blob, technically called the *Airy disk* (named in honor of its discoverer, Sir George Airy, Britain's Astronomer Royal from 1835 to 1892). Because light is composed of waves, rays from different parts of a telescope's prime optic (be it a mirror or lens) alternately interfere with and enhance each other, producing a series of dark and bright concentric rings around the Airy disk (Figure 1.2a). The whole display is known as a *diffraction pattern*. Ideally, through a telescope without a central obstruction (that is, without a secondary mirror), 84% of the starlight remains concentrated in the central disk, 7% in the first bright ring, and 3% in the second bright ring, with the rest distributed among progressively fainter rings.

Figure 1.2b graphically presents a typical diffraction pattern. The central peak represents the bright central disk, and the smaller humps show the successively fainter rings.

The apparent diameter of the Airy disk plays a direct role in determining an instrument's resolving power. This becomes especially critical for observations of close-set double stars. How large an Airy disk will a given telescope produce? Table 1.2 summarizes the results for most common amateur-size telescopes.

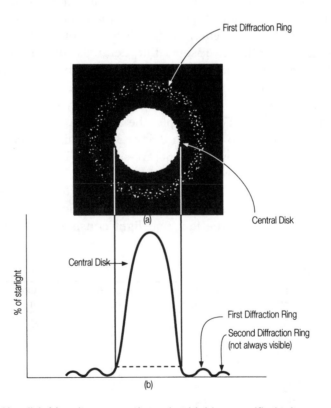

Figure 1.2 *The Airy disk (a) as it appears through a highly-magnified telescope and (b) graphically showing the distribution of light.*

Table 1.2 **Resolving Power**

Telescope In.	Aperture mm	Diameter of Airy Disk (theoretical) arc-seconds	Telescope In.	Aperture mm	Diameter of Airy Disk (theoretical) arc-seconds
2	51	5.5	14	356	0.78
3	76	3.7	16	406	0.68
4	102	2.8	18	457	0.60
6	152	1.7	20	508	0.54
8	203	1.4	24	610	0.46
10	254	1.1	30	762	0.26
12.5	318	0.88			

Although these values would appear to indicate the resolving power of the given apertures, some telescopes can actually exceed these bounds. The nineteenth-century English astronomer William Dawes found experimentally that the closest a pair of 6th-magnitude yellow stars can be to each other and still be distinguishable as two points can be estimated by dividing 4.56 by the telescope's aperture. This is called *Dawes' Limit* (Figure 1.3). Table 1.3 lists Dawes' Limit for some common telescope sizes.

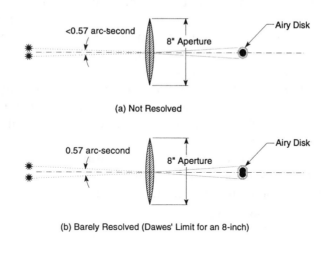

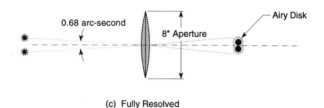

Figure 1.3 *The resolving power of an 8-inch telescope: (a) not resolved; (b) barely resolved, and the Dawes' Limit for the aperture; (c) fully resolved.*

Table 1.3 *Dawes' Limit*

Telescope in.	Aperture mm	Limit of Resolution arc-seconds	Telescope in.	Aperture mm	Limit of Resolution arc-seconds
2	51	4.6	14	356	0.66
3	76	3.0	16	406	0.58
4	102	2.2	18	457	0.50
6	152	1.5	20	508	0.46
8	203	1.1	24	610	0.38
10	254	0.92	30	762	0.30
12.5	318	0.72			

When using telescopes of less than 6-inch aperture, some amateurs can readily exceed Dawes' Limit, while others will never reach it. Does this mean that they are doomed to be failures as observers? Not at all! Remember that Dawes' Limit was developed under very precise conditions that may have been far different than your own. Just as with limiting magnitude, reaching Dawes' Limit can be adversely affected by many factors, such as turbulence in our atmosphere, a great disparity in the test stars' colors and/or magnitudes, misaligned or poor-quality optics, and the observer's visual acuity.

Rarely will a large-aperture telescope—that is, one greater than about ten inches—resolve to its Dawes' Limit. Even the largest backyard instruments can almost never show detail finer than between 0.5 arc-seconds (abbreviated 0.5″) and 1 arc-second (1″). In other words, a 16- to 18-inch telescope will offer little additional detail compared to an 8- to 10-inch one when used under most observing conditions. Interpret Dawes' Limit as a telescope's equivalent to the projected gas mileage of an automobile: "These are test results only—your actual numbers may vary."

We have just begun to digest a few of the multitude of telescope terms that are out there. Others will be introduced in the succeeding chapters as they come along, but for now, the ones we have learned will provide enough of a foundation for us to begin our journey.

2

In the Beginning . . .

To appreciate the grandeur of the modern telescope, we must first understand its history and development. It is a rich history, indeed. Since its invention, the telescope has captured the curiosity and commanded the respect of princes and paupers, scientists and laypersons. Peering through a telescope renews the sense of wonder we all had as children. In short, it is a tool that sparks the imagination in us all.

Who is responsible for this marvelous creation? Ask this question of most people and they probably will answer, "Galileo." Galileo Galilei did, in fact, usher in the age of telescopic astronomy when he first turned his telescope, illustrated in Figure 2.1, toward the night sky. With it, he became the first person in human history to witness craters on the Moon, the phases of Venus, four of the moons orbiting Jupiter, and many other hitherto unknown heavenly sights. Though he was ridiculed by his contemporaries and persecuted for heresy, Galileo's observations changed humankind's view of the universe as no single individual's ever had before or has since. But he did not make the first telescope.

So who did? The truth is that no one knows for certain just who came up with the idea, or even when. Many knowledgeable historians tell us that it was Jan Lippershey, a spectacle maker from Middelburg, Holland. Records indicate that in 1608 he first held two lenses in line and noticed that they seemed to bring distant scenes closer. Subsequently, Lippershey sold many of his telescopes to his government, which recognized the military importance of such a tool. In fact, many of his instruments were sold in pairs, thus creating the first field glasses.

Other evidence may imply a much earlier origin for the telescope. Archaeologists have unearthed glass in Egypt that dates to about 3500 B.C., while

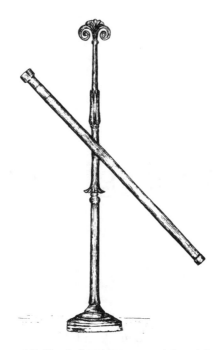

Figure 2.1   *Artist's rendition of Galileo's first telescope. Artwork by David Gallup.*

primitive lenses have been found in Turkey and Crete that are thought to be 4,000 years old! In the third century B.C., Euclid wrote about the reflection and refraction of light. Four hundred years later, the Roman writer Seneca referred to the magnifying power of a glass sphere filled with water.

Although it is unknown if any of these independent works led to the creation of a telescope, the English scientist Roger Bacon wrote of an amazing observation made in the thirteenth century: "... Thus from an incredible distance we may read the smallest letters ... the Sun, Moon and stars may be made to descend hither in appearance ..." Might he have been referring to the view through a telescope? We may never know.

Refracting Telescopes

Though its inventor may be lost to history, this early kind of telescope is called a *Galilean* or *simple* refractor. The Galilean refractor consists of two lenses: a convex (curved outward) lens held in front of a concave (curved inward) lens a certain distance away. As you know, the telescope's front lens is called the objective, while the other is referred to as the eyepiece, or *ocular*. The Galilean refractor placed the concave eyepiece *before* the objective's prime focus; this produced an upright, extremely narrow field of view, like today's inexpensive opera glasses.

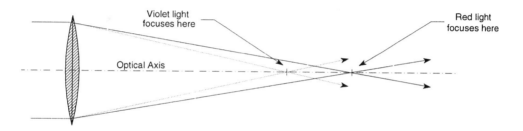

Figure 2.2 *Chromatic aberration, the result of a simple lens focusing different wavelengths of light at different distances.*

Not long after Galileo made his first telescope, Johannes Kepler improved on the idea by simply swapping the concave eyepiece for a double convex lens, placing it behind the prime focus. The *Keplerian refractor* proved to be far superior to Galileo's instrument. The modern refracting telescope continues to be based on Kepler's design. The fact that the view is upside down is of little consequence to astronomers because there is no up and down in space; for terrestrial viewing, extra lenses may be added to flip the image a second time, reinverting the scene.

Unfortunately, both the Galilean and the Keplerian designs have several optical deficiencies. Chief among these is *chromatic aberration* (Figure 2.2). As you may know, when we look at any white-light source, we are not actually looking at a single wavelength of light but rather a collection of wavelengths mixed together. To prove this for yourself, shine sunlight through a prism. The light going in is refracted within the prism, exiting not as a unit but instead broken up, forming a rainbow-like spectrum. Each color of the spectrum has its own unique wavelength.

If you use a lens instead of a prism, each color will focus at a slightly different point. The net result is a zone of focus, rather than a point. Through such a telescope, everything appears blurry and surrounded by halos of color. This effect is called chromatic aberration.

Another problem of simple refractors is *spherical aberration* (Figure 2.3). In this instance, the curvature of the objective lens causes the rays of light entering around its edges to focus at a slightly different place than those striking the center. Once again, the light focuses within a range rather than at a single point, making the telescope incapable of producing a clear, razor-sharp image.

Modifying the inner and outer curves of the lens proved somewhat helpful. Experiments showed that both defects could be reduced (but not totally eliminated) by increasing the focal length—that is, decreasing the curvature—of the objective lens. And so, in an effort to improve image quality, the refractor became longer . . . and longer . . . and even longer! The longest refractor on record was constructed by Johannes Hevelius in Denmark during the latter part of the seventeenth century; it measured about one hundred and fifty feet from objective to eyepiece and required a complex sling system suspended

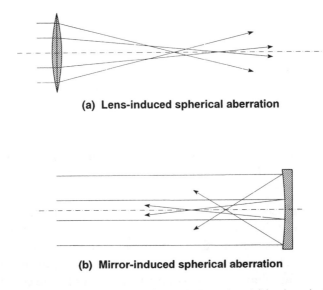

(a) Lens-induced spherical aberration

(b) Mirror-induced spherical aberration

Figure 2.3 *Spherical aberration. Both (a) lens-induced and (b) mirror-induced spherical aberration are caused by incorrectly figured optics.*

high above the ground on a wooden mast to hold it in place! Can you imagine the effort it must have taken to swing around such a monster just to look at the Moon or a bright planet? Surely, there had to be a better way.

In an effort to combat these imperfections, Chester Hall developed a two-element *achromatic lens* in 1733. Hall learned that by using two matching lenses made of different types of glass, aberrations could be greatly reduced. In an achromatic lens, the outer element is usually made of crown glass, while the inner element is typically flint glass. Crown glass has a lower dispersion effect and therefore bends light rays less than flint glass, which has a higher dispersion. The convergence of light passing through the crown-glass lens is compensated by its divergence through the flint-glass lens, resulting in greatly dampened aberrations. Ironically, though Hall made several telescopes using this arrangement, the idea of an achromatic objective did not catch on for another quarter century.

In 1758, John Dollond reacquainted the scientific community with Hall's idea when he was granted a patent for a two-element aberration-suppressing lens. Though quality glass was hard to come by for both of these pioneers, it appears that Dollond was more successful at producing a high-quality instrument. Perhaps that is why history records John Dollond, rather than Chester Hall, as the father of the modern refractor.

Regardless of who first devised it, this new and improved design has come to be called the *achromatic refractor* (Figure 2.4a), with the compound objective simply labelled an *achromat*. Though the methodology for improving the refractor was now known, the problem of getting high-quality glass (especially flint glass) persisted. In 1780, Pierre Louis Guinard, a Swiss bell maker, began experimenting with various casting techniques in an attempt to improve the glass-making process. It took him close to twenty years, but Guinard's efforts ultimately paid off, for he learned the secret of producing flawless optical disks as big as roughly six inches in diameter.

Later, Guinard was to team up with Joseph von Fraunhofer, inventor of the spectroscope. While studying under Guinard's guidance, Fraunhofer experimented by slightly modifying the lens curves suggested by Dollond,

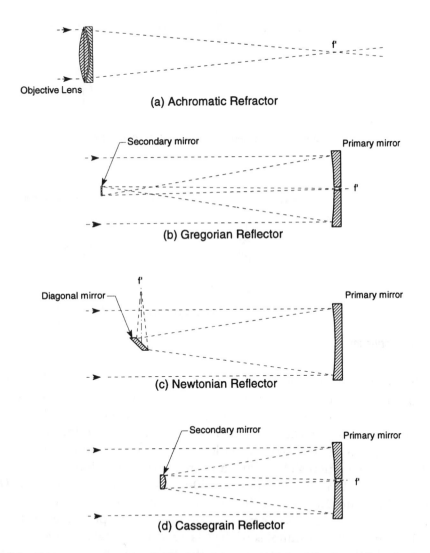

Figure 2.4 *Telescopes come in all different shapes and sizes: (a) achromatic refractor, (b) Gregorian reflector, (c) Newtonian reflector, (d) Cassegrain reflector, (e) Schmidt cata-dioptric telescope, (f) Maksutov-Cassegrain telescope, and (g) Schmidt-Cassegrain telescope.*

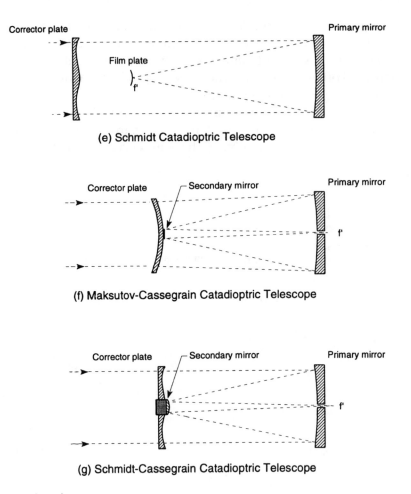

(e) Schmidt Catadioptric Telescope

(f) Maksutov-Cassegrain Catadioptric Telescope

(g) Schmidt-Cassegrain Catadioptric Telescope

Figure 2.4 continued

which resulted in the highest-quality objective yet created. In Fraunhofer's design, the front surface is strongly convex. The two central surfaces differ slightly from each other, requiring a narrow air space between the elements, while the innermost surface is almost perfectly flat. These innovations bring two wavelengths of light across the lens's full diameter to a common focus, thereby greatly reducing chromatic and spherical aberration.

The world's largest refractor is the 40-inch f/19 telescope at Yerkes Observatory in Williams Bay, Wisconsin. This mighty instrument was constructed by Alvan Clark and Sons, Inc., America's premier telescope maker of the nineteenth century. Other examples of the Clarks' exceptional skill include the 36-inch at Lick Observatory in California, the 26-inch at the U.S. Naval Observatory in Washington, D.C., and many smaller refractors at universities and colleges worldwide. Even today, Clark refractors are considered to be among the finest available.

The most advanced modern refractors offer features that the Clarks could not have imagined. *Apochromatic refractors* effectively eliminate just about all aberrations common to their Galilean, Keplerian, and achromatic cousins. More about these when we examine consumer considerations in chapter 3.

Reflecting Telescopes

But there is more than one way to skin a cat. The second general type of telescope utilizes a large mirror, rather than a lens, to focus light to a point—not just any mirror, mind you, but a mirror with a precisely figured surface. To understand how a mirror-based telescope works, we must first reflect on how mirrors work (sorry about that). Take a look at a mirror in your home. Chances are it is flat, as shown in Figure 2.5a. Light that is cast onto the mirror's polished surface in parallel rays is reflected back in parallel rays. If the mirror is

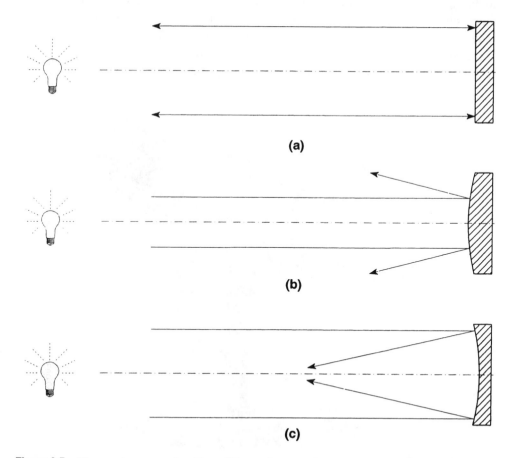

Figure 2.5 *Three mirrors, each with a different front-surface curve, reflect light differently. A flat mirror (a) reflects light straight back to the source, a convex mirror (b) causes light to diffuse, and a concave mirror (c) focuses light.*

convex (Figure 2.5b), the light diverges after it strikes the surface. But if the mirror is concave (Figure 2.5c), then the rays converge toward a common point, or focus. (It should be pointed out here that household mirrors are *second-surface* mirrors; that is, their reflective coating is applied onto the back surface. Reflecting telescopes use *front-surface* mirrors, coated on the front.)

The first reflecting telescope was designed by James Gregory in 1663. His system centered around a concave mirror (called the *primary mirror*). The primary mirror reflected light to a smaller concave *secondary mirror,* which, in turn, bounced the light back through a central hole in the primary and out to the eyepiece. The *Gregorian reflector* (Figure 2.4b) had the benefit of yielding an upright image, but its optical curves proved difficult for Gregory and his contemporaries to fabricate.

A second design was later conceived by Sir Isaac Newton in 1672 (Figure 2.6). Like Gregory, Newton realized that a concave mirror would reflect and focus light back along the optical axis to a point called the prime focus. Here an observer could view a magnified image through an eyepiece. Quickly realizing that his head got in the way, Newton inserted a flat mirror at a 45° angle some distance in front of the primary. The secondary, or *diagonal,* mirror acted to bounce the light at a 90° angle out through a hole in the side of the tele-

Figure 2.6 *Newton's first reflecting telescope. From* Great Astronomers *by Sir Robert S. Ball, London, 1912.*

scope's tube. This arrangement has since become known as the *Newtonian reflector* (Figure 2.4c).

That same year, the French sculptor Sieur Cassegrain announced a third variation of the reflecting telescope. His system is strongly reminiscent of Gregory's original design. The biggest difference between a *Cassegrain reflector* (Figure 2.4d) and a Gregorian reflector is the curve of the secondary mirror's surface. The Gregorian design uses a concave secondary mirror positioned outside the main focus, whereas the Cassegrain uses a convex secondary mirror inside the main focus.

Both Newton and Cassegrain received acclaim for their independent inventions, but neither telescope saw further development for more than half a century. It seems that good mirrors were just too difficult to come by. One of the greatest difficulties to overcome was the lack of information on suitable materials for their construction. Newton, for instance, made his mirrors out of bell metal whitened with arsenic. Others chose speculum metal, an amalgam consisting of copper, tin, and arsenic.

Another complication faced by makers of early reflecting telescopes was generating accurately figured mirrors. In order for all the light striking its surface to focus to a point, a primary mirror's concave surface must be a parabola precisely shaped to within a few millionths of an inch—a fraction of the wavelength of light. Unfortunately, the first reflectors were outfitted with spherically figured mirrors. In this case, rays striking the mirror's edge come to a different focus than the rays striking its center. The net result: spherical aberration.

The first reflector to use a parabolic mirror was constructed by Englishman John Hadley in 1722. The primary mirror of his Newtonian measured about six inches across and had a focal length of $62^{5}/_{8}$ inches. But whereas Newton and the others had failed to generate mirrors with accurate parabolic concave curves, Hadley succeeded. Extensive tests were performed on Hadley's reflector after he presented it to the Royal Society. In direct comparison between it and the society's 123-foot-focal-length refractor of the same diameter, the reflector performed equally well and was immeasurably simpler to use.

A second success story for the early reflecting telescope was that of James Short, another English craftsman. Short created several fine Newtonian and Gregorian instruments in his optical shop from the 1730s through the 1760s. He placed many of his telescopes on a special type of support that permitted easier tracking of sky objects (what is today termed an *equatorial mount*—see chapter 3). Today, the popularity of the Gregorian reflector has long since faded away, although it is interesting to note that NASA chose that design for its highly successful Solar Max mission of 1980.

Sir William Herschel, a musician who became interested in astronomy when he was given a telescope in 1722, ground some of the finest mirrors of his day. As his interest in telescopes grew, Herschel continued to refine the reflector by devising his own system. The *Herschelian* design called for the primary mirror to be tilted slightly, thereby casting the reflection toward the front rim of the oversized tube, where the eyepiece would be mounted. The biggest advantage to this arrangement is that with no secondary mirror to block the incoming light, the telescope's aperture is unobstructed by a second mirror; disadvantages included image distortion due to the tilted optics and heat from the observer's head. Herschel's largest telescope was completed in 1789. The metal speculum around which it was based measured 48 inches across and had a focal length of 40 feet. Records indicate that it weighed something in excess of one ton.

Even this great instrument was to be eclipsed in 1845, when Lord Rosse completed the largest speculum ever made. It measured 72 inches in diameter and weighed in at an incredible 8,380 pounds. This telescope (Figure 2.7),

Figure 2.7 *Lord Rosse's 72-inch reflecting telescope. From* Elements of Descriptive Astronomy *by Herbert A. Howe, New York, 1897.*

mounted in Parsonstown, Ireland, is famous in the annals of astronomical history as the first to reveal spiral structure in what were then thought to be nebulae and are now known to be spiral galaxies.

The poor reflective qualities of speculum metal, coupled with its rapid tarnishing, made it imperative to develop a new mirror-making process. That evolutionary step was taken in the following decade. The first reflector to use a glass mirror instead of a metal speculum was constructed in 1856 by Dr. Karl Steinheil of Germany. The mirror, which measured 4 inches across, was coated with a very thin layer of silver; the procedure for chemically bonding silver to glass had been developed by Justus von Liebig about 1840. Although it apparently produced a very good image, Steinheil's attempt received very little attention from the scientific community. The following year, Jean Foucault (creator of the Foucault pendulum and the Foucault mirror test procedure, among others) independently developed a silvered mirror for his astronomical telescope. He brought his instrument before the French Academy of Sciences, which immediately made his findings known to all. Foucault's methods of working glass and testing the results elevated the reflector to new heights of excellence and availability.

Although silver-on-glass specula proved far superior to the earlier metal versions, this new development was still not without flaws. For one thing, silver tarnished quite rapidly, although not as fast as speculum metal. The twentieth century dawned with experiments aimed to remedy the situation, which ultimately led to the process used today of evaporating a thin film of aluminum onto glass in a vacuum chamber. Even though aluminum is not quite as highly reflective as silver, its longer useful lifespan more than makes up for that slight difference.

Although reflectors do not suffer from the refractor's chromatic aberration, they are anything but flawless. We have already seen how spherical aberration can destroy image integrity, but other problems must be dealt with as well. These include *coma,* which describes objects away from the center of view appearing like tiny comets, with their tails aimed outward from the center; *astigmatism,* resulting in star images that focus to crosses rather than points; and *light loss,* which is caused by obstruction by the secondary mirror and the fact that no reflective surface returns 100% of the light striking it.

Today, there exist many variations of the reflecting telescope's design. While the venerable Newtonian has remained popular among amateur astronomers, the Gregorian is all but forgotten. In addition to the classical Cassegrain, we find two modified versions: the Dall-Kirkham and the Ritchey-Chretien. The former employs simpler mirror curves than a true Cassegrain and is therefore favored by amateur telescope makers. The latter is the best of the three at correcting aberrations but is quite difficult to make. Finally, for the true student of the reflector, there are several lesser-known instruments, such as the Tri-Schiefspiegler (a three-mirror telescope with tilted optics).

Like the refractor, today's reflectors enjoy the benefit of advanced materials and optical coatings. Although they are a far cry from the first telescopes of

Newton, Gregory, and Cassegrain, we must still pause a moment to consider how different our understanding of the universe might be if it were not for these and other early optical pioneers.

Catadioptric Telescopes

Earlier this century, some comparative newcomers launched a whole new breed of telescope: the *catadioptric*. These telescopes combine attributes of both refractors and reflectors into one instrument. They can produce wide fields with few aberrations. Many declare that this genre is (at least potentially) the perfect telescope; others see it as a collection of compromises.

The first catadioptric was devised in 1930 by German astronomer Bernhard Schmidt. The *Schmidt telescope* (Figure 2.4e) passes starlight through a corrector plate *before* it strikes the spherical primary mirror. The curves of the corrector plate eliminate the spherical aberration that would result if the mirror were used alone. One of the chief advantages of the Schmidt is its fast f-ratio, typically f/1.5 or less. However, owing to the fast optics, the Schmidt's prime focus point is inaccessible to an eyepiece, restricting the instrument to photographic applications only. To photograph through a Schmidt, film is placed in a special curved holder (to accommodate a slightly curved focal plane) at the instrument's prime focus, not far in front of the main mirror.

The second type of catadioptric instrument to be developed was the *Maksutov telescope.* By rights, the Maksutov telescope should probably be called the Bouwers telescope, after A. Bouwers of Amsterdam, Holland. Bouwers developed the idea for a photovisual catadioptric telescope in February 1941. Eight months later, D. Maksutov, an optical scientist working independently in Moscow, came up with the exact same design. Like the Schmidt, the Maksutov combines features of both refractors and reflectors. The most distinctive trait of the Maksutov is its deep-dish front corrector plate, or *meniscus,* which is placed inside the spherical primary mirror's radius of curvature. Light passes through the corrector plate to the primary and then to a convex secondary mirror.

Most Maksutovs resemble a Cassegrain in design and are therefore referred to as *Maksutov-Cassegrains* (Figure 2.4f). In these, the secondary mirror returns the light toward the primary mirror, passing through a central hole and out to the eyepiece. This layout allows a long focal length to be crammed into the shortest tube possible.

In 1957, John Gregory, an optical engineer working for Perkin-Elmer Corporation in Connecticut, modified the original Maksutov-Cassegrain scheme to improve its overall performance. The main difference in the *Gregory-Maksutov* telescope is that instead of a separate secondary mirror, a small central spot on the interior of the corrector is aluminized to reflect light to the eyepiece.

Though not as common, a Maksutov telescope may also be constructed in a Newtonian configuration. In this scheme, the secondary mirror is tilted at

45°. As in the classical Newtonian reflector, light from the target then passes through a hole in the side of the telescope's tube to the waiting eyepiece. The greatest advantage of the Maksutov-Newtonian over the traditional Newtonian is the availability of a short focal length (and therefore a wide field of view) with greatly reduced coma and astigmatism.

Finally, two hybrids of the Schmidt camera have also been developed: the *Schmidt-Newtonian* and the *Schmidt-Cassegrain* (Figure 2.4g). The Newtonian hybrid remains mostly in the realm of the amateur telescope maker, but since its introduction in the 1960s, the Schmidt-Cassegrain has grown to become the most popular type of telescope sold today. It combines a short-focal-length spherical mirror with an elliptical-figured secondary mirror and a Schmidt-like corrector plate. The net result is a large-aperture telescope that fits into a comparatively small package. Is the Schmidt-Cassegrain the right telescope for you? Only you can answer that question—with a little help from the next chapter, that is.

The telescope has certainly come a long way in its nearly four-hundred-year history, but that history is by no means finished. The age of orbiting observatories, such as the Hubble Space Telescope, has just dawned with untold possibilities. Back here on the ground, newly designed giant telescopes, like the Keck reflector in Hawaii, using segmented mirrors—and even some whose exact curves are controlled and varied by computers to compensate for atmospheric conditions (so-called *adaptive optics*)—are now being aimed toward the universe. New materials, construction techniques, and accessories are coming into use. All this means that the future will see even more diversity in this already diverse field. Stay tuned!

3

So You Want
to Buy a Telescope

So you want to buy a telescope? That's wonderful! A telescope will let you visit places that most people are not even aware exist. With it, you can soar over the stark surface of the Moon, travel to the other worlds in our solar system, and plunge into the dark void of deep space to survey clusters of jewel-like stars, huge interstellar clouds, and remote galaxies. You will witness firsthand exciting celestial objects that were unknown to astronomers only a generation ago. You can become a citizen of the universe without ever leaving your backyard.

Just as a pilot needs the right aircraft to fly from one point to another, so too must an amateur astronomer have the right instrument to journey into the cosmos. As we have seen already, many different types of telescopes have been devised in the past four centuries. Some remain popular today, while others are of interest from a historical viewpoint only.

Which telescope is right for you? Had I written this book back in the 1950s or 1960s, there would have been one answer: a 6-inch f/8 Newtonian reflector. Just about every amateur either owned one or knew someone who did. Though many different companies made this type of instrument, the most popular model was the RV6 Dynascope by Criterion Manufacturing Company of Hartford, Connecticut, which for years retailed for $194.95. The RV6 was to telescopes what the Volkswagen Beetle was to cars—a triumph of simplicity and durability at a great price!

Times have changed; the world has grown more complicated, and so has the hobby of amateur astronomy. The venerable RV6 is no longer manufactured, although some can still be found in classified advertisements. Today, looking through astronomical product literature, we find sophisticated Schmidt-Cassegrains, mammoth Newtonian reflectors, and state-of-the art refractors. With such a variety from which to choose, it is hard to know where to begin.

Optical Quality

Before examining specific types of telescopes, a few terms used to rate the caliber of telescope lenses and mirrors must be defined and discussed. In the everyday world, when we want to express the accuracy of something, we usually write it in fractions of an inch, centimeter, or millimeter. For instance, when building a house, a carpenter might call for a piece of wood that is, say, 4 feet long plus or minus $1/8$ inch. In other words, as long as the piece of wood is within an eighth of an inch of 4 feet, it is close enough to be used.

In the optical world, however, close is not always close enough. Because the curves of a lens or mirror must be made to such tight tolerances, it is not practical to refer to optical quality in everyday measurements. Instead, it is usually expressed in fractions of the wavelength of light. Each color in the spectrum has a different wavelength, so opticians use the color to which the human eye is most sensitive: yellow-green. Yellow-green, in the middle of the visible spectrum, has a wavelength of 550 nanometers (that's 0.00055 mm, or 0.00002 inch).

For a lens or a mirror to be accurate to, say, $1/8$ (0.125) wave (a value frequently quoted by telescope manufacturers), its surface shape cannot deviate from perfection by more than 0.000069 mm, or 0.000003 inch! This means that none of the little irregularities (commonly called *hills* and *valleys*) on the optical surface exceed a height or depth of $1/8$ of the wavelength of yellow light. As you can see, the smaller the fraction, the better the optics. Given the same aperture and conditions, telescope A with a $1/8$-wave prime optic (lens or mirror) should outperform telescope B with a $1/4$-wave lens or mirror, while both should be exceeded by telescope C with a $1/20$-wave prime optic.

Stop right there. Companies are quick to boast about the quality of their primary mirrors and objective lenses, but in reality, we should be concerned with the *final wavefront* reaching the observer's eye, which is double the wave error of the prime optic alone. For instance, a reflecting telescope with a $1/8$-wave mirror has a final wavefront of $1/4$ wave. This value is known as *Rayleigh's Criterion* and is usually considered the lowest quality level that will produce acceptable images. Clearly, an instrument with a $1/8$ to $1/10$ final wavefront is very good. However, even these figures must be taken loosely because there is no industrywide method of testing.

Owing to increasing consumer dissatisfaction with the quality of commercial telescopes, both *Sky & Telescope* and *Astronomy* magazines have begun to purchase instruments for testing and evaluation, subsequently publishing the results. Talk about a shot heard around the world! Both organizations quickly found out that the claims made by some manufacturers (particularly a few producers of Newtonian reflectors and Schmidt-Cassegrains) were a bit, shall I say, inflated.

In light of this shake-up, many companies have dropped claims of their optics' wavefront, referring to them instead as being *diffraction limited,* meaning that the optics are so good that performance is limited only by the wave properties of light itself and not by any flaws in optical accuracy. In general, to be diffraction limited, an instrument's final wavefront must be at least 1/4 wave, the Rayleigh Criterion. Once again, however, this can prove to be a subjective claim.

Telescope Point-Counterpoint

So which telescope would I recommend for you? None of them ... or all of them! Actually, the answer is that there is no one answer anymore. It all depends on what you want to use the telescope for, how much money you can afford to spend, and many other considerations. To help sort all this out, you and I are about to go telescope hunting together. We will begin by looking at each type of telescope that is commercially available. The chapter's second section will examine the many different mounting systems used to hold a telescope in place. Finally, all considerations will be weighed together to let you decide which telescope is right for you.

Binoculars

What are binoculars doing in a book about telescopes? The fact of the matter is that every amateur astronomer should own a pair of quality binoculars regardless of his or her other telescopic equipment. In fact, if you are limited in budget or are just starting out in the hobby, do NOT even consider buying an inexpensive telescope. Spend your money wisely by purchasing a good pair of binoculars plus a star atlas and a few of the books listed in chapter 6.

Binoculars may be thought of as two refracting telescopes strapped together. Light from a target enters a pair of objective lenses, bounces through two identical sets of prisms, and exits through the eyepieces. Modern binoculars are available in two basic styles depending on the type of prisms used: *roof-prism* and *porro-prism*. Which should you consider? All other things being equal, porro-prism binoculars will yield brighter, sharper images than roof-prism glasses.

Roof prisms require one internal surface to be aluminized in order to bounce light through the binoculars. A bit of the light is lost as it strikes the aluminized surface, resulting in a dimmed image. By contrast, porro prisms (at least those made from high-quality glass) totally reflect the incoming light without the need for a mirrored surface, allowing for a brighter final image. Another disadvantage to roof-prism binoculars is their high cost, part of which results from the precision required for their assembly. Unless constructed with great care, the image quality of roof-prism binoculars will be unacceptable, but for viewing the night sky, even the best roof-prism glasses will be inferior to well-made porro-prism models.

Porro-prism binoculars come in two body styles: German and American. The German style consists of a two-piece body, with the objective lenses held in one set of barrels and the prisms and eyepieces held in another. Because of the joint between the two pieces, German glasses may be knocked out of alignment with a slight jolt. In addition, these construction joints provide poor seals against dust and moisture. It is not unusual for German-style glasses to suffer from internal fogging, especially during times of high humidity. American-style binoculars, illustrated in Figure 3.1, consist of a one-piece body that offers a sturdier method of holding the optics, maintaining alignment, and sealing out water and dust.

All binoculars are labeled with two numbers, such as 7 × 35 or 10 × 50. The first refers to the pair's magnification, while the second specifies the diameter (in millimeters) of the two front lenses. Typically, values range from 7 power (7×) to 20 power (20×), with objectives measuring between 35 mm (1.5 inches) and 125 mm (5 inches). Choosing the right combination of magnification and diameter is most important when selecting a new pair of binoculars. Before an intelligent choice can be made, ask yourself how you plan on holding them. Most likely, you will answer, "by hand." If so, then limit your search to magnifications of 8× or less. In practice, models at 9× or higher prove too heavy to hold by hand for long periods without fatigue setting in. I know that some readers will say that they can hold higher-powered binoculars without tiring, but truth be known, most are exaggerating their physical prowess. Then there are the so-called giant binoculars, with magnifications of 10× and apertures of 70 mm and larger. These are great for comet hunting and deep-sky observing, but a sturdy external mounting for them is a *must*.

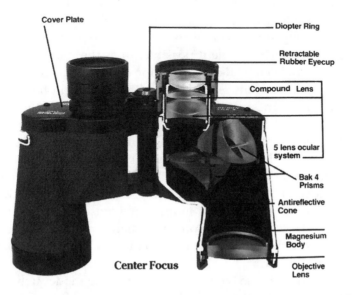

Figure 3.1   *Cross-sectional view of porro-prism binoculars. Photo courtesy of Swift, Inc.*

Many binocular manufacturers offer *zoom binoculars,* which permit the user to instantly double or even triple the power of the instrument. Unfortunately, nothing in this life is free, and zoom binoculars exact quite a toll. To perform this feat, zoom models require far more complex optical systems than fixed-power glasses, thus increasing the risk of imperfections and inferior performance. Additionally, image brightness suffers terribly at the high end of their magnification range, causing faint objects to vanish.

Just as important is the binoculars' objective lens diameter. When it comes to selecting a telescope, amateur astronomers always assume that bigger is better. Although this is generally true for telescopes, it is not always the case for binoculars. For optimum performance, the objective diameters should be matched to both your eyes and your observing site. To be perfectly matched for wide-field nighttime skywatching, the diameter of the beam of light leaving the binoculars' eyepieces (the *exit pupil*) should match the clear diameter of the observer's pupils. Although the diameter of everybody's pupil can vary (especially with age), values typically range from about 2.5 mm in the brightest lighting conditions to about 7 mm under the dimmest conditions. If the binoculars' exit pupil is much smaller, then we lose the binoculars' *rich-field* viewing capability, while too large an exit pupil will waste some of the light the binoculars collect. (For a complete discussion on what exactly an exit pupil is, fast-forward to chapter 5. Go ahead, I'll wait.)

To find the exit pupil for any pair of binoculars, simply divide the size of the aperture by the magnification. If you are young and plan on doing most of your observing from a rural setting, then you would do best with a pair of binoculars that yield a 7-mm exit pupil (such as 7 × 50 or 10 × 70). However, if you are older or are a captive of a light-polluted city or suburb, then you may do better with binoculars yielding a 4-mm or 5-mm exit pupil (7 × 35, 10 × 50, and so on).

When shopping for a pair of astronomical binoculars, look at several brands and models side by side if possible. Does the manufacturer state that the lenses are coated? Optical coatings improve light transmission and reduce scattering. An uncoated lens reflects about 4% of the light hitting it. By applying a thin layer of magnesium fluoride onto both surfaces of the lens, reflection is reduced to 1.5%. A lens with the proper thickness of magnesium fluoride looks purplish when held at a narrow angle toward a light. If the coating is too thin, it will appear pinkish; if too thick, it looks greenish. Uncoated lenses have a whitish glint. Reject any that do not have the correct coating thickness. Top-of-the-line binoculars receive multiple antireflection coatings, which reduce reflection to less than 0.5%. Most multicoated lenses show a greenish reflection when turned toward a light. Some less-than-reputable companies may try to pass off a lens with too thick a single coating as being multicoated, as both can appear very similar.

A warning: The phrase "coated optics" on some moderately priced binoculars should be interpreted as "only the exposed faces of the objectives and eye lenses are coated"; the internal optics are probably not. Others may state that all air-to-glass optics are coated, which implies all internal surfaces as well. The manufacturer, however, may intentionally forget to mention that some of the internal lenses are made of plastic. Chances are those nonglass elements are uncoated. Be sure to insist on fully coated, or better yet, fully multicoated optics.

Another feature to look for is the type of glass used to make the porro prisms. Better binoculars use prisms made from BaK-4 (barium crown) glass, while less expensive binoculars utilize prisms of BK-7 (borosilicate). BaK-4 glass yields slightly brighter, sharper images because it passes all of the light that enters (what optical experts call *total internal reflection*). BK-7 prisms do not have total internal reflection, causing light falloff and, consequently, somewhat dimmer images. Manufacturers aluminize one face of these prisms to boost their reflectivity.

Most manufacturers will state "BaK-4 prisms" right on the binoculars, but if not, you can always check for yourself. Hold the binoculars at arm's length and look at the circle of light floating, as it were, inside each eyepiece. This is the previously mentioned exit pupil. Are they both clear and uniform, or do you notice a diamond shape within? Clear circles indicate that the prisms are made of BaK-4 glass, while the diamond effect is caused by the light falloff from using cheaper, BK-7 glass prisms.

It is also important to know the binoculars' *field of view*. This value is usually found stamped on the barrels' tail stock and may be expressed in two different ways. Typically, manufacturers will either note the area of view in degrees or express it as a separation in feet at 1,000 yards (e.g., 325 feet at 1,000 yards). This second method simply means that if a 325-foot-long ruler were viewed through these glasses from 1,000 yards away, it would just squeeze into the eyepieces' field of view. If your binoculars use this second method of specifying the field, you may determine their true angular field by dividing the number of feet by 52.5. Therefore, the view through the 325-feet binoculars actually covers a little over six degrees of sky.

The ultimate in portability, binoculars offer unparalleled views of rich Milky Way starfields thanks to their low power and wide fields of view. As much as this is an advantage to the deep-sky observer, it is a serious drawback to those interested in looking for fine detail on the planets, where higher powers are required. In these cases, the hobbyist has no choice but to purchase a telescope.

Refracting Telescopes

After many years of being all but ignored by the amateur community, the astronomical refractor (Figure 3.2) is making a strong comeback. Hobbyists are rediscovering the exquisite images seen through well-made refractors.

Figure 3.2 *The 4-inch Tele Vue Genesis-SDF apochromatic refractor, one of the finest refractors for the amateur astronomer. Photo courtesy of Tele Vue, Inc.*

Crisp views of the Moon, razor-sharp planetary vistas, and pinpoint stars are all possible through the refracting telescope.

Achromatic refractors. As mentioned in chapter 2, many refractors of yester-year were plagued with a wide and varied assortment of aberrations and image imperfections. The most difficult of these faults to correct are chromatic aberration and spherical aberration.

Achromatic objective lenses, in which a convex crown lens is paired with a concave flint element, go a long way in suppressing chromatic aberration. Indeed, at f/15 or greater, chromatic aberration is effectively eliminated. Even at focal ratios down to f/10, chromatic aberration is frequently not too offensive if the objective elements are *well made*. High-quality achromatic refractors sold today range in size from 2.4 inch (6 cm) up to 6 inch (15 cm). Even though chromatic aberration can be dealt with effectively, a lingering bluish or purplish glow will frequently be seen around brighter stars and planets. This glow is known as *secondary spectrum* and is almost always present in achromatic refractors.

Another point in favor of the refractor is that its aperture is *clear*—that is, nothing blocks any part of the light as it travels from the objective to the eyepiece. As you can tell from looking at the diagrams in chapter 2, this is not the case for reflector and catadioptric instruments. As soon as a secondary mirror interferes with the path of the light, some loss of contrast and image degradation are inevitable.

In addition to sharp images, the achromatic refractor is also famous for its portability and ruggedness. If constructed properly, a refractor should deliver years of service without its optics needing to be realigned (recollimated). The sealed-tube design means that dust and dirt are prevented from infiltrating the optical system, and contaminants can be kept off the objective's exterior simply by using a lens cap.

On the minus side of the achromatic refractor is its small aperture. Although this is of less concern to lunar, solar, and planetary observers, the instrument's small light-gathering area means that faint objects such as nebulae and galaxies will appear dimmer than with larger-but-cheaper reflectors. In addition, the long tubes of achromatic refractors can make them difficult to store and transport to dark, rural skies.

Another problem common to refractors is their inability to provide comfortable viewing angles at all elevations above the horizon. The long tube and short tripod typically provided can work against the observer in some cases. For instance, if the mounting is set at the proper height to view near the zenith, the eyepiece will swing high off the ground as soon as the telescope is aimed toward the horizon. This disadvantage can be partially offset by using a star diagonal between the telescope's drawtube and eyepiece, but doing so has disadvantages of its own. Most star diagonals use either a prism or a flat mirror to bounce the light at a right angle, flipping the field left to right and making it difficult to compare the view with star charts.

Apochromatic refractors. For the true connoisseur who will settle for nothing but the best, there are apochromatic refractors. While an achromat brings two wavelengths of light at opposite ends of the spectrum to a common focus, it still leaves a secondary spectrum along the optical axis. Though not as distracting as chromatic aberration from a single lens, secondary spectrum can still contaminate critical viewing and photography.

Apos, as they are affectionately known to many owners, greatly reduce chromatic aberration and secondary spectrum, allowing manufacturers to increase aperture and decrease focal length. First popularized in the 1980s, apochromatic refractors use either two-, three-, or four-element objective lenses with one or more elements of an unusual glass type—often fluorite or ED (short for *extra-low-dispersion*) glass. All apochromats minimize dispersion of light by bringing all wavelengths to just about the same focus, reducing chromatic aberration and secondary spectrum dramatically and thereby permitting shorter, more manageable focal lengths.

Much has been written about the pros and cons of fluorite (monocrystalline calcium fluorite) lenses. The most popular myth is that they do not stand the test of time. Some so-called experts claim that fluorite absorbs moisture and/or fractures more easily than other types of glass. This is simply not true. Fluorite objectives work very well and are just as durable as conventional lenses. Like all lenses, they will last a lifetime if given a little care. (Besides, fluorite is not normally used as an outer element.)

Although durability is not a problem, there are a couple of hitches to fluorite refractors. One problem that is not popularly known is fluorite's high thermal expansion. This means that the fluorite element will require relatively more time to adjust to ambient temperature; telescope optics change shape slightly as they cool or warm, and this tendency is more pronounced in fluorite than in other materials.

Other hindrances are shared by all apos. Like most commercially sold achromatic refractors, apochromatic refractors are limited to smaller apertures, usually somewhere in the range from 3 to 7 inches or so. This is not because of unleashed aberrations at larger apertures; it's simply a question of economics, which brings us to their second (and biggest) stumbling block: Apochromats are not cheap! When we compare dollars per inch of aperture, it soon becomes apparent that apochromatic refractors are *the* most expensive telescopes. Given the same type of mounting and accessories, an apochromatic refractor can retail for more than twice the price of a comparable achromat. That's a big difference, but the difference in image quality can be even bigger. For a first telescope, an achromatic refractor is just fine, but if this is going to be your ultimate dream telescope, then you ought to consider an apochromat.

Reflecting Telescopes

Reflectors offer an alternative to the small apertures and big prices of refractors. Let's compare. Each of the two or more elements in a refractor's objective lens must be accurately figured and made of high-quality, homogeneous glass. By contrast, the single optical surfaces of a reflector's primary and secondary mirrors favor construction of large apertures at comparatively modest prices.

Another big advantage that reflectors enjoy over refractors is complete freedom from chromatic aberration, which is a property of light refraction but not reflection. This means that only the true colors of whatever a reflecting telescope is aiming at will come shining through. Of course, the eyepieces used to magnify the image for our eyes use lenses, so we are not completely out of the woods.

These two important pluses are frequently enough to sway amateurs in favor of a reflector. They feel that although there are drawbacks to the designs, these are outweighed by the many strong points. But just what are the problems with reflecting telescopes? Some are peculiar to certain breeds, while others affect them all.

One problem common to all telescopes of this genre is the simple fact that mirrors do not reflect all the light that strikes them. Just how much light is lost depends on the kind of reflective coating used. For instance, most telescope mirrors are coated with a thin layer of aluminum and overcoated with a clear layer of silicon monoxide for added protection against scratches and pitting. This combination reflects about 89% of visible light. But consider this: Given primary and secondary mirrors with standard aluminum coatings, the combined reflectivity is only 79% of the light striking the primary! That's why special enhanced coatings have become so popular in recent years. Enhanced coatings increase overall system reflectivity to about 90%—a noticeable improvement.

Reflectors also lose some light and image contrast because of obstruction by the secondary mirror. The degree of blockage depends on the size of the secondary, which in turn depends on the focal length of the primary mirror. Generally speaking, the shorter the focal length of the primary, the larger its secondary must be to bounce all of the light toward the eyepiece. For primaries with very short focal lengths, this value can exceed 10% of the primary's total light-gathering area. The only reflectors that do not suffer from this ailment are Herschelians and members of the Schiefspiegler family of instruments. Nevertheless, their availability is severely limited because of economics and practicality.

Since the idea of a telescope that uses mirrors to focus light was first conceived in 1663, different schemes have come and gone. Today, two designs continue to stand the test of time: the Newtonian reflector and the Cassegrain reflector. Each shall be examined separately.

Newtonian reflectors. For sheer brute-force light-gathering ability, Newtonian reflectors rate as a best buy. No other type of telescope will give you as large an aperture for the money. Given a similar style mounting, we could buy an 8-inch Newtonian reflector for the same amount of money needed for a 4-inch achromatic refractor.

Newtonians (Figure 3.3) are famous for their panoramic views of star fields, making them especially attractive to deep-sky fans, but they also can be equally adept at providing moderate-to-high-powered glimpses of the Moon and planets. These highly versatile instruments come in a wide variety of styles. Commercial models range from 3 inches to more than two *feet* in diameter, with focal ratios stretching between f/3.5 and about f/10. Of course, not all apertures are available at all focal ratios. Can you imagine climbing more than twenty feet to the eyepiece of a 24-inch f/10?

For the sake of discussion, I have divided Newtonians into two groups based on focal ratio. Those less than f/6 have been broadly classified as *rich-field* telescopes, or RFTs for short. Newtonian reflectors with focal lengths f/6 and greater here will be called *normal-field* telescopes, or simply NFTs.

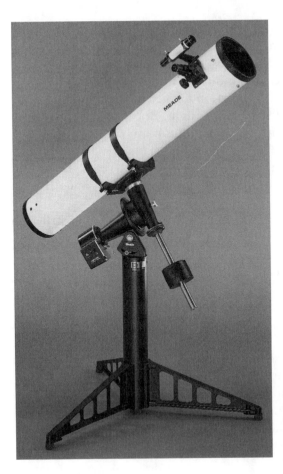

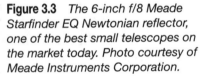

Figure 3.3 *The 6-inch f/8 Meade Starfinder EQ Newtonian reflector, one of the best small telescopes on the market today. Photo courtesy of Meade Instruments Corporation.*

Let's examine normal-field telescopes first. Pardon my bias, but NFTs have always been my favorite type of telescope. They are capable of delivering clear views of the Moon, Sun, and other members of the solar system as well as thousands of deep-sky objects. NFTs with apertures between 3 inches (8 cm) and 8 inches (20 cm) are usually small enough to be moved from home to observing site and set up quickly with little trouble. Once the viewing starts, most amateurs happily find that looking through both the eyepiece and small finderscope is effortless because the telescope's height closely matches their eye level.

A 6-inch f/8 Newtonian is still one of the best all-around telescopes for those new to astronomy. It is compact enough so as not to be a burden to transport and assemble, yet it is large enough to provide years of fascination and is reasonably priced—maybe not $194.95 like the old RV6, but it's still not a bad deal. Better yet is an 8-inch f/6 to f/9 Newtonian. The increased aperture permits even finer views of nighttime targets. Keep in mind, however, that as aperture grows, so grows a telescope's size and weight. Unless you live in the country and can store your telescope where it is easily accessible, an NFT larger than an 8-inch might be difficult to manage.

Most experienced visual observers agree that NFT Newtonian reflectors are tough to beat. In fact, an optimized Newtonian reflector can deliver views of the Moon and planets that eclipse those possible through a Schmidt-Cassegrain telescope and compare favorably with a refractor of similar size, but at a fraction of the refractor's cost. Though the commercial telescope market now offers a wide range of superb refractors, it has yet to embrace the long-focus reflector fully. Why? NFT Newtonians were quite popular back in the 1950s and 1960s; we old-timers still remember 12.5-inch f/8s.

The dawn of the 1970s saw both reflectors and refractors taking a back seat to the incredibly popular Schmidt-Cassegrain telescope. The reason was simple. Long-focus reflectors and refractors can be much more difficult to store and transport, so manufacturers, fearing the loss of customers, dropped them from their lines. But with today's widespread renewal of interest in observing and photographing the planets, amateurs are rediscovering the virtues of these fine instruments. Several companies have reintroduced 6- and 8-inch Newtonians in the f/6 to f/8 range that are hits among amateurs.

While NFTs provide fine views of the planets and deep-sky objects up to about half a degree across, rich-field telescopes (RFTs) return outstanding vistas of extended deep-sky objects such as widespread open clusters and diffuse nebulae. When combined with one of the wide-field eyepieces described in chapter 5, these instruments give panoramic views of Milky Way starfields that are beyond written description. A few companies sell small-aperture RFT Newtonians, but most sold today are 10 inches across and larger (in some cases, much larger).

Most large RFTs use thin-section primary mirrors of short focal length. Traditionally, primary mirrors have a diameter-to-thickness ratio of 6 : 1. This means that a 12-inch mirror measures a full 2 inches thick. That is one heavy piece of glass to support. Thin-section mirrors cut this ratio to 12 : 1 or 13 : 1, slashing the weight by 50%. This sounds good at first, but practice shows that large, thin mirrors tend to sag under their own weight (thicker mirrors are more rigid), thereby distorting the parabolic curve, when held in a conventional three-point mirror cell. To prevent mirror sag, a new support system was devised to support the primary at nine (or more) evenly spaced points across its back surface. These cells are frequently called *mirror flotation systems,* as they do not clamp down around the mirror's rim, thereby preventing possible edge distortions by pinching.

If big apertures mean bright images, why not buy the biggest aperture available? Actually, there are several reasons not to. For one thing, unless they are made very well, Newtonians (especially those with short focal lengths) are susceptible to a number of optical aberrations, including *spherical aberration* and *astigmatism.* Spherical aberration results when light rays near the edge of an improperly made mirror (or lens) focus to a slightly different point than those from the optic's center. Astigmatism is due to a mirror (or lens, once again) that was not symmetrically ground around its center. The result: elongated star images that appear to flip their orientation by 90° when the eyepiece is brought from one side of the focus point to the other. Coma, especially apparent in short-focal-length RFTs, is evident when stars near the edge of the field of view distort into tiny blobs resembling comets, while stars at the center appear as sharp points. With any or all of these present, resolution suffers greatly. (Note that coma can be, for all purposes, eliminated using a *coma corrector*—see the discussion in chapter 5.)

Furthermore, if you observe from a light-polluted area, large apertures will likely produce results inferior to instruments with smaller apertures. While they gather more starlight, larger mirrors also gather more sky glow, washing out the field of view. In these cases, you probably will do best by sticking with a telescope no larger than 8 to 10 inches in aperture.

Both NFT and RFT Newtonians share many other pitfalls as well. One of the more troublesome is that of all the different types of telescopes, Newtonians are among the most susceptible to collimation problems. If either or both of the mirrors are not aligned correctly, image quality will suffer greatly, possibly to the point of making the telescope worthless. Sadly, many commercial reflectors are delivered with misaligned mirrors. The new owner, perhaps not knowing better, immediately condemns his or her telescope's poor performance as a case of bad optics. In reality, however, the optics may be fine, just a little out of alignment. Chapter 8 details how to examine and adjust a telescope's collimation, a procedure that should be repeated frequently. The need for precise alignment grows more critical as the primary's focal ratio shrinks, making it especially important to double-check collimation at the start of every observing session if your telescope is f/6 or less.

There are cases where no matter how well aligned the optics are, image quality is still lacking. If this is the case, then the fault undoubtedly lies with one or both of the mirrors themselves. As the saying goes, you get what you pay for, and that is as true with telescopes as with anything else. Clearly, manufacturers of low-cost models must cut their expenses somewhere in order to underbid their competition. These cuts are usually found in the nominal-quality standard equipment supplied with the instrument but may also sometimes affect optical testing procedures and quality control.

Cassegrain reflectors. Though they have never attained the widespread following among amateur astronomers that Newtonians continue to enjoy, Cassegrain reflectors (Figure 3.4) have always been considered highly competent instruments. Cassegrains are characterized by long focal lengths, making them ideally suited for high-power, high-resolution applications such as solar, lunar, and

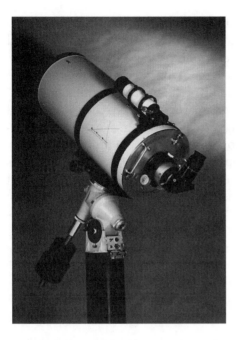

Figure 3.4 *The RC-10, 10-inch Ritchey-Chretien Cassegrain reflector. Photo courtesy of Optical Guidance Systems.*

planetary studies. Although Newtonians also may be constructed with these focal ratios, observers would have to go to great lengths to reach their eyepieces! This is not the case with the Cassegrain, where the eyepiece is conveniently located along the optical axis behind the backside of the primary mirror.

The Cassegrain's long focal length is created not by the primary mirror (which typically ranges around f/4) but rather by the convex, hyperbolic secondary mirror. As it reflects the light from the primary back toward the eyepiece, the convex secondary actually magnifies the image, thereby stretching the telescope's effective focal ratio to between f/10 and f/15. The net result is a telescope that is much more compact and easier to manage than a Newtonian of equivalent aperture and focal length.

Unfortunately, while the convex secondary mirror gives the Cassegrain its great compactness, it also contributes to many of the telescope's biggest disadvantages. First, in order to reflect all the light from the primary back toward the eyepiece, the secondary mirror must be placed quite close to the primary. This requires its diameter to be quite large, noticeably larger than the flat diagonal of a Newtonian. With the secondary blocking more light, image brightness, clarity, and contrast all suffer. Secondly, the convex secondary combined with the short-focus primary mirror makes alignment critical to the Cassegrain's proper function and at the same time causes the telescope to be more difficult to collimate than a similar Newtonian. Finally, Cassegrains are prone to coma just like RFT Newtonians, making it impossible to achieve sharp focus around the edge of the field of view.

The advantage of the eyepiece's placement along the optical axis can also work against the instrument's performance. The most obvious objection will become painfully apparent the first time an observer aims a Cassegrain near the zenith and tries to look through the eyepiece. That can be a real pain in the neck, although the use of a star diagonal will help alleviate the problem. Another problem that may not be quite as apparent involves a very localized case of light pollution, caused by extraneous light passing around the secondary and flooding the field of view. To combat this, manufacturers invariably install a long baffle tube protruding in front of the primary. The size of the baffle is critical, as it must shield the eyepiece field from all sources of incidental light while allowing the full intensity of the target to shine through.

Though Cassegrains remain the most common type of telescope in professional observatories, their popularity among today's amateur astronomers is low. So it should come as no surprise to find that so few companies offer complete Cassegrain systems for the hobbyist.

Catadioptric Telescopes

Most amateur astronomers who desire a compact telescope now favor hybrid designs that combine some of the best attributes of the reflector with some from the refractor, creating a completely different kind of beast: the catadioptric. Catadioptric telescopes (also known as *compound telescopes*) are comparative johnny-come-latelies on the amateur scene. Yet in only a few decades, they have developed a loyal following of backyard astronomers who staunchly defend them as the ultimate telescopes.

Most lovers of catadioptrics fall into one, two, or possibly all three of the following categories:

1. They are urban or suburban astronomers who prefer to travel to remote observing sites.
2. They enjoy astrophotography (or aspire to at least try it).
3. They just love gadgets.

If any or all of these profiles fit you, then a catadioptric telescope might just be the one for you.

Catadioptric telescopes for visual use may be constructed in either Newtonian or Cassegrain configurations. Only two catadioptrics have made lasting impacts on the world of amateur astronomy: the Schmidt-Cassegrain and the Maksutov-Cassegrain. For our purposes here, the discussion will be confined to these two designs.

Schmidt-Cassegrain telescopes. Take a look through practically any astronomy magazine published just about anywhere in the world and you are bound to find at least one advertisement for a Schmidt-Cassegrain telescope (also known as a *Schmidt-Cas* or an *SCT*). As your eyes digest the ads chock-full of mouth-watering celestial photographs that have been taken through these instruments, you suddenly get the irresistible urge to run right out and buy one. Don't worry—you would not be the first to find these telescopes so appealing. In the last few decades, sales of Schmidt-Cassegrains have outpaced both refractors and reflectors to make them the most popular serious telescopes among amateur astronomers. Though SCTs are available in apertures from 4 to 16 inches, the favorite size of all is the 8-inch model.

Is the Schmidt-Cassegrain (Figure 3.5) the perfect telescope? Admittedly, it can be attractive. By far, its greatest asset has to be the compact design. No other telescope can fit as large an aperture and as long a focal length into such a short tube assembly as a Schmidt-Cas; they are usually only about twice as

Figure 3.5 *The Meade 8-inch LX200 Schmidt-Cassegrain telescope, one of the most sophisticated instruments on the market today. Photo courtesy of Meade Instruments Corporation.*

long as the aperture. If storing and transporting the telescope are major concerns for you, then this will be an especially important benefit.

Here is another point in their favor. Nothing can end an observing session quicker than a fatigued observer. For instance, owning a Newtonian reflector, with its eyepiece positioned at the front end of the tube, usually means having to remain standing—sometimes even on a stool or a ladder—just to take a peek. Compare this to a Schmidt-Cassegrain telescope, which allows the observer to enjoy comfortable, seated viewing of just about all points in the sky. Your back and legs will certainly thank you! The eyepiece is difficult to reach only when the telescope is aimed close to the zenith. As with a refractor and Cassegrain, a right-angle star diagonal placed between the telescope and eyepiece will help a little, but these have their drawbacks, too. Most annoying of all is that a diagonal will flip everything right-to-left, creating a mirror image that makes the view difficult to compare with star charts.

All commercially made Schmidt-Cassegrain telescopes look pretty much the same *at a quick glance*, but then again, so do many products to the uninitiated. Only after closer scrutiny will the features unique to individual models come shining through. Standard-equipment levels vary greatly, as reflected in the wide price range of SCTs. Some basic models come with an undersized finderscope, one eyepiece, maybe a couple of other bare-bones accessories, and some cardboard boxes for storage, whereas top-of-the-line instruments are supplied with foam-lined footlockers, advanced eyepieces, large finders, and a multitude of electronic gadgets. (As I mentioned before, if you love widgets and whatchamacallits, then the Schmidt-Cassegrain will certainly appeal to you.) Most amateurs can find happiness with a model somewhere between these two extremes.

Another big plus of the Schmidt-Cassegrain is its sealed tube. The front corrector plate acts as a shield to keep dirt, dust, and other foreign contaminants off the primary and secondary mirrors. This is especially handy if you travel a lot with your telescope and are constantly taking it in and out of its carrying case. A sealed tube can also help extend the useful life of the mirrors' aluminized coatings by sealing well against the elements. (Always make sure the mirrors are dry before storing the telescope to prevent the onset of mold and mildew.)

While the corrector seals the two mirrors against dust contamination, it can also act as a dew collector when you are observing. Depending on local weather conditions, correctors can fog over in a matter of hours or even minutes, or they may remain clear all night. To help fight the onslaught of dew, manufacturers sell *dew caps* or *dew shields*. Dew caps are a must-have accessory for all Cassegrain-based catadioptrics. Consult chapter 6 for more information.

Many of the accessories for SCTs revolve around astrophotography, an activity enjoyed by many amateur astronomers. Here again, the SCT pulls ahead of the crowd. Because of their comparatively short, lightweight tubes,

Schmidt-Cassegrains permit easy tracking of the night sky. Just about all are held on fork-style equatorial mounts complete with motorized clock drives. Once the equatorial mount is properly aligned to the celestial pole (a tedious activity at times—see chapter 9), you can turn on the drive motor, and the telescope will track the stars by compensating for Earth's rotation. With various accessories (many of which are intended to be used only with SCTs), the amateur is now ready to photograph the universe.

What about optical performance? Here is where the Schmidt-Cassegrain telescope begins to teeter. Due to the comparatively large secondary mirrors required to reflect light back toward their eyepieces, SCTs produce images that are fainter and show less contrast than other telescope designs of the same aperture size. This can prove especially critical when searching for fine planetary detail or hunting for faint deep-sky objects at the threshold of visibility. One way to help the situation is to use enhanced optical coatings. As mentioned earlier in this chapter, these coatings improve light transmission and reduce scattering. They can make the difference between seeing a marginally visible object and missing it, and they are an absolute must for all Schmidt-Cassegrains.

Most 8-inch SCTs operate at f/10, while a few work at f/6.3. What's the difference? On the outside, they both look the same, the only difference being in the secondary mirrors. Are there pluses to using one over the other? Yes and no. If the telescopes are used visually (that is, if you are just going to look through them), then there should be negligible difference between the performance of an f/10 telescope and an f/6.3 telescope *when operated at the same magnification.* Image brightness is controlled by clear aperture, not by f-ratio.

The faster focal ratio may actually work against the observer. To achieve an f/6.3 instrument, a larger secondary mirror is required (3.5 inches across, compared to between 2.75 to 3 inches across in an f/10 instrument). The larger central obstruction in f/6.3 SCTs causes a decrease in contrast, making them less useful for planetary observation than their f/10 brethren. If you really want to split hairs, there is also a slight difference in image brightness—only about 5%. (Of course, in 8-inch f/6 Newtonian reflectors, the secondary blocks only 1.8 inches of the full aperture, which helps to explain their superior performance.)

The biggest advantage to using an f/6.3 SCT is enjoyed by astrophotographers. When set up for prime-focus photography, with the camera body coupled directly to the eyepieceless telescope, exposure time can be cut by a factor of 2.5 to get the same image brightness as an f/10. Of course, image size is going to be reduced at the same time, but for many deep-sky objects, this is usually not a problem. (See also the discussion in chapter 5 about focal-length reducers for SCTs.)

Image sharpness in a Schmidt-Cassegrain is not as precise as that obtained through a refractor or a reflector. Perhaps this is due to the loss of contrast mentioned above or because of optical misalignment, another problem of the Schmidt-Cassegrain. In any telescope, optical misalignment will play havoc with image quality. What should you do if the optics of a Schmidt-Cassegrain are out of alignment? If only the secondary is off, then you may follow the procedure outlined in chapter 8, but if the primary is out, then manufacturers suggest that the telescope be returned to the factory. That is good advice. Remember—although just about anyone can take a telescope apart, not everyone can put it back together!

Finally, aiming an SCT can sometimes prove to be a frustrating experience. This is not the fault of the telescope but is due instead to the low position of the *finderscope*, a small auxiliary telescope mounted sidesaddle and used to aim the main instrument. Traditionally, most SCTs are supplied with right-angle finders. Although these greatly reduce back fatigue, they introduce a whole cauldron of problems of their own. See chapter 6 for more about right-angle finders and why you shouldn't use them.

In general, Schmidt-Cassegrain telescopes represent good values for the money. They offer acceptable views of the Sun, Moon, planets, and deep-sky objects and work reasonably well for astrophotography. But for exacting views of celestial objects, SCTs are outperformed by other types of telescopes. For observations of solar system members, it is hard to beat an NFT Newtonian (especially f/10 or higher) or an apochromatic refractor, while the myriad faint deep-sky objects are best seen with large-aperture Newtonians. You might think of Schmidt-Cassegrain telescopes as jack-of-all-trades-but-master-of-none telescopes.

Maksutov-Cassegrain telescopes. The final stop on our telescope world tour is the Maksutov-Cassegrain catadioptric. Many people feel that Maksutovs are the finest telescopes of all. And why not? They offer all the advantages inherent in the Cassegrain and Schmidt-Cassegrain in an even smaller parcel while effectively eliminating coma and other aberrations. Maks, as they are called by some, provide views of the Moon, Sun, and planets that rival those possible with the best refractors and long-focus reflectors, and they are easily adaptable for astrophotography (though their high focal ratios mean longer exposures than needed with other telescopes of similar aperture). And traveling with them is a breeze.

Is there a downside to the Maksutov? Unfortunately, yes—a big one. Unlike Schmidt-Cassegrains, for which manufacturers have developed methods of mass-producing optical components of consistent quality while holding prices down, Maksutovs require precise handcrafting. In other words, they cost a lot. A second restriction of Maksutovs is aperture—or lack thereof. Even the smallest Maks cost more per inch of aperture than nearly any other type of telescope.

To help digest all this, take a look at Table 3.1, which summarizes all the pros and cons mentioned above. Use it to compare the good points and the bad among the more popular types of telescopes sold today.

Table 3.1   ***Telescope Point-Counterpoint: A Summary***

	Point	Counterpoint
A. Binoculars Typically 1.4″ to 4″ apertures	• Most are comparatively inexpensive • Extremely portable • Wide field makes them ideal for scanning	• Low power makes them unsuitable for objects requiring high magnification • Small aperture restricts magnitude limit
B. Achromatic Refractors Typically 2.4″ to 5″ aperture, f/10 and above	• Portable in smaller apertures • Sharp images • Moderate price vs. aperture • Good for Moon, Sun, planets, double stars, and bright astrophotography	• Small apertures • Mounts may be shaky (attention, department-store shoppers!) • Possible chromatic aberration
C. Apochromatic Refractors Typically 3″ to 8″ aperture, f/5 and above	• Portable in smaller apertures • Very sharp images of high contrast • Excellent for Moon, Sun, planets, double stars, and astrophotography	• Very high cost vs. aperture
D. Normal-field Newtonian Reflectors (NFTs) Typically 4″ and larger apertures, f/6 and above	• Best all-around telescope • Low cost vs. aperture • Easy to collimate • Good for Moon, Sun, planets (especially at f/10 and above), deep-sky objects, and astrophotography	• Bulky/heavy, over 8″ aperture • Collimation must be checked often • Open tube end permits dirt and dust contamination
E. Rich-field Newtonian Reflectors (RFTs) Typically 4″ and larger apertures, below f/6	• Very low cost vs. aperture • Wide fields of view • Large apertures mean maximum magnitude penetration • Easy to collimate • Good for both bright and faint deep-sky objects (solar system objects okay, but usually inferior to longer-focal-length instruments)	• Heavy • Larger apertures may require a ladder to reach the eyepiece • Low cost may indicate compromise in quality • Mounting does not track the stars • Collimation is critical (must be checked before each use) • Open tube end permits dirt and dust contamination • Susceptible to light pollution
F. Cassegrain reflector Typically 6″ and larger apertures, f/12 and above	• Portability • Convenient eyepiece position • Good for Moon, Sun, planets, and smaller deep-sky objects (e.g., double stars, planetary nebulae, and some galaxies)	• Large secondary • Moderate-to-high price vs. aperture • Narrow fields • Offered by few companies

Continued on next page

Table 3.1 continued

	Point	Counterpoint
G. Schmidt-Cassegrain (Catadioptric) Typically 4″ to 16″ apertures, f/6.3 or f/10	• Moderate cost vs. aperture • Portability • Convenient eyepiece position • Wide range of accessories • Easily adaptable to astrophotography • Good for viewing Moon, Sun, planets, bright deep-sky objects, and especially astrophotography	• Large secondary • Image quality not as good as with refractors or reflectors (though optics with enhanced coatings help) • Slow f-ratio means longer exposures than faster Newtonians and refractors • Corrector plates are prone to dewing over • Potentially difficult to find objects without using an auxiliary finder or setting circles • Mirror shift (see chapter 4)
H. Maksutov-Cassegrain (Catadioptric) 3.5″ to 12″ apertures, f/12 to f/15	• Sharp images • Convenient eyepiece position • Easily adaptable for astrophotography • Good for Moon, Sun, planets, and bright deep-sky photography	• Very high price vs. aperture • Some models use threaded eyepieces, making an adapter necessary to use other brand oculars. • Difficult to collimate • Slow f-ratio means longer exposures than faster Newtonians and refractors

Support Your Local Telescope

The telescope itself is only half of the story. Can you imagine trying to hold a telescope *by hand* while trying to look through it? If the instrument's weight did not get you first, surely every little shake would be magnified into a visual earthquake! To use a true astronomical telescope, we have no choice but to support it on some kind of external mounting. For small spotting scopes, this might be a simple tabletop tripod, whereas the most elaborate telescopes come equipped with equally elaborate support systems.

Selecting the proper mount is *just as important* as picking the right telescope. A good mount must be strong enough to support the telescope's weight while minimizing any vibrations induced by the observer (such as during focusing) and the environment (from wind gusts or even nearby road traffic). Indeed, without a sturdy mount to support the telescope, even the finest instrument will produce only blurry, wobbly images. A mounting also must allow for smooth motions when moving the telescope from one object to the next and permit easy access to any part of the sky.

Though Figure 3.6 shows many different types of telescope-mounting systems, all fall into one of two broad categories based on their construction: *altitude-azimuth* and *equatorial*. We shall examine both.

Altitude-Azimuth Mounts

Frequently referred to as either *alt-azimuth* or *alt-az*, these are the simplest types of telescope support available. As their name implies, alt-az systems move both in azimuth (horizontally) and in altitude (vertically). All camera-tripod heads, for instance, are alt-az systems.

This is the type of mounting most frequently supplied with smaller, less expensive refractors and Newtonian reflectors. It allows the instrument to be aimed with ease toward any part of the sky. Once pointed in the proper direction, the mount's two axes (that is, the altitude axis and the azimuth axis) can be locked in place. Better alt-azimuth mounts are outfitted with *slow-motion controls,* one for each axis; together, they permit fine adjustment of the telescope's aiming simply by twisting one or both of the control knobs.

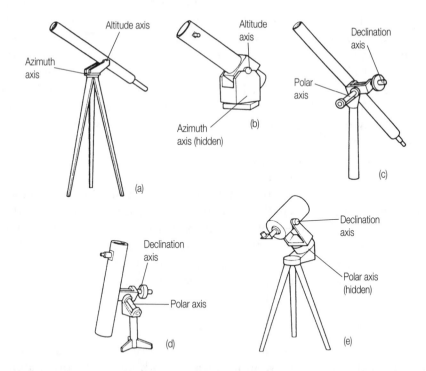

Figure 3.6 *A variety of modern telescope mountings. (a) A simple alt-azimuth mounting for a small refractor, (b) a Dobsonian alt-azimuth mounting for a Newtonian, (c) a German equatorial mounting for a refractor, (d) a German equatorial for a Newtonian, and (e) a fork equatorial mounting for a catadioptric telescope. Illustration from* Norton's 2000.0 Star Atlas and Reference Handbook, *edited by Ian Ridpath, Longman Scientific & Technical.*

In the past 20 years, a variation of the alt-azimuth mount called the *Dobsonian mount* has become extremely popular among hobbyists. Dobsonian mounts are named for John Dobson, an amateur telescope maker and astronomy popularizer from the San Francisco area. Back in the 1970s, Dobson began to build large-aperture Newtonian reflectors in order to see the "real universe," as he put it. With the optical assembly complete, he faced the difficult challenge of designing a mount strong enough to support the instrument's girth yet simple enough to be constructed from common materials using hand tools. What resulted was an offshoot of the alt-az mount.

Using plywood, Formica, and Teflon, along with some glue and nails, Dobson devised a telescope mount that was capable of holding steady his huge Newtonians. Plywood is an ideal material for a telescope mount, as it has incredible strength as well as terrific vibration-damping capability. Formica and Teflon together create smooth bearing surfaces, allowing the telescope to flow across the sky. No wonder Dobsonian mounts have become so popular.

Though both traditional alt-az mounts as well as Dobsonian mounts are wonderfully simple to use, they also possess some drawbacks. Perhaps the most obvious is caused not by the mounts but by the Earth itself! If an alt-azimuth mount is used to support a terrestrial spotting scope, then the fact that it moves horizontally and vertically plays in its favor. However, the sky is always moving owing to the Earth's rotation. Therefore, to study or photograph celestial objects for extended periods without interruption, our telescopes have to move right along with them. If we were located exactly at either the North or South Pole, the stars would appear to trace arcs parallel to the horizon as they move around the sky. In these two cases, tracking the stars would be a simple matter with an alt-az mounting; one would simply tilt the telescope up at the desired target, lock the altitude axis in place, and slowly move the azimuth axis with the sky.

Once we leave the poles, however, the tilt of the Earth's axis causes the stars to follow long, curved paths in the sky, causing most to rise diagonally in the east and set diagonally in the west. With an alt-azimuth mount, it now becomes necessary to nudge the telescope both horizontally and vertically in a steplike fashion to keep up with the sky. This is decidedly less convenient than the single motion enjoyed by an equatorial mount, a second way of supporting a telescope.

Equatorial Mounts

"If you can't raise the bridge, lower the river," so the saying goes. This is the philosophy of the equatorial mount. Because nothing can be done about the stars' apparent motion across the sky, the telescope's mounting method must accommodate it. An equatorial mount may be thought of as an altitude-azimuth mount tilted at an angle that matches your location's latitude.

Like its simpler sibling, an equatorial mount is made up of two perpendicular axes, but instead of referring to them as altitude and azimuth axes, we use the terms *right-ascension* (or *polar*) axis and *declination* axis. (No doubt you have encountered these terms before, but to refresh your memory, you might want to turn to the discussion of *celestial coordinates* found in chapter 9.) In order for an equatorial mount to track the stars, its polar axis must be aligned with the celestial pole, a procedure also detailed in chapter 9.

There are many benefits to using an equatorial mount, the greatest of which is the ability to attach a motor drive onto the right-ascension axis so the telescope follows the sky automatically and (almost) effortlessly. But there are more reasons favoring an equatorial mount. One is that once aligned to the pole, an equatorial will make finding objects in the sky much easier by simplifying hopping from one object to the next using a star chart, as well as by permitting the use of setting circles.

On the minus side of equatorial mounts, however, is that they are almost always larger, heavier, more expensive, and more cumbersome than alt-azimuth mounts. This is why the simple Dobsonian alt-az design is so popular for supporting large Newtonians. An equatorial large enough to support, say, a 12- to 14-inch f/4 reflector would probably tip the scales at close to two hundred pounds, while a plywood Dobsonian mount would weigh under fifty pounds.

Just as there are many kinds of telescopes, so too are there many kinds of equatorial mounts. Some are quite extravagant, while others are simple to use and understand. We will examine the two most common styles.

German equatorial mounts. For years, this was the most popular type of mount among amateur astronomers. The German equatorial is shaped like a tilted letter **T** with the polar axis representing the long leg and the declination axis marking the letter's crossbar. The telescope is mounted to one end of the crossbar, while a weight for counterbalance is secured to the opposite end.

The simplicity and sturdiness of German equatorials have made them the perennial favorite for supporting refractors and reflectors as well as some catadioptrics. They allow free access to just about any part of the sky (as with all equatorial mounts, things get a little tough around the poles), are easily outfitted with a clock drive, and may be held by either a tripod or a pedestal base. To help make polar alignment easier, some German equatorials have small alignment scopes built right into their right-ascension axes—a big hit among astrophotographers.

Of course, as with everything in life, there are some flaws to the German mount as well. One strike against the design is that it cannot sweep continuously from east to west. Instead, when the telescope nears the meridian, the user must move it away from whatever was in view, swing the instrument around to the other side of the mounting, and re-aim it back to the target. Inconvenient as this is for the visual observer, it is disastrous for the astrophotographer caught in the middle of a long exposure because as the telescope is spun around, the orientation of the field of view is also rotated.

A second burden imposed by the German-style mount is a heavy one to bear: They can weigh a lot, especially for telescopes larger than 8 inches. Most of their weight comes from the axes (typically made of solid steel) as well as the counterweight used to offset the telescope. At the same time, I must quickly point out that weight does *not* necessarily beget sturdiness. For instance, some heavy German equatorials are so poorly designed that they could not even steadily support telescopes half as large as those with which they are sold. (More about checking a mount's rigidity later in this chapter.)

If you are looking at a telescope that comes with a German mount, pay especially close attention to the diameter of the right-ascension and declination shafts. On well-designed mounts, each shaft is *at least* 1/8 of the telescope's aperture. For additional solidity, superior mounts use tapered shafts instead of straight shafts for the polar and declination axes. The latter carry the weight of the telescope more uniformly, thereby giving steadier support. It is easy to tell at a glance if a mount has tapered axes or not by looking at an equatorial's T-housing. If the mount has tapered shafts, then the housing will look like two truncated cones joined together at right angles; otherwise, it will look like two long, thin cylinders.

Finally, if you must travel with your telescope to a dark-sky site, moving a large German equatorial mount can be a tiring exercise. First the telescope must be disconnected from the mounting. Next, the equatorial mount (or *head*) must frequently be separated from its tripod or pedestal. Last, all three pieces (along with all eyepieces, charts, and other accessories) must be carefully stored away. The reverse sequence occurs when setting up at the site, and the whole thing happens all over again when it is time to go home. All this can add to the burden of observing, something we always try to minimize.

Fork equatorial mounts. Although German mounts are preferred for telescopes with long tubes, fork equatorial mounts are usually supplied with more compact instruments such as Schmidt-Cassegrains and Maksutov-Cassegrains. Fork mounts support their telescopes on bearings set between two short *tines*, or *prongs*, that permit full movement in declination. The tines typically extend from a rotatable circular base, which, in turn, acts as the right-ascension axis when tilted at the proper angle.

Perhaps the biggest plus to the fork mount is its light weight. Unlike its Bavarian cousin, a fork equatorial usually does not require counterweighting to achieve balance; instead, the telescope is balanced by placing its center of gravity within the prongs, sort of like a seesaw. This is an especially nice feature for Cassegrain-style telescopes, as it permits convenient access to the eye-piece regardless of where the telescope is aimed ... that is, except when it is aimed near the celestial pole. In this position, the eyepiece can be notoriously difficult to get to, usually requiring the observer to lean over the mounting without bumping into it.

The fork mounts that come with SCTs and Maksutovs are designed for maximum convenience and portability. They are compact enough to remain attached to their telescopes and fit together into their cases for easy transport-ing. Once at the observing site, the fork quickly secures to its tripod using thumbscrews. It can't get much better than that, especially when compared with the German alternative.

Fork mounts quickly become impractical, however, for long-tubed tele-scopes such as Newtonians and refractors. In order for a fork-mounted tele-scope to be able to point toward any spot in the sky, the mount's two prongs must be long enough to let the ends of the instrument swing through without colliding with any other part of the mounting. To satisfy this requirement, the prongs must grow in length as the telescope becomes longer. At the same time, the fork tines must also grow in girth to maintain rigidity; otherwise, if the fork arms are undersized, they will transmit every little vibration to the telescope. (In those cases, maybe they ought to be called tuning fork mounts.)

One way around the need for longer fork prongs is to shift the telescope's center of gravity by adding counterweights onto the tube. Either way, however, the total weight will increase. In fact, in the end the fork-mounted telescope might weigh more than if it was held on an equally strong German equatorial.

A Telescope Pop Quiz

Now let's put all this newly acquired knowledge of yours to work. Here are eight questions to help you focus on which telescope might be best for you. Respond to each question as honestly and realistically as you can; remember, there is no right or wrong answer. Once completed, add up the scores that are listed in brackets after each response. By comparing your total score with those found in Table 3.2 at the end of the quiz, you will get a good idea of which telescopes are best suited for your needs, but use the results only as a guide, not as an absolute. And no fair peeking at your neighbor's answers!

1. Which statement best describes your level of astronomical expertise?
   a. Casual observer [1]
   b. Enthusiastic beginner [4]
   c. Intermediate space cadet [6]
   d. Advanced amateur [10]
2. Will this be your first telescope or binoculars?
   a. Yes [4]
   b. No [8]

3. If not, what other instrument(s) do you already own? (If you own more than one, select only the one that you use most often.)
   a. Binoculars [1]
   b. Achromatic refractor [4]
   c. Apochromatic refractor [11]
   d. Newtonian reflector (2″ to 4″ aperture) [4]
   e. Newtonian reflector (6″ to 10″ aperture on equatorial mount) [6]
   f. Newtonian reflector (>10″ aperture on equatorial mount) [11]
   g. Newtonian reflector (<12″ aperture on Dobsonian mount) [5]
   h. Newtonian reflector (12″ to 14″ aperture on Dobsonian mount) [7]
   i. Newtonian reflector (16″ + aperture on Dobsonian mount) [10]
   j. Cassegrain reflector (<10″ aperture) [7]
   k. Cassegrain reflector (10″ and larger aperture) [11]
   l. Schmidt-Cassegrain Catadioptric [9]
   m. Maksutov [11]
4. For what do you want to use the telescope primarily? Choose only one.
   a. Casual scan of the sky [1]
   b. Informal lunar/solar/planetary observing [4]
   c. Estimating magnitudes of variable stars [6]
   d. Comet hunting [5]
   e. Detailed study of solar system objects [10]
   f. Bright deep-sky objects (star clusters, nebulae, galaxies) [6]
   g. Faint deep-sky objects [8]
   h. Astrophotography of bright objects (Moon, Sun, etc.) [4]
   i. Astrophotography of faint objects (deep-sky, etc.) [9]
5. How much money can you afford to spend on this telescope? (Be conservative; remember that you might want to buy some accessories for it—see chapters 5 and 6.)
   a. $100 or less [1]
   b. $200 to $400 [3]
   c. $400 to $800 [5]
   d. $800 to $1,200 [7]
   e. $1,200 to $1,600 [9]
   f. $1,600 to $2,000 [11]
   g. $2,000 to $3,000 [13]
   h. As much as it takes (are you looking to adopt an older son?) [15]
6. Which of the following scenarios best describes your particular situation?
   a. I live in the city and will use my telescope in the city [1]
   b. I live in the city but will use my telescope in the suburbs [6]
   c. I live in the city but will use my telescope in the country [8]
   d. I live in the suburbs and will use my telescope in the suburbs [9]
   e. I live in the suburbs but will use my telescope in the country [10]
   f. I live in the country and will use my telesceope in the country [12]
   g. I live in the country but will use my telescope in the city (Just kidding) [−5]
7. Which of the following best describes your observing site?
   a. A beach [4]
   b. A rural open field or meadow away from any body of water [8]
   c. A suburban park near a lake or river [5]

d. A suburban site with a few trees and a few lights but away from any water   [6]
e. An urban yard with a few trees and a lot of lights   [4]
f. A rural hilltop far from all civilization   [13]
g. A desert   [11]
h. A rural yard with a few trees and no lights   [9]
8. Where will you store your telescope?
a. In a room on the ground floor of my house   [6]
b. In a room in my ground-floor apartment/co-op/condominium   [5]
c. Upstairs   [4]
d. In my (sometimes damp) basement   [4]
e. In a closet on the ground floor   [5]
f. I'm not sure; I have very little extra room   [4]
g. In a garden/tool shed outside   [9]
h. In my garage (protected from car exhaust and other potential damage)   [7]
i. In an observatory   [11]

Now add up your results and compare them to Table 3.2.

Table 3.2 **The Results Are In . . .**

If your score is between . . .	Then a good telescope for you might be . . .
17 and 30	Binoculars (7 × 35 or 10 × 50 for city dwellers, 7 × 50 or 10 × 70 for country residents)
25 and 45	Achromatic Refractor (3″ aperture on a *sturdy* alt-azimuth mount)
35 and 55	Newtonian Reflector (4″ to 8″ aperture on an alt-azimuth or Dobsonian mount)
45 and 65	Newtonian Reflector (6″ to 8″ aperture on an equatorial mount) -OR- Achromatic Refractor (3″ to 4″ aperture on alt-azimuth or equatorial mount)
55 and 75	Newtonian Reflector (12″ to 14″ aperture on a Dobsonian mount) -OR- Schmidt-Cassegrain Catadioptric (8″ aperture)—for astronomers who own small cars and must travel to dark skies
65 and 85	Schmidt-Cassegrain Catadioptric (8″ to 11″ aperture)—for astronomers who own small cars and must travel to dark skies -OR- Newtonian Reflector (10″ or larger aperture on an equatorial mount)
75 and 90	Cassegrain Reflector (8″ or larger aperture) -OR- Apochromatic Refractor (4″ or larger aperture) -OR- Newtonian Reflector (16″ or larger aperture on a Dobsonian or equatorial mount) -OR- Maksutov (3.5″ or larger aperture)

The results of this test are based on buying a new, complete telescope from a retail outlet and should be used *only as a guide.* The range of choices for each score is purposely broad to give the reader the greatest selection. For instance, if your score was 52, then you can select from either the 45 to 65 or the 55 to 75 score ranges. These indicate that good telescopes for you to consider include a 3- to 4-inch achromatic refractor or a 6- to 8-inch Newtonian reflector on an equatorial mount, a 12- to 14-inch Newtonian on a Dobsonian mount, or an 8-inch SCT. You must then look at your particular situation to see which is best for you based on what you have read up to now. If astrophotography is an interest, then a good choice would be an 8-inch SCT. If your primary interest is in observing faint deep-sky objects, then I would suggest the Dobsonian, while someone interested in viewing the planets would do better with the refractor. If money (or, rather, lack thereof) is your strongest concern, then choose either the 6- or 8-inch Newtonian. Of course, you might also consider selecting from a lower-score category if one of those suggestions best fits your needs.

Some readers may find the final result inconsistent with their responses. For instance, if you answer that you live and observe in the country, you already own a Schmidt-Cassegrain telescope, and you want a telescope to photograph faint deep-sky objects but are willing to spend $100 or less, then your total score would correspond to, perhaps, a Newtonian reflector. That would be the right answer, except that the price range is inconsistent with the other responses, because such an instrument would cost anywhere from $600 on up. Sorry, that's just simple economics, but there are always alternatives.

Just what are your alternatives? Who makes the best telescopes for the money? For a survey of today's telescope marketplace, as well as a review of some of yesteryear's best amateur instruments, take a look at chapter 4.

4

Attention, Shoppers!

It is time to lay all the cards on the table. From the discussion in chapter 3, you ought to have a pretty good idea of what type of telescope you want. But we have only just begun! There is an entire universe of brands and models from which to choose. Which is the best one for you, and where can you buy it? These are not simple questions to answer.

Let's first consider where you should *not* go to buy a telescope. This is an easy one! Never buy a telescope from a department store, consumer-club warehouse, toy store, hobby shop, or any other mass-market retail outlet (yes, this includes those 24-hour consumer television channels) that advertises a "400 × 60 telescope." First, what exactly does that mean anyway? These confusing numbers are specifying the telescope in a manner similar to a pair of binoculars; that is, magnification (400 power, or 400×) and the aperture of the primary optic (60 mm, or 2.4 inches). This brings us to a very important rule: **Never buy a telescope that is marketed by its maximum magnification.** Remember, you can make any telescope operate at any magnification just by changing the eyepiece. Telescopes that are sold under this ploy almost always suffer from mediocre optics, flimsy mounts, and poor eyepieces. They should be avoided!

Many of the telescopes mentioned throughout this chapter, and itemized in appendix A, come outfitted with finderscopes, eyepieces, and other accessories. If you come across an item with which you are unfamiliar, chances are good that it is defined in a later chapter. A discussion of various types of eyepieces can be found in chapter 5, while finderscopes and a plethora of other accessories are discussed in detail in chapter 6.

So where *should* you go to buy a telescope? To help shed a little light on this all-consuming question, let us take a look at the current offerings of the more popular and reputable telescope manufacturers from around the world. All of these companies are a big cut above those department-store brands, and you can buy from them with confidence.

I have chosen to organize the reviews by telescope type, with manufacturers listed in alphabetical order. Department-store brands and models have been omitted, because they really fall more under the rubric of toys than of scientific instruments. Most toy telescopes come with plastic tubes, plastic lenses, weak mountings, and horrible eyepieces. It is disconcerting to see that some of the reputable companies mentioned in this chapter are beginning to sell such models with their brand names emblazoned across the telescope tubes. Don't be fooled; these are no better than the others being purposely left out of this review. Those small-aperture refractors and reflectors are universally inferior to the better-made telescopes detailed in this chapter.

In the course of writing this book, I have looked through more than my fair share of telescopes, but it takes more than one person to put together an accurate overview of today's marketplace. To compile this chapter, as well as the chapters to come on eyepieces and accessories, I solicited the help and opinions of amateur astronomers everywhere. Surveys were distributed in print and on-line. I was quickly flooded with hundreds of interesting and enlightening replies and have incorporated many of those comments and opinions into this discussion. (You can help me prepare for the third edition of *Star Ware* by completing the survey form found in appendix J.)

Binoculars

Bausch & Lomb. Dozens of different binocular models are offered in this company's lineup. Some are sold under the Bausch & Lomb trademark, while others are available with the Bushnell label. But for all of this diversity, few Bausch & Lomb binoculars pass the astronomical litmus test because most of them are designed with other purposes in mind, such as sporting and theater events and other terrestrial pursuits.

Of all the binoculars wearing the Bausch & Lomb name, only some of the glasses in its Discoverer, Custom, and Legacy lines are suited to an astronomer's needs, with the Legacy 7 × 42 or 10 × 42 models being the best choice; unfortunately, neither includes a built-in tripod adapter. The Custom 8 × 36 and 10 × 40 binoculars, however, do include tripod sockets. Legacy binoculars range in size from 8 × 22 up to 10 × 50, including two zoom variations, but only the 7 × 35, 8 × 40, and 10 × 50 models offer useful combinations of aperture and magnification. All have wider-than-average fields of view and built-in tripod sockets but suffer from very short eye relief.

Bushnell binoculars are a step below all of the Bausch & Lomb binoculars mentioned above in terms of both price and quality. Most of these models are aimed toward the low end of the market, although a couple in the Natureview series rise to the occasion. While they suffer from cut corners—not too surprising, given their comparatively low cost—they remain good choices if you are working on a tight budget.

Celestron International. This company currently offers five lines of binoculars suitable for astronomical viewing: the Enduro, Pro, Ultima, ED, and Giant. The least expensive are the Enduro models, which are easily held by hand, although eye relief is rather short for eyeglass wearers. Enduros lack the image contrast and brightness of more expensive glasses because they use BK-7 prisms, but they still make good first binoculars for budding stargazers.

Both the Pro and Ultima series feature center focus, premium components, and long eye relief. Images are both bright and sharp, with edge definition a little sharper in the Ultima models. In addition, Ultimas are exceptionally lightweight, making them ideal for support by hand. The Pros are good, but the Ultimas are among the finest moderately priced binoculars on the market today.

A cut above the Pros and Ultimas are Celestron's ED binoculars, so named because their objective lenses are made from extra-low-dispersion glass to virtually eliminate chromatic aberration. Available in 6.5 × 44 and 9.5 × 44 versions, ED glasses are exceptional performers on terrestrial targets. Astronomically speaking, however, the 6.5 × 44s are very low powered for stargazing, while the 9.5 × 44s have a small, 4.6 mm-exit pupil, again not ideal for nighttime viewing.

Last but not least, Celestron's 20 × 80 Standard and Deluxe Giant binoculars produce sharp, clear images that compare favorably to those of other glasses of similar aperture and magnification, such as Orion and Swift. These binoculars are too large and heavy to hold by hand, so you will need some sort of external support. For this, both have built-in tripod sockets, but only the new Deluxe version includes a bar between the objectives and prisms for added stability and structural reinforcement.

Edmund Scientific. In addition to selling other brands, Edmund offers its own Japanese-made 7 × 50 binoculars that feature fully coated optics (a plus), BK-7 glass prisms (a minus), and an average field of view. The aluminum German-style body is sheathed with rubber to help absorb shock. Edmund 7 × 50s come with all the standard fare, such as case, strap, and lens caps. Although adequate for astronomical purposes, they are priced higher than some comparable binoculars from other manufacturers.

Fujinon. A name that has come to be known among photographers for excellent film and camera products over the years, Fujinon is now synonymous with the finest in binoculars as well. Fujinon binoculars are famous for sharp, clear images that snap into focus. They suffer from little or no astigmatism, chromatic aberration, or other optical faults that plague glasses of lesser design. Binoculars just don't come much better.

Not all Fujinon binoculars are designed for astronomical viewing, but its Polaris series (also known as the FMT-SX series; see Figure 4.1) is made with the stargazer in mind. Ranging between 6 × 30 (too small for astronomical viewing) to 16 × 70, all models are waterproofed and sealed, purged of air, and filled with dry nitrogen to minimize the chance of internal lens fogging in high-humidity environments. To maintain their airtight integrity, eyepieces must be individually focused, which is inconvenient when supporting by hand. Another inconvenience is their high price, but many are willing to pay for the outstanding optical quality. The FMT-SX 7 × 50s and 10 × 70s are especially popular among amateur astronomers.

If you're working within a tighter budget but still want Fujinon quality, consider the Poseidon MT-SX series. Available in 7 × 50 and 10 × 50 models, these retail for about two hundred dollars less than their FMT-SX counterparts. Like their more expensive brethren, MT-SX binoculars are waterproof and nitrogen filled, again requiring individually focused eyepieces. So what's the difference? For one, their eye relief measures only about half that of the FMT-SX binoculars, which may make viewing difficult, especially for eyeglass wearers. FMT-SX binoculars also use flat-field optics to reduce edge-of-field distortion, while MT-SX models do not. But the amount of distortion present in any Fujinon binoculars is so small that few complain.

Fujinon also makes what it calls high-power fixed-mount binoculars. Featuring the same quality construction as the Polaris line, these giants are available in either 25 × 150 or 40 × 150 versions. The 25 × 150s come in three models: the least expensive MTs, featuring straight-through eyepieces; EDMTs utilizing ED (extra-low-dispersion) glass in their objective lenses for better

Figure 4.1 *Fujinon 10 ×70 FMT-SX binoculars. Photograph by Eric Hilton.*

color correction; and EM-SX models, including ED objectives and eyepieces mounted in 45° prisms. The 40 × 150 EDMT-SX binoculars also have ED objectives and 45° eyepieces. Fujinon is justifiably proud of these binoculars; after all, Japanese amateur Yuji Hyakutake used 25 × 150 MTs to discover Comet Hyakutake (C/1996 B2), while the 40 × 150s are the world's largest commercially produced binoculars (and the world's heaviest, too!).

Meade Instruments. Several lines of binoculars, many of which are suitable for astronomical viewing, were recently introduced by this company. Least expensive is its Infinity series, with the 7 × 35, 8 × 40, and 10 × 50 being the most attractive. All bodies are soft-rubber covered and include fully coated optics but have only BK-7 crown-glass prisms. The fields of view in both the Infinity and Travel View series are wider than average, but eye relief is short. Still, they represent very good values for budget-conscious astronomers.

If you must wear glasses while viewing, Meade's 7 × 36, 8 × 42, and 10 × 50 Safari Pro porro-prism binoculars offer longer eye relief. BaK-4 prisms and multicoated optics, combined with smooth mechanicals and wider-than-average fields of view, make these good midpriced glasses. Images are sharp, though with slightly less contrast than fully multicoated models.

Minolta. Though best known for its cameras, Minolta also manufactures several lines of binoculars. Stargazers will find some of the XL family most attractive, notably the 7 × 35, 7 × 50, 10 × 50, and 12 × 50s. Perhaps the nicest thing about them is their wider-than-average fields of view. Edge sharpness is good, although some coma is evident. Moderately priced, Minolta Standard XL binoculars represent some of the best buys in binoculars today.

Miyauchi. Not exactly a brand name that rolls off the tongue easily, Miyauchi nonetheless manufactures binoculars that are sure to make mouths water. The smallest Miyauchi is the 20 × 77 BS-77. With an exit pupil of less than 4 mm, this model is marginally useful for astronomy, although it does come with "semi-apochromatic" objective lenses and a beautifully executed one-piece aluminum die-cast body. Each binocular barrel is filled with dry nitrogen to minimize internal fogging. Their eyepieces are tilted at a 45° angle for easier viewing and focus individually, with click-stop detentes for precise setting.

Miyauchi also makes three 20 × 100 binoculars. The BJ100iA and BJ100iB are identical to one another except that the latter features interchangeable 1.25-inch eyepieces. Like the BS-77, both offer tilted eyepieces and aluminum construction.

The most expensive Miyauchi binocular, the BJ100iBF, includes all of these features plus apochromatic objective lenses for near-perfect color correction. Although they cost more than some large telescopes, the BJ100iBF glasses are hard to beat.

Nikon. Like Fujinon, the name Nikon may be more familiar to photographers than astronomers, but all that is changing fast. Nikon produces several lines of binoculars, although many are based on the roof-prism design, which is more appropriate for terrestrial, rather than celestial, study.

Nikon's Sky & Earth series includes ergonomically designed barrels that fit nicely in the observer's hand. Performance is very good, although their fields of view are a little narrower than others in the same size classes.

The finest Nikon binoculars wear the Criterion badge. Of these, Nikon's 7 × 50 IF SP Prostars have almost all the right stuff for handheld use. Not surprisingly, optical performance is outstanding thanks to fully multicoated objective lenses made of extra-low-dispersion (ED) glass that effectively eliminates chromatic and spherical aberrations as well as coma. Like Fujinon's Polaris FMT binoculars, Nikon Criterion Prostars come with individually focused eyepieces and are sealed and waterproofed to prevent internal fogging. Their field of view is comparable to the Fujinons, but eye relief is not as good.

The Nikon Criterion 10 × 70 IF SP Astroluxes are remarkable giant binoculars. Sharing all the features of their smaller Prostar cousins, these glasses have a 5.1° field of view, nearly identical to their Fujinon 10 × 70 rivals. Again, their optical performance is exceptional.

There are only two drawbacks to Nikon binoculars. One is that none are designed with a built-in tripod socket to accept an industry-standard 90° bracket. Nikon, however, does offer a special, custom-designed (and expensive) tripod adapter. Although 7 × 50 binoculars can be supported by hand, 10 × 70s absolutely need the help of a tripod to hold them still. Another problem is price. Nikon binoculars are several hundred dollars more expensive than their Fujinon counterparts. If there is any difference at all in performance between the two brands, it is negligible.

Orion. Few astronomy-related companies are into binoculars in as big a way as Orion, now doing business as the Telescope and Binocular Center. Besides offering models from Bausch & Lomb, Celestron, Fujinon, Nikon, Pentax, Swift, and Zeiss, Orion also features a wide selection of models marketed under its own name.

In the popular 7×-to-10× range, Orion has four different grades of porro-prism binoculars, the least expensive of which are the Scenix models. They come with fully coated optics and built-in tripod sockets, but, not surprisingly, their prisms are made of BK-7 glass.

Orion's Explorer series is the next step up. These center-focus glasses produce bright, sharp images with good contrast, although their fields of view are a little narrower than those of comparably sized Scenix binoculars as well as those from some other manufacturers. Explorers represent good value for the price, offering features that are absent from some more costly glasses.

Orion UltraView binoculars, another step up on the quality ladder, offer bright images across the field of view. I have owned a pair of 10 × 50 Ultra-Views for more than a year now and am very happy with them. Focusing is smooth, with no binding or rough spots throughout the ample travel range. Eyeglass wearers especially should note the UltraViews' exceptional eye relief. Some observers, however, complain about the UltraViews' limited amount of eyepiece travel for focusing.

Orion Vista binoculars compare favorably to other moderately priced models. Images are very clear and crisp, with little or no edge distortion detectable. As with the UltraViews, focusing is very smooth. The only real drawback to Vistas is their slightly narrower than average fields of view, which is why I ended up with the wider-field UltraViews instead.

Orion also markets 8 × 56 and 9 × 63 Mini Giant binoculars. These are ideal for those who want a little extra magnification but still want the option of holding binoculars by hand for a quick view of the night sky out the back door before bedtime. Views through the Mini Giants compare favorably with glasses costing much more, and I highly recommend them.

Looking for *big* binoculars? Orion sells 11 × 70, 15 × 70, and 20 × 70 Little Giant IIs; 11 × 80, 16 × 80, and 20 × 80 Giants; as well as mammoth SuperGiant 14 × 100s and 25 × 100s. Images in the center of their fields are sharp, but like many giant binoculars, all suffer from some edge curvature. This was quite evident through an older pair of Orion 10 × 70s that I tested but perhaps has been lessened in the Little Giant II design. The Little Giant IIs and Giants come with center focusing and include adapters for attaching the glasses to camera tripods, while the SuperGiants come with individually focused eyepieces and are mounted on an integral platform-type tripod mounting plate.

Finally, as this was being written, Orion just introduced three giant binoculars manufactured in Japan by Vixen Optical Industries. All feature 45° prisms for easier viewing when the binoculars are tilted skyward. The Vixen BT80M-A offers owners the unique ability to use any 1.25-inch diameter eyepiece, thus permitting a wide range of magnifications. The binoculars come supplied with a pair of 25-mm orthoscopic eyepieces, yielding 36×. To support these, Orion offers an alt-azimuth mount made by Vixen specifically for this "binocular telescope." Although the mounting is heavily constructed, one owner complained that it still shakes under the weight of the binocular telescope. Additionally, the tripod does not have a center post for adjusting height; instead, the only way to change the height of the binoculars is by extending the tripod's legs, judged an inconvenience. Orion has also added Vixen's enormous 20 × 125 and 30 × 125 binoculars to their lineup. These are, quite literally, a pair of 4.9-inch refractors strapped together! Each comes with an alt-azimuth fork mount; a 7 × 50 finderscope is sold separately. Like Fujinon's monstrous 150-mm binoculars, these also weigh pounds, not ounces! Although I have not heard any firsthand reports on any of these, Vixen's reputation would indicate that they are well made, though very expensive.

Pentax. Best known for its cameras, Pentax offers some exciting porro-prism binoculars that are well suited for astronomical viewing. Least expensive are the seven models that comprise the PCF III series, with glasses ranging from 7× to 20×. Each has a center-focus thumbwheel as well as a locking mechanism and a click-stop diopter adjustment, a very useful feature! All Pentax PCF III binoculars use superb optics that yield good-quality views of the night sky, although their fields of view are narrower than those of some competitors' models. Like many binoculars, their housings are rubber coated to help absorb the inevitable bumps.

The PIF series is Pentax's top of the line. PIF binoculars compare favorably with the most expensive binoculars sold, including those offered by Fujinon and Nikon. Each features field-flattener lenses that suppress edge distortion, curvature of field, and other aberrations. The rubberized waterproof housings are purged of air and filled with dry nitrogen to enhance optical performance further, mandating individual-focus eyepieces. The only downside to Pentax PIF binoculars is their weight, but thankfully, tripod sockets are also built into the center hinge of each model.

Swift Instruments. Established in 1926, Swift offers one of the world's largest and most varied selections of binoculars. Most are designed for earthly pursuits such as birding and yachting, but a few are also appropriate for more ethereal viewing.

Looking first at binoculars that can be held by hand, several Swift models utilize prisms of BK-7 glass, making them somewhat less desirable from an astronomical vantage point. Standing above the crowd, however, are their Ultra-Lite glasses. All three offer multicoated optics, BaK-4 porro prisms, built-in tripod sockets, and retractable rubber eyecups. The images they produce are sharp, and their fields of view are about average for the size and magnification. Eye relief on the 7 × 42s is very good but grows shorter as magnification rises.

Three Swift porro-prism birding binoculars are also well suited for astronomical viewing. The Swift 8.5 × 44 Audubon and 10 × 50 Kestrel glasses offer fields of view that are among the widest available because of their five-lens ocular system. Both are among the finest moderately priced binoculars in their respective size classes, with their only real shortcoming being a lack of eye relief.

The Audubon ED 8.5 × 44 binoculars have all of the features found on the non-ED Audubons but with the added benefit of objective lenses made from (as you might have guessed) ED glass for exceptional color correction. Their field of view also measures 8.2°, a big plus, but their eye relief is slightly shorter.

For those who want something more substantial, Swift has the 11 × 80 Observer, 15 × 60 Vanguard, and 20 × 80 Satellite models. All three come with built-in tripod sockets and smooth center-focusing thumbwheels, which are all pluses. Images obtained through the Observer and Satellite are clear and crisp, roughly comparable to those of similarly priced giants from Celestron and Orion. The 15 × 60s, however, are somewhat inferior performers because they use BK-7 prisms.

Zeiss (Carl). Long known as one of the world's premier sources for fine optics, Zeiss continues the tradition by offering some of the finest, and most expensive, binoculars of all. For instance, the Zeiss 7 × 42 and 10 × 40 B/GA T Dialyt roof-prism binoculars include conventional center focus but no tripod socket. Eye relief is a little shorter than some other brands, but image quality is second to none.

Zeiss B/GA Design Selection roof-prism binoculars also offer outstanding performance. Both the 8 × 56 and 10 × 56 models in this series come with fully multicoated optics, individually focused eyepieces, and are sealed with o-rings to prevent internal lens fogging. Each features a wider-than-most field of view, a heavier-than-most weight, and the highest price of any similar-sized binoculars mentioned here! They are too rich for my blood but perhaps not for yours.

Refracting Telescopes

Although many small refractors continue to flood the telescope market, the discussion here is limited to only those with apertures of 3 inches and larger, with two notable exceptions (see Tele Vue). I used to own a couple of 2.4-inch refractors, but I found that I quickly outgrew those instruments' capabilities and hungered for more. Why? At the risk of possibly offending some readers, let me be brutally honest: A 2.4-inch long-focus telescope will produce marginal views of the Moon and maybe Venus, Jupiter's satellites, Saturn, and a few of the brightest deep-sky objects but nothing else. Most are supplied on shaky mountings that only add to an owner's frustration. If a 2.4-inch refractor is all your budget will permit, then I strongly urge you to buy a good pair of binoculars instead.

Astro-Physics. This name is immediately recognizable to the connoisseur of fine refractors on rock-steady mounts. Owners Roland and Marjorie Christen introduced their first high-performance instruments in the early 1980s and effectively revived what then was sagging interest in refractors among amateur astronomers. Now, nearly twenty years later, Astro-Physics refractors (Figure 4.2) remain unsurpassed by any other apochromat sold today.

Currently, four refractors make up the Astro-Physics line. The heart of each is the superb three-element apochromatic objective lens. Two crown meniscus elements combine with a third lens made of an advanced fluorite ED glass to eliminate all hints of false color (secondary spectrum) and other optical aberrations, both visually and photographically.

Figure 4.2 *Astro-Physics Starfire apochromatic refractor.*

Least expensive is the 105EDFS, a 4.1-inch f/6 refractor measuring only 19 inches long, aptly named the Traveler. As with the larger Starfires, the performance of the Traveler is outstanding! I can vividly recall the view I once had of the Veil Nebula complex in Cygnus through a Traveler owned by some friends. With a 35-mm Panoptic eyepiece, the three main pieces of the Veil, spanning nearly 3°, fit into a single eyepiece field! The nebula's resolution and brightness looked just like a photograph, but I was actually seeing this "live." Now, years later, that single observation remains one of the highlights of my astronomical career.

The Traveler is as superb mechanically as it is optically. Especially impressive is its oversized rack-and-pinion focuser of aluminum and brass. Its inside diameter measures 2.7 inches, designed to let advanced astrophotographers use medium-format cameras with minimal vignetting. Adapters are supplied for both 1.25 and 2-inch eyepieces.

For those looking for a little larger aperture and longer focal length, the Starfire series should be given serious consideration. Like its smaller sibling the Traveler, the 130EDFS 5.1-inch f/6 Starfire is an exceptional instrument both optically and mechanically. The largest refractor in Astro-Physics's corral is the 155EDFS 6.1-inch f/7 Starfire, the pinnacle of its aperture class. Because of its fast focal ratio, the 155EDFS measures only 41 inches long with the dewcap fully retracted, making it a truly portable telescope. No trace of false color is evident in either of these instruments, even on such critical objects as the planet Venus. Stars show textbook Airy disks, planetary contrast is sharp as can be, and resolution is crisp and clear. Both use the same oversized rack-and-pinion focuser as the Traveler for effortless focusing without any hint of binding or shift, even in subfreezing temperatures. All machining is done by Astro-Physics in its factory, ensuring top-notch quality control on all components.

Finally, there is the 155EDF 6.1-inch f/7 astrograph, a variation of the 155EDFS. This instrument uses the same three-element objective lens as the 155EDFS but also includes a two-element, 4-inch field-flattener lens in the rear of the optical tube assembly to eliminate astigmatism and field curvature in photographs.

Notice that I have not mentioned anything about Astro-Physics mountings. That is because none of the telescopes come with mountings; instead, you may select from their line of homegrown German equatorial mounts to suit your needs. The Astro-Physics 400 German equatorial mount was designed with the Traveler and 130EDFS instruments in mind and can be attached to either a metal or hardwood tripod. Many choose the Astro-Physics 600 German equatorial mount for the 155EDFS as it combines solidity and light weight, with the equatorial head weighing in at 27 pounds. For those contemplating long-exposure astrophotography with either the 155EDFS or the 155EDF, the Astro-Physics 900 and 1200 mountings are more appropriate. These are *serious* telescope mounts designed to attach directly to large-diameter metal piers rather than tripods. All Astro-Physics mounts include built-in, dual-axis 12-volt DC motor drives and are designed to accept a wide variety of accessories, camera-mounting plates, and digital setting circles.

For either visual or photographic use, Astro-Physics apochromatic refractors are the finest of their kind, but be forewarned that because of the high demand and limited production, delivery may take several months or even a year.

Celestron International. Within the last 15 years, Celestron, best known for its Schmidt-Cassegrain telescopes, has branched out into other telescope markets. In the refractor category, Celestron features imported instruments that range in aperture from 2.4 to 4 inches.

Celestron's Firstscope 80 comes mounted on either an altitude-azimuth or German equatorial mount. Both feature the same 3.1-inch f/11.4 optical tube assembly and come with a mediocre 25-mm SMA (Kellner; see chapter 5) eyepiece. Overall optical performance is adequate, although their small aperture limits these instruments to prominent sky objects such as the Moon, planets, and brighter deep-sky objects. The alt-azimuth version includes a 45° erect-image diagonal for upright terrestrial viewing, while the equatorial version includes a 90° star diagonal. Be aware that the 45° diagonal produces dimmer images and proves less comfortable for night-sky viewing than the 90° diagonal. Mechanically, however, the alt-az version is sturdier than the equatorial mounting.

Those interested in a superportable terrestrial spotting scope that can do double duty as a low-power, rich-field instrument should consider the Celestron Rich Field 80. This 3.1-inch f/5 achromatic refractor measures only 13.5 inches long and is designed to accept standard 1.25-inch eyepieces and accessories. The air-spaced doublet objective lens is fully multicoated to yield high contrast and good resolution, but keep in mind that at f/5, false color (secondary spectrum) will detract from views of bright objects such as the Moon. Also, the short 400-mm focal length is meant to operate at low power only; this is not intended as a planetary telescope. The Rich Field 80 does not come with an eyepiece, star diagonal, finder, or mounting, although a built-in adapter plate allows easy mounting to a camera tripod.

Two excellent refractors imported by Celestron are the 4-inch f/9.8 GP-C102 and the 4-inch f/9 GP-C102ED from Vixen Optical Industries of Japan.* Both appear nearly identical on the outside, with each held in a pair of metal rotating rings atop a Great Polaris German equatorial mount. The Great Polaris (or GP) mount, successor to Vixen's popular Super Polaris (SP) mount, is executed nicely and features a handy polar alignment scope built into the right ascension axis. A motorized clock drive is available for an additional charge.

Both the GP-C102 and GP-C102ED come with a nice 20-mm Plössl eyepiece and a 1.25-inch star diagonal but only a 6 × 30 finder. The finder is best augmented with a lightweight, 1× aiming device. The rack-and-pinion focusing mount takes 1.25-inch eyepieces but will also accept 2-inch eyepieces with an optional adapter.

The biggest difference between the GP-C102 and the GP-C102ED is in the optics. The objective lens of the C102 is a classic achromat, while the C102ED uses a two-element multicoated objective made of ED glass. As the owner of an older C102, I can attest to its optical excellence with enthusiasm. Color correction is good for an f/10 achromat (though not as good as a slower ratio), with images that are both sharp and clear.

As good as the C102 is, the C102ED is in a whole other class. Although it costs considerably more than the C102, instrument's higher price is offset by its exceptional quality. Contrast, false color suppression, and sharpness are all very good, although personally I feel that Vixen's fluorite version, now sold here in the United States by Orion Telescopes (described later in this chapter), has better color correction.

D&G Optical Company. This small company specializes in superb-quality achromatic refractors ranging in aperture from 5 to 10 inches, all available at either f/12 or f/15. D&G also makes larger refractors on a custom-order basis. All optics are handmade and tested to exacting standards, producing the finest results. The 5- and 6-inch objectives receive coatings of magnesium fluoride, while coatings on the 8-inch and larger objectives are available at extra cost.

Unlike most refractors, whose objective lenses are mounted in nonadjustable cells, D&G objectives come with a "push-pull" cell. This allows the

*See note on page 110.

owner to align the lenses' optical axis precisely, resulting in better images and a telescope that stays in collimation even with constant transport. All D&G refractors also come with oversized rack-and-pinion focusing mounts complete with adapters for 1.25- and 2-inch eyepieces. Owners of this company's telescopes consistently rave about their performance, pointing to how secondary spectrum is all but nonexistent while contrast and sharpness are both outstanding.

D&G refractors are sold without mounts, although they may be easily supported by many commercial mountings. For its 5- and 6-inch instruments, D&G offers its own DG-150 German equatorial mount, which features 1.5-inch stainless steel shafts and an AC-powered clock drive, made by Edward R. Byers Company, on the right-ascension axis. The DG-150 is adequately strong for the task at hand and has good rigidity with little flexure. Other mounts worth considering are those by Astro-Physics, Hollywood General machining (Losmandy), and others cited in chapter 6.

Meade Instruments. To celebrate its twentieth anniversary in 1992, Meade introduced a line of new and exciting state-of-the-art 4- to 7-inch f/9 apochromatic refractors. Years later, the Meade apochromats remain among the most sophisticated refractors on the market today. All come mounted on tripod-supported German equatorial mounts that offer a good mix of stability, portability, and convenience. Both fast-speed slewing and slow-motion control are smooth, with little or no backlash. The only mechanical problem evident in the overall design is in the focusing mount, which is not as smooth as in some of the competition's models. This shortcoming was especially noticeable in earlier production scopes but has since been improved.

Like many other premium refractors today, Meade apochromats use ED glass in their objectives to minimize false color. The resulting images yield pinpoint stars and good field contrast, two characteristics of better refractors. When viewing brighter targets, however, some minor blue fringing becomes evident, as does some slight astigmatism and spherical aberration. This led one respondent of my survey to describe these as semi-apochromatic telescopes—perhaps not as good as some other apochromats but better than achromatic refractors with similar-size apertures.

Meade also offers an imported 3.5-inch f/11 achromatic refractor for the budget-minded hobbyist. Available as the Model 390 on an alt-azimuth mount or as the Model 395 on a German equatorial mount, this instrument comes with a 6 × 30 finderscope and a 25-mm eyepiece and will accept standard 1.25-inch diameter eyepieces. Both mountings are adequate, but the alt-azimuth unit is sturdier. The equatorially mounted version is nearly identical to Orion's SkyView 90 (see page 64), except that the latter comes with two Kellner eyepieces.

Orion Telescopes (U.S.). In addition to marketing telescopes by several major manufacturers, Orion Telescopes (now also known as the Telescope and

Binocular Center, as previously mentioned) of California imports several refractors and sells them under the Orion name.

The Orion ShortTube 3.1-inch achromatic refractor operates at f/5 to yield a 400-mm focal length and is a very interesting instrument. Images are surprisingly good, with some spectacularly wide fields of Milky Way star clouds possible, although a fair amount of color fringing is evident around brighter stars because of its fast focal ratio. Potential purchasers should also note that the instrument is not really suitable for high-powered planetary views. Given that it comes with a built-in camera-tripod adapter plate, a detachable 6 × 30 finderscope, and two Kellner eyepieces, the Orion Short-Tube refractor is one of the best buys in small rich-field achromatic telescopes today. But readers should note that it falls short of the performance offered by pricier semi-apochromatic refractors such as Tele Vue's Ranger and Pronto.

Orion also offers the more conventionally styled Observer 80 EQ Ultra, a 3.1-inch f/11.3 achromatic refractor that is a clone of the Celestron Firstscope 80 EQ. The Observer 80 features fully coated optics and comes on a German equatorial mount equipped with manual slow-motion controls, with an AC-powered clock drive available as an option. Also included are a small 6 × 30 finder and two 1.25-inch Kellner eyepieces. As with the Celestron, the mounting is the weak link with the Observer 80, which will be especially noticeable in windy conditions.

Orion's SkyView 90 3.5-inch f/11.1 achromatic refractor is very similarly styled to Meade's Model 395, although the Orion comes with two 1.25-inch Kellner eyepieces as opposed to Model 395's one, and is equipped with a 6 × 30 finder. The Orion optics, like Meade's, perform quite well for the aperture, with good star images and fairly small amounts of secondary spectrum. (Better eyepieces will go a long way to improving the instrument's performance, especially at moderate and high magnification.) The SkyView 90 is supported by a German equatorial mount on an aluminum tripod. A pair of aluminum rings grasps the telescope's tube assembly, allowing it to be rotated to a more convenient viewing angle if necessary. Manual slow-motion controls adorn the mounting, with a battery-powered motor drive available as an option. Although beefier than that of the Observer 80, the mounting is still the SkyView's greatest weakness. All things considered, the Orion SkyView 90 is competent optically and passable mechanically, although I still favor the Meade 390 with its simpler, albeit sturdier, alt-azimuth mount.

Recently, Orion began marketing the VX80, a 3.1-inch f/11.4 achromatic refractor made by Vixen Optical Industries of Japan and in a completely different league than its telescopes mentioned previously. Although I have not received any reviews from owners, I expect optical quality should be tops for the aperture, as with all Vixen instruments. Its use is limited to the Moon, planets, and brighter sky objects, however, because of the comparatively small aperture. The VX80 comes with a well-made alt-azimuth mounting complete with slow-motion controls and an adjustable wooden tripod as well as a 1.25-

inch 26-mm Plössl eyepiece (the first measurement refers to the size of the barrel; the second, the focal length), a small 6 × 30 finderscope, and a 45° correct-image prism diagonal. The latter is far more appropriate for terrestrial than astronomical viewing, as users will quickly discover when they first aim the instrument near the zenith. If you are looking for a high-quality small refractor, the Orion/Vixen VX80 represents a very good choice.

At the same time, Orion added a pair of 4-inch refractors from Vixen to its lineup. The Orion/Vixen VX102-GP is essentially the same instrument as the Celestron GP-C102 achromatic refractor. Both feature the acclaimed Great Polaris (GP) German equatorial mount, with the only apparent difference being the tripod. The Celestron uses a telescoping wooden tripod, while Orion's extendable version comes with extruded aluminum legs; either is adequate. Refer to the previous comments about the GP-C102 for additional thoughts on performance.

The Orion/Vixen VX102-GP Fluorite is a slightly repackaged version of the no-longer-offered Celestron C102F 4-inch f/9 instrument, one of the finest 4-inch apochromatic telescopes ever offered. Although Celestron now sells a version that features an objective lens using ED glass (the GP-C102ED, reviewed previously), Orion offers the original fluorite model. Again, mechanically, both instruments are identical, so I'll refer you to the comments made on page 62. But optically, although the GP-C102ED is a terrific instrument, the VX102-GP Fluorite version has it beat when it comes to the suppression of false color and chromatic aberration. Add to this the fact that the VX102-GP Fluorite is typically less expensive than the GP-C102ED, and the choice is clear.

Takahashi. This company's apochromatic refractors have a well-deserved reputation for their excellent performance. In fact, to many they represent the pinnacle of the refractor world—they just don't come much better. The only drawback is that this superior quality does not come cheaply!

The largest Takahashi refractors belong to its older FC and FCT lines. The 6-inch f/8 FC-150 is characterized by doublet objectives with the internal element made of fluorite, whereas the FCT telescopes are designed around triplet objectives that include a fluorite element to yield uncompromising optical performance. Astrophotographers should be especially interested in the photographically fast 6-inch f/7 FCT-150. And if you are in the market for the ultimate large-aperture refractor, then the 8-inch f/10 FCT-200, the largest apochromatic refractor in production today, is for you. Both it and the FCT-150 feature huge 11 × 70 finderscopes (the FC-150 includes a 7 × 50 finder) and come mounted on one of the steadiest equatorial mounts available.

The new FS line of instruments includes the 3-inch f/8 FS-78, the 4-inch f/8 FS-102, and the 5-inch f/8.1 FS-128. As in the FC refractors, Takahashi FS doublet objectives include one element made from fluorite for exceptional color correction; the difference is that the outer element of FS objectives is made of fluorite rather than the inner. This new design claims to produce even better results than before. Many optical experts feel this may be little more than an advertising scheme, but there is no denying that image clarity and contrast are amazing (as they always have been with Takahashi apochromats).

All Takahashi refractors yield aberration-free images that can only be described as perfect. Planetary detail is simply stunning. Given steady atmospheric conditions, the optical quality of Takahashi instruments routinely permits the use of magnifications in excess of 100× per inch (see chapter 5 for a discussion of magnification and its limits).

All FS, FC, and FCT refractors ride on unyielding German equatorial mounts that are among the best in the world. An illuminated polar-alignment scope is built right into the right-ascension axis on all Takahashi mounts. A single-axis motor drive is included on the EM-2 mounts supplied as standard with the 3- and 4-inch models. The optional EM-10 mount has dual-axis drives, which are also standard on the EM-200 mounts that are supplied with the FS-128 and optional on the FS-102. The FCT-150 and FCT-200 refractors come with the massive EM-200 mounting. The EM-2, EM-10, and EM-200 come with adjustable wooden tripods, while an observatory-style metal pier is available for the EM-200 and standard on the EM-500.

Tele Vue, Inc. This outfit manufactures several outstanding short-focus refractors for the amateur. The views they provide are among the sharpest and clearest produced by any telescope in their class, with excellent aberration correction.

Least expensive is the Ranger, a compact 2.8-inch (70-mm) f/6.8 instrument. Though technically below the 3-inch threshold I set earlier in this section, this instrument's sharp, wide-field performance warrants its mention. The Ranger features a two-element objective lens, with the front element made of crown ED glass and the rear made of high-index flint. All surfaces are coated to reduce internal reflections. The net result is proclaimed by the manufacturer to be semi-apochromatic, that is, not as completely free of chromatic aberration as a true apochromatic refractor but better than a standard achromat. From the optical testing I've done, the Ranger stands up well, even when compared to short-focal-length eyepieces. Images are crisp, and stray-light reflections are nonexistent, even (surprisingly) in the absence of internal light baffles. The 1.25-inch helical focuser and tube are made from machined aluminum, the latter coming nicely finished with either ivory-white paint or brass coating. Although I am not a proponent of helical focusing mounts, Tele Vue attacks the problem intelligently by including a sliding drawtube for coarse adjustments. The Ranger includes a 20-mm Tele Vue Plössl eyepiece, a 90° mirror diagonal, and a mounting plate for attaching the instrument to a camera tripod. No finder is provided, although the Ranger's wide field at low power makes one unnecessary.

The Ranger's more affluent older sibling is the Pronto. Both the Ranger and Pronto use the same semi-apochromatic objective lens, but the appointments offered with the Pronto are more lavish. Chief among these is its excellent 2-inch rack-and-pinion focuser, one of the smoothest available. Another nice feature is the mounting that attaches the instrument to any standard camera tripod or to Tele Vue's Panoramic alt-azimuth mounting (sold separately). Unlike the Ranger's sliding mounting ring, which is fixed in place, the Pronto's allows the user to balance the instrument after attaching a camera or other accessory. Also included with the Pronto are a 2-inch mirror star diagonal (a 1.25-inch adapter is included) and a 20-mm Tele Vue Plössl.

If glimpses of exceptional planetary detail are what you crave, then consider Tele Vue's Genesis-SDF 4-inch f/5.4 apochromatic refractor. Although the short focal length keeps the optical tube assembly to a toss-it-over-your-shoulder length, the four-element special-dispersion-plus-fluorite (SDF) objective permits the use of a broad range of eyepiece focal lengths. The four lens elements are coupled in two groups of two: A pair make up the front objective, and another two mounted toward the rear of the tube serve as a corrector. The result is a compact instrument of outstanding optical quality that is just as useful for close-up planetary work as it is for rich-field, deep-sky observing.

Tele Vue recently introduced a further refinement to the Genesis-SDF. The new 4-inch f/5.4 Tele Vue 101 looks just like the Genesis-SDF except that it incorporates an internal fluorite lens and two special-dispersion glasses into its optical tube assembly. The optical assembly produces absolutely textbook images, high in contrast and resolution and completely free of extraneous color.

Both the Genesis-SDF and TV-101 are packaged in aluminum tubes painted "vanilla white," although the Genesis is also available with a polished brass tube under the name Renaissance-SDF. All three include 2-inch focusing mounts and star diagonals, 1.25-inch eyepiece adapters, and 20-mm Tele Vue Plössl eyepieces but no finderscopes. And although all three are sold mountless, Tele Vue's sturdy Gibraltar alt-azimuth mounting was specifically designed with each in mind. If you are looking for an equatorial mount, consider one of the Astro-Physics or Takahashi mounts (see the previous discussions of them in this chapter) or the Losmandy GM-8 (see chapter 6).

Finally, Tele Vue's new flagship is the Tele Vue-140, a 5.5-inch f/5 apochromatic refractor built around a larger version of the exotic four-element lens system used in the TV-101. My views through an early version were breathtaking, with an especially spectacular view of Perseus's Double Cluster being perhaps the most memorable. The short 700-mm focal length means that wide expanses of the sky can squeeze into the eyepiece's field. For instance, a 27-mm Tele Vue Panoptic will take in 2.6°, more than adequate to show the Pleiades or other broad sky objects. Like I said, simply breathtaking! The TV-140 comes with a 2-inch focusing mount, a 2-inch star diagonal, and a 1.25-inch eyepiece adapter but no eyepieces, mount, or finderscope. An optional special adapter will fit the TV-140 to either the Gibraltar or Losmandy GM-8 mounts. Readers who are considering a TV-140 should note that it is not a production telescope but rather produced only on a per-order basis. But for those who are willing to pay the $5,000-plus price and wait for a custom-built telescope, the TV-140 is hard to beat!

Right about now, you might be thinking, "Okay, so which is better: Takahashi, Astro-Physics, or Tele Vue?" That's an awfully close call. Astro-Physics owners swear their instruments are the best, while owners of Takahashi and Tele Vue telescopes are equally loyal to theirs. The important thing to note is that all received glowing reviews with few, if any, complaints. It's the same as asking which car is better: Mercedes-Benz, BMW, or Lexus? Chances are, they are all pretty good if you can afford them!

Reflecting Telescopes

Celestron International. Long known for its Schmidt-Cassegrain instruments, Celestron introduced a line of Dobsonian-mounted Newtonian reflectors in 1995 called Star Hoppers. Star Hoppers are aimed at entry-level amateurs, designed to go head-to-head against Meade Starfinder Dobsonians and Orion Deep-Space Explorers. As this book was going to press, Celestron redesigned the mountings on the 6- and 8-inch Star Hoppers and added three larger siblings. Currently, Star Hoppers come in six apertures, ranging from 4.25 to 17.5 inches.

The 6- and 8-inch Star Hoppers continue to attract wide attention among amateur astronomers in search of that perfect first telescope. Like their competition from Meade Instruments and Orion Telescopes, Star Hoppers feature cardboard tubes, although the 8-inch is available with an aluminum tube. They also come with a 25-mm modified achromat eyepiece (a very similar design to a Kellner) and are designed to accept standard 1.25-inch eyepieces. They do not come with a finderscope, but Celestron offers a 6×30 finder as an option.

The 6- and 8-inch Star Hoppers feature good-quality 1.25-inch rack-and-pinion focusing mounts; the 11-, 14-, and 17.5-inch models are sold only with 2-inch helical focusing mounts, which have been judged an inconvenience. Once focus is achieved, the Pyrex primary mirrors in Star Hoppers perform quite well. Image clarity and contrast are both excellent, which led one respondent to note that a friend's 8-inch Star Hopper outperformed his own 8-inch Schmidt-Cassegrain telescope: "In side-by-side tests, I liked the images in his telescope better than in my 8″ SCT in moderately light-polluted skies and fair seeing conditions. Viewing M51 and M13 at about 120×, his telescope had better contrast."

One of the more unusual features of the 6- and 8-inch Star Hoppers is their tapered primary mirrors that are supported in the center and mounted on an excellent mirror cell. The cell's open and allows very good ventilation, while three large knobs on its back make for easy adjustment without tools. Likewise, the diagonal mirrors in the 6- and 8-inch Star Hoppers, mounted on a thick single-vane holder, are easily collimated by turning three nylon thumbscrews while watching through the focuser.

The diagonal mirrors in the 11-, 14-, and 17.5-inch Star Hoppers are held in place by conventional spider mounts, while their primaries are attached to their mountings with silicone adhesive.

The newly redesigned Star Hopper Dobsonian mountings closely resemble those supplied with Meade or Orion telescopes, with a white laminate material covering particle board. Celestron's designers were thinking when working on the altitude bearings on the 6- and 8-inch models. Instead of sedentary plastic trunnions, the altitude bearings are made of machined aluminum and ride on dovetailed tracks. This innovative design allows the user to customize the telescope's balance point. Although adding a larger finder or heavier eyepiece to a Dobsonian may normally throw balance off, the Star Hopper altitude bearings can be moved up and down the tube to compensate. Unfortunately, this useful feature was not carried over to the larger models, perhaps because a new eyepiece or finder would add only a small percentage to the tube's weight or perhaps because of economics. Instead, the three largest Star Hoppers are supported on plastic bearings that appear identical to those found on Orion's Premium Deep-Space Explorers. Protruding handles from the altitude bearings are a great assistance when moving the tube from storage to observing site, though bear in mind that the larger models should be carried by only two people.

Overall, Star Hoppers are good performers that feature some nice touches seemingly forgotten by the competition. If I had to single out their biggest weakness, it would have to be their motion in azimuth, although the smoothness has been greatly improved since the 6- and 8-inch models were first introduced. The larger scopes' plastic altitude bearings might also prove a little sticky when aimed toward high altitudes.

The Star Hopper 4.5 is a different sort of beast altogether. It features the same optical tube assembly as the Firstscope 114 series (see the following paragraph) but comes on an unusual alt-azimuth mounting, described by Celestron as a "unique, small-size Dobsonian." The tube is carried in a wide rotating ring and mounted on a one-armed "half-Dobsonian" wooden mount, the altitude axis being held with a large bolt. The azimuth bearing is similar to those of the other Star Hoppers.

As noted previously, the Star Hopper 4.5 uses the same Chinese-made optical-tube assembly as the Firstscope 114, Firstscope 114 Deluxe, and Firstscope 114 Premium. All three feature a 4.5-inch f/8 mirror mounted in an aluminum tube. Although the standard Firstscope 114 comes with poor 4-mm and 25-mm Huygens subdiameter (0.965-inch) eyepieces, its 1.25-inch rack-and-pinion focuser will accept better-quality eyepieces that may be purchased later on. The Deluxe and Premium versions also come with a 1.25-inch focuser and 1.25-inch 10-mm and 25-mm modified-achromat (Kellner) eyepieces. The standard and Deluxe include inadequate 5×24 finderscopes, as does the Star Hopper 4.5, while the Premium uses a 6×30 finder. Optically, the Firstscope instruments are good, although the mountings are shaky.

Finally, Celestron offers the C4.5, a 4.5-inch f/7.9 Newtonian reflector made in Japan by Vixen Optical Industries.* The C4.5 is secured to a German equatorial mount and an adjustable wooden tripod that are the sturdiest in its class. It also comes with a high-quality 26-mm Plössl eyepiece, a 6 × 30 finderscope, and a 1.25-inch rack-and-pinion focuser.

Overall performance has been improved recently now that the C4.5 comes with a larger (29.8-mm minor axis) secondary mirror. As mentioned in the first edition of *Star Ware,* image brightness used to be less than expected because the secondary mirror was smaller than the cone of light reaching it from the primary, effectively reducing the telescope's aperture. With this problem corrected, the C4.5 proves to be the best 4.5-inch reflector sold today. But be forewarned that the C4.5 costs more than many larger, Dobsonian-mounted reflectors. For most observers, they would prove the better choice.

Coulter Optical Company. This name has long been associated with Dobsonian-mounted Newtonian reflectors. Indeed, it was Coulter's introduction of the 13.1-inch f/4.5 Odyssey 1 in 1980 that started the whole revolution in commercial Dobsonian telescopes. They were an immediate success, with orders coming in faster than could be satisfied. Sadly, illness closed the family-owned Coulter Optical Company in the mid-1990s, but like the Phoenix rising from the ashes, the name Coulter and the trademark line of Odyssey telescopes have been resurrected by Murnaghan Instruments of West Palm Beach, Florida. Five instruments, from a 6-inch f/8 to a 13.1-inch f/4.5, make up the new Coulter lineup.

At first glance, the new Coulters look just like the old. All sport the same red-tube-and-black-base paint job while retaining their famous low price. Looking a little deeper, we find an improved helical focusing mount that slides in and out for coarse focusing and twists for fine adjustment, a far superior primary-mirror mount, and a smoother moving Dobsonian mount. Options available for an additional charge include finderscopes, lift handles, and cooling fans.

So how do the new Coulters stack up against the competition? To help answer that question, I put a 10.1-inch f/4.5 Coulter Odyssey Compact through its paces. The optical tube assembly and rocker box arrived fully assembled, although in two separate boxes.

My test Odyssey moved smoothly in altitude but was a little sticky in azimuth. Once the supplied Telrad aiming device was attached to the instrument, however, the tube's balance was thrown off enough for it to sink to the horizon whenever the telescope was aimed below an altitude of about 30°. The situation only became worse when I substituted a heavier eyepiece for the standard 27-mm ocular. I must stress that this is not a problem unique to Odysseys, however. Meade's and Orion's Dobsonians may also suffer from this balancing act; only Celestron's Star Hoppers, with their sliding altitude bearings, are immune. Coulter offers an add-on counterbalance accessory, as does the Telescope and Binocular Center (Orion).

*See note on page 110.

The plate-glass primary mirror is attached to its wooden mounting cell by several blobs of silicone adhesive. (See more about plate-glass versus Pyrex mirrors under "Orion Telescopes (US)" later in this section.) Three pairs of adjustment thumbscrews on the back of the mirror cell make collimation easy. The Pyrex diagonal mirror is held in place by the same bar-style mounting used in the originals. Although sturdier than a thin single-vane mount, the design makes it difficult to collimate the mirror or to remove it for realuminizing.

At first, I was very impressed with how the Odyssey performed optically. Images were both sharp and clear with good contrast, even with the inexpensive 27-mm eyepiece that came with the instrument (which I would recommend replacing with better eyepieces as soon as economically feasible). Once the instrument was properly collimated, stars focused well even at 165× (with a 7-mm eyepiece).

But after leaving the scope in a warm car for an afternoon, performance fell off noticeably. I do not know for sure what caused the degradation, but I suspect that either the chip-board mirror mount or the silicone adhesive somehow warped or sagged under the heat. This is a big problem, because telescopes are often transported by day to dark-sky sites for a night's observing. Coulter has said that it plans to look into the problem and to redesign whatever caused the distortion, but until that is done, I cannot recommend these telescopes.

As of this writing, only a few Coulter Odysseys have been delivered as Murnaghan continues to work out the bugs and holdups that are inevitable with starting up a new production line. Tardiness was a problem faced by countless amateurs who ordered the original Odysseys, and it seems to follow the name even today.

Dark Star Telescopes. Comparatively unknown in the United States and Canada but well regarded in the United Kingdom and Europe, Dark Star Telescopes offers six Dobsonian-mounted Newtonian reflectors in its Custom line, ranging in aperture from 6.25 to 16 inches and all with Pyrex optics. Company literature states that all optics are 1/8 wave, although whether this is the final wavefront measurement and how this number was determined are not specified. All optics are made specifically for Dark Star by British mirror manufacturer Jon Owen and then aluminized in Dark Star's own plant. (Mirrors and telescope kits are also sold separately, should you wish to make your own instrument.)

According to several respondents to my survey, Dark Star telescopes perform very well optically, with images that are both sharp and contrasty. All come with low-profile rack-and-pinion focusing mounts and small diagonal mirrors to minimize central obstruction. In addition, all feature open primary-mirror cells for quick mirror cooling, a big plus when moving from a warm room to cooler, outdoor conditions. The diagonals are held steadily in place with four-vane spider mounts that are readily adjustable, if necessary. All come with finderscopes; the 6.25- and 8.75-inch telescopes come only with small 6 × 30 finders, while the others feature 10 × 50s.

All Dark Star Custom telescopes use solid tubes made of blue polyvinyl chloride (PVC), a plastic derivative often used in piping and plumbing fixtures. Although fine for the smaller apertures, solid tubes prove quite heavy in the 10-inch-and-larger range. Each instrument rides on a Dobsonian mount made from 18-mm (about 0.7-inch) marine plywood, with altitude and azimuth bearings made from a combination of Teflon, nylon, and Formica. Owners report that motions in both axes are smooth, with little binding evident. None of the tube assemblies are supplied with carrying handles, although the Dobsonian rocker boxes include a pair.

The 10- through 16-inch Dark Star telescopes are also available outfitted with either JMI's NGC-miniMAX or NGC-MAX digital setting circles, both detailed in chapter 6. Called CompuDobs, these instruments come from the factory wired and ready to go.

Finally, if you are looking for a BIG computerized telescope, Dark Star offers the CompuDob 500, a 20-inch f/4 Newtonian reflector. Unlike its smaller siblings, the CompuDob 500 sports an open-truss optical assembly that breaks this huge instrument down to a size small enough to fit into many cars. The CompuDob 500 comes either uncomputerized or with a choice of JMI's NGC-miniMAX or NGC-MAX digital setting circles. Regardless of computerization, all include a 2-inch rack-and-pinion focuser, cloth light shroud, and 10 × 50 finderscope.

D&G Optical Company. In addition to its line of fine refractors, D&G offers a 20-inch f/15 classical Cassegrain reflector. This is *not* a portable telescope by any means; it is meant to be placed in a small observatory owned by a college, museum, or possibly an amateur. Optics are all hand figured and guaranteed to be diffraction limited. Each features a fully baffled aluminum tube, a large rack-and-pinion focuser with adapters for 2-inch and 1.25-inch eyepieces, and an 8 × 50 finderscope. Mountings are available separately.

Edmund Scientific Company. This outfit was once one of the most often mentioned sources for telescopes and accessories, but Edmund's share of the astronomical marketplace has dwindled over the last quarter-century. Many of their once-popular products are no longer sold, having been replaced by a new, limited assortment of astronomical gear. Most popular of all the Edmund telescopes around today is the 4.25-inch f/4 Astroscan 2001 rich-field Newtonian. The Astroscan is immediately recognizable by its unique design, which resembles a bowling ball with a cylinder growing out of one side. The primary mirror is held inside the 10-inch ball opposite the tube extension that supports the diagonal mirror and eyepiece holder. The telescope is supported in a three-point tabletop base that may also be attached to a camera tripod.

The primary mirror, advertised as being accurate to $1/8$ wave at the mirror surface, yields good star images when used with the provided 28-mm RKE eyepiece. As magnification increases, however, image quality degrades. This should come as no great shock, because small RFT Newtonians are really not suitable for high-power applications. Although the Astroscan does not come with a finder in the classic sense, it does come with a peep sight that permits easy aiming of the telescope.

Edmund also offers a 3-inch f/6 Newtonian designed to meet the needs of today's young astronomers. The heart of the instrument is a 3-inch f/6 primary mirror that is aluminized and overcoated. The manufacturer assures that the optics will perform "to the theoretical limits possible," stating that the primary and secondary mirrors are each figured to $1/8$ wave at the mirror surface. Like the Astroscan, these optics are permanently mounted and collimated, although the primary can be removed with its cell for cleaning. The 15-inch-long telescope tube is made from ABS plastic tubing, bright red on the outside and flat black to reduce stray light on the inside. It comes with a good-quality 1.25-inch RKE 15-mm eyepiece. Aiming the instrument is accomplished by sighting through a pair of tube-mounted eyelets, a more practical solution than the typically poor-quality finders supplied on most telescopes in this price range.

The telescope comes mounted on a fork equatorial mount and an adjustable metal tripod. The mounting's polar-axis angle is permanently set at 40° and is useful from midnorthern and midsouthern latitudes. The fork arms are a little flimsy but adequate for the weight.

Jim's Mobile, Incorporated (JMI). This outfit produces some of the most innovative large-aperture Newtonian reflectors on the market today. As a schooled mechanical engineer, I appreciate a product that emphasizes excellence in design and workmanship. That is why, when I first laid eyes on JMI's Next Generation Telescopes (NGT, for short), my heart skipped a beat! Of all the giant Newtonians on the market today, no other has the innovation or craftsmanship of the NGT series (Figure 4.3). But be forewarned: This innovation does not come cheaply. These are some of the most expensive telescopes in their respective size classes.

Two models currently make up the NGT line of telescopes: a 12.5-inch f/4.5 and an 18-inch f/4.5. Sibling resemblance is unmistakable, with each instrument sporting an open-truss tube mounted on a sturdy split-ring equatorial mount similar in design to the 200-inch Hale telescope atop Mount Palomar. All components are assembled using large knobs, making extra tools unnecessary. Unfortunately, the black knobs are not tethered to the telescopes, which increases the possibility of losing them at night.

The image quality obtained through the NGTs is consistently excellent thanks to their fine optics. Of those I have used, all sported mirrors with resolution limited only by seeing conditions and not by mirror imperfections. The

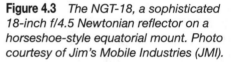

Figure 4.3 *The NGT-18, a sophisticated 18-inch f/4.5 Newtonian reflector on a horseshoe-style equatorial mount. Photo courtesy of Jim's Mobile Industries (JMI).*

NGTs are graced by all sorts of neat little gadgets, such as a rotatable nose assembly that permits comfortable viewing of any part of the sky. Bear in mind, however, that collimation may wander a bit when rotating the nose from one side to another. The NGT-18 also features a rotating azimuth base and polar-alignment scope to make setup easier. The polar scope is a recommended option for the NGT-12.5.

Two other options offered for NGTs are absolutely required if you observe from light-polluted or damp environs. One is a black shroud that wraps around the telescope's skeletal structure to prevent light and wind from crossing the instrument's optical path. It also slows dewing of the optics, a problem common to all open-truss telescopes, not just the NGTs. The other must is a tube extension that sticks out beyond the telescope's front and serves to slow fogging of the diagonal mirror, which lies *very* close to the end of the nose assembly, as well as to prevent stray light from scattering into the diagonal and washing out the scene.

JMI also offers the NTT-25, which is short for New Technology Telescope. The NTT-25 is a 25-inch f/5 folded Newtonian, the only telescope of its kind on the market. (Larger versions up to 40 inches in aperture are available on a custom-order basis.) Light entering the NTT bounces from the primary to an optically flat mirror mounted at the front end of the tube and parallel to the primary. The light then bounces from it back toward the primary but is redirected out the side of the tube by a flat diagonal mirror. All this cuts the instrument's overall length dramatically. Now instead of having to climb to towering heights to look through the eyepiece (as is the case with all 25-inch unfolded Newtonians), most observers need only a stepstool. The NTT-25 comes on a state-of-the-art clock-driven alt-azimuth mounting, while the deluxe NTT-25C incorporates something called a "field-rotation drive" that actually turns the eyepiece holder during long-exposure photography. Without this, the field of view will rotate as the alt-azimuth–mounted telescope tracks the sky.

The folded Newtonian design is not without its flaws, the biggest of which is the huge flat secondary mirror at the front of the tube that is required to redirect the light back toward the diagonal. In the case of the NTT-25, it measures 9 inches across—larger than some primaries! This translates to a central obstruction equal to 36% of the primary's diameter (13% by area), much greater than in any other Newtonian. Not surprisingly, the large central obstruction dramatically reduces image clarity and contrast when compared to a conventional Newtonian. The net result is a telescope with excellent optics delivering only a lukewarm optical performance.

Mag One Instruments. Looking for something different in a Newtonian reflector? Consider a Mag One telescope. Owner Peter Smytka has created the PortaBall 12.5-inch f/4.8 and newly introduced 8-inch f/6, two of the most unique telescopes to come along in years.

The PortaBall looks like the big brother of Edmund's Astroscan. Rather than employing a conventional design, the PortaBall centers around a hollow sphere that houses the primary mirror. The sphere, made from 7-gauge aluminum reinforced with a central flange, is hand sanded and finished with a smooth powder coat of white paint.

Mirrors for PortaBalls come from Wavefront Consultants and are acclaimed by owners for their excellence. All are coated with standard aluminum (88% reflectivity), although enhanced (96%) coatings are available at extra cost. Because the spherical mounting completely encloses the back of the primary, a conventional mirror cell is impractical. Instead, each primary mirror is glued in its mounting with small blobs of silicone adhesive, a technique that will prove troublesome when the mirror needs to be realuminized. A pair of black-anodized metal rings and a short tube segment of a composite material make up the telescope's upper-end assembly, which holds the focuser and secondary mirror. Six (as opposed to the usual eight) aluminum truss tubes join the upper assembly to the primary-mirror ball.

Are there drawbacks to the PortaBall? Owners could find only a few. One is that digital setting circles cannot be adapted to the unusual mounting design. Another is the issue of balance. Although the telescope is custom designed to maintain its aim with its owner's heaviest eyepieces (or no eyepiece) in the focuser, very heavy eyepieces, such as some from Tele Vue, Meade, and Pentax, might throw the balance off, especially when aimed toward the horizon. Finally, the price of PortaBalls, which is higher than conventional Dobsonians, may turn some people off. But most who have purchased a PortaBall say they are well worth it.

Meade Instruments. When Meade introduced its apochromatic refractors in 1992, it also unveiled a revamped line of Newtonian reflectors called Starfinders. Consumers can select models on clock-driven German equatorial mounts or Dobsonian-style alt-azimuth mounts. All represent excellent values.

Starfinders come in five aperture sizes: 6-inch f/8, 8-inch f/6, 10-inch f/4.5, 12.5-inch f/4.8 (only available on the Dobsonian mount), and 16-inch f/4.5, all with optical components made from fine-annealed Pyrex glass. Optically, Starfinders perform very well. Images are sharp and clear even at relatively high magnification.

At first glance, the Starfinder optical tube assemblies appear identical regardless of mounting. All have tubes made of spiral-wound cardboard and share the same 1.25-inch rack-and-pinion focusing mounts made from metal and plastic (except the 16-inch equatorial model, which has a higher quality 2-inch focuser), but subtle differences lurk beneath the surface.

One of the biggest differences is the primary-mirror mount. Equatorial Starfinders use superior open-design mirror mounts that allow the passage of air through the tube, promoting temperature equilibrium of the optics while minimizing image-degrading heat currents. The Dobsonians use mirror mounts fashioned from particle board that seal off the bottom of the tube. While providing good protection against dust, this design stops any air passage through the tube and as a result will retard the optics' reaching the ambient outdoor temperature, thereby degrading the image.

The Starfinder Dobsonian mounting uses Teflon pads against smooth plastic laminate for the azimuth axis and a pair of nylon pads against each hard-plastic altitude trunnion. Both up-and-down and left-to-right motions are smooth, with little binding. The mountings themselves are made of particle board covered with a plastic countertop-style laminate.

The 6-, 8-, and 10-inch equatorially mounted models are each held by a pair of flexible straps on German equatorial mounts atop 4-inch-diameter piers. The straps make it possible to rotate the tube for easier access to the eyepiece. Although the equatorial mount is fine for the 6-inch and adequate for the 8-inch, it begins to teeter under the weight of the 10-inch. Starfinders come outfitted with a recently redesigned DC spur-gear drive, passable for visual use as well as some simple astrophotography. Meade states that the drive will run for 50 hours on six AA (penlight) batteries.

The 16-inch equatorial Starfinder is a little different than its smaller siblings. Included with the package are a 2-inch focusing mount (an adapter for 1.25-inch eyepieces is included), an 8 × 50 finder, and an inexpensive 25-mm eyepiece. Optical quality is very good, equal to that obtained from many other, more expensive telescopes, although Meade's tall eyepiece holder requires a large (4-inch) secondary mirror, which tends to lower image contrast. The 16-inch comes on a German equatorial mount that rides low to the ground. Recently, Meade has incorporated a pair of plastic straps to secure the tube to the mounting's cradle. As mentioned previously, these let the user rotate the tube for easier access to the eyepiece. The straps also hold the tube more securely than if it were bolted directly to the mount, as was the practice with older 16-inch Starfinders.

To sum everything up, Starfinders are very good choices for amateurs who want well-made instruments without spending a fortune. If finances permit, I would strongly recommend the Starfinder equatorials over the Dobsonians. But beware that the 12.5- and 16-inch Starfinders, like all solid-tubed, large-aperture telescopes, are very heavy. Their weight can make transport and setup very difficult, especially for one person.

One other reflector belongs to the Meade stable of Newtonians. The Meade 4500, an imported 4.5-inch f/8 instrument, accepts standard 1.25-inch diameter eyepieces and comes with a 25-mm modified-achromat eyepiece (similar to a Kellner). This telescope is also supplied with a 6 × 30 finder, better than the 5 × 24 finders found on some competitors. Optically, the instrument is surprisingly good, although the mounting is wobbly. Readers should note that, for the same price as the Meade 4500, you can purchase a 6-inch Newtonian reflector on a Dobsonian mount, which is far more satisfying to use.

MorningStar Telescope Works. This new telescope manufacturer makes its debut in this edition. Like many other small-volume telescope makers, MorningStar handcrafts all of its Newtonian telescopes on a per-order basis. This has its pros and cons. On the pro side, you can customize your instrument to fit your particular needs and preferences, which allows a flexibility that is not possible with mass-produced take-it-or-leave-it instruments. But on the con side, delivery is liable to take several weeks, possibly months, and prices are apt to be higher.

MorningStar's smallest telescope is the MS-8, an 8-inch reflector available with either an f/6 or f/8 focal ratio. The MS-8 uses Pyrex mirrors from Orion Telescopes coated with 94% enhanced aluminizing. The mirror is secured with silicone pads to its oak mount, which in turn is attached to an outer rear ring. The focuser, a Crayford-style mount from AstroSystems, accepts both 1.25- and 2-inch eyepieces. The optics are held in a cardboard Sonotube that is reinforced with oak end rings and painted in your choice of colors. The Dobsonian mount is made from oak-veneer plywood, stained and varnished. The telescope can be rotated within its cradle for the best eyepiece angle, a nice feature that will be appreciated by short and tall observers alike, and motions in both altitude and azimuth are smooth thanks to the use of Teflon-on-Formica bearing surfaces. Overall, the MS-8 is an excellent choice for observers on a fairly limited budget.

MorningStar's flagship telescope is the MS-10, a 10-inch Newtonian that is unique in many ways. For one, its f/8 focal length is the longest of any 10-inch

Newtonian marketed today. The increased focal length shrinks the diagonal mirror's minor axis to only 1.3 inches, minimizing central obstruction (only 13% by diameter, or 1.7% by area) and maximizing image contrast. The optics in the MS-10, like those in the MS-8, come from Nova Optics.

Realizing that a 10-inch f/8 telescope is rather long, MorningStar owner Bob Duke chose to mount the optics in an open-truss tube rather than a more traditional solid tube. This way, even though the assembled telescope measures close to seven feet in length, it can still break down into a 2-foot cube (excluding the eight truss tubes, of course). Unfortunately, loose wing nuts are used to secure the truss tubes to the mirror box and eyepiece cage, which can prove troublesome when assembling the telescope in a dark field, especially when the temperature makes gloves a must. It's too bad captive hardware is not available for this model.

The MS-10 is mounted on a nicely crafted Dobsonian mount of Baltic birch and oak plywood. Stain and four coats of clear varnish serve both to preserve the wood as well as to let its richness show through. Bearings of Teflon and Ebony Star Formica are used throughout for smooth motions in both altitude and azimuth.

MorningStar also offers a 12.5-inch f/6 and 16-inch f/6, both larger versions of the MS-10, as well as an innovative clock-driven 10-inch reflector, the MS-10D. This latter incarnation incorporates the Dob Driver II computerized clock drive system by Tech2000 (see chapter 6). To couple the Dob Driver to the MS-10, Duke had to completely redesign the optical tube assembly; the end result is a telescope that is half tube and half truss. One advantage that the MS-10D has over the MS-10 is that the tube of the MS-10D can be rotated and balanced by moving the tube assembly in its wooden cradle. It may also be broken down for easy storage and transport, although it is not as compact as its undriven partner (still, no part is longer than 32 inches).

North Star Systems. The folks at North Star Systems offer three inexpensive Newtonian reflectors, a 6-inch f/8, an 8-inch f/6, and a 10-inch f/6. All are designed to compete with instruments of similar aperture sold by Meade, Celestron, and Orion, but they sell for about one hundred dollars less. Indeed, in many ways, they are less expensive versions of Orion's Deep-Space Explorers, as all North Star telescopes use Orion's plate-glass mirrors, although Pyrex mirrors (also from Orion) are available at extra cost. (For more thoughts on this, see the discussion under "Orion Telescopes (US)" on page 82.) The primaries are held in place with plywood mirror mounts that seal the ends of their tubes. While good at keeping dust out, a sealed tube also slows mirror cooling. All come with inexpensive-but-inconvenient helical focusers and single-vane diagonal-mirror holders.

Each optical tube assembly is held on a conventional Dobsonian mount with nylon-on-plastic bearing surfaces for movement in altitude and nylon-on-

Teflon for movement in azimuth. North Star mounts are made of plywood, which is less prone to chipping than the laminated particle board used by some companies but causes the mounts to lack the refined countertop finish of their laminated counterparts sold by other companies.

Obsession Telescopes. This company makes some of the finest large-aperture alt-azimuth Newtonians sold today. They are famous for their sharp optics, clever design, fine workmanship, and ease of assembly and use. All add up to a winning combination.

Five models constitute the Obsession telescope line: 15-, 18-, 20-, 25-, and 30-inchers! (Sorry, the 32- and 36-inch models are no longer available.) Customers may choose optics made by either Nova Optics or Galaxy Optics in the 18- through 25-inch models, while only Nova Optics are used in the 15- and 30-inch instruments. The secondary mirrors feature a nonaluminum dielectric coating that the manufacturer states has 99% reflectivity. The primaries come with either standard aluminizing (89% reflectivity) or, for an extra charge, enhanced aluminizing (96% reflectivity).

The overall design of the Obsessions (Figure 4.4) makes them some of the most user-friendly large-aperture light buckets around. Their open-truss tube design allows them to be broken down for easy transport to and from dark-sky sites (keep in mind, we're still talking about *big* telescopes that are much more unwieldy than smaller instruments). Once at the site, the 15-, 18-, and 20-inch

Figure 4.4 *The Obsession-20, an outstanding 20-inch f/5 Newtonian reflector on a state-of-the-art Dobsonian-style alt-azimuth mounting. Photo courtesy of Obsession Telescopes.*

Obsessions can be set up by one person in about fifteen minutes without any tools; the bigger scopes really require two people but are still quick to assemble. This feature is especially appreciated by those of us who know what it is like to drive to a remote site only to find out that the telescope cannot be set up without a certain tool inadvertently left at home. Even nicer is the fact that all hardware remains attached to the telescope, so it is impossible to misplace anything. The only way to lose Obsession hardware is to lose the telescope!

Another plus is the Obsession's low-profile Crayford focusing mount that is supplied by JMI. With the eyepiece held closer to the tube, a smaller secondary mirror can be used to keep the size of the diagonal (and the central obstruction) to a minimum.

Many conveniences are included with Obsessions. For instance, a small 12-volt DC pancake fan is built into the primary mirror mount to help the optics reach thermal equilibrium with the outside air more quickly. Another advantage is that each instrument comes with a pair of removable, metal wheelbarrow-style handles attached to a set of rubber wheels (except for the 15-inch, which is supplied with hand grips). The handles quickly attach to either side of the telescope's rocker box, letting the observer wheel the scope around like a wheelbarrow. Of course, these handles do not solve the problem of getting the scope in and out of a car. If you own a van, hatchback, or station wagon with a low rear tailgate, the scope may be rolled out using a makeshift ramp, but if it has to be lifted, two people will be required.

It should come as no surprise, given the aperture of these monsters, that their eyepieces ride high off the ground. When aimed toward the zenith, the eyepiece of the 20-inch, for instance, is at an altitude of about ninety-five inches. Unless you are as tall as a basketball player, you will probably have to climb a stepladder to enjoy the view. A tall ladder is a must for looking through the bigger Obsessions; the eyepiece of the 30-inch scope (when aimed at the zenith) is 11 feet off the ground. Like all open tubes, a cloth shroud is recommended when the telescope is used in light-polluted locations; Obsession offers optional custom-fit black shrouds for all of its instruments.

Nothing is small about the Obsession scopes. Besides large apertures, considerable weight, and substantial girth, their cost is quite heavy, too. Yes, they are expensive, but after all, they are a lot of telescope—and they remain the standard by which all other Dobsonian-style Newtonians are judged! For the observer who enjoys looking at the beauty and intrigue of the universe and has ready access to a dark sky, using Obsession telescopes can really become obsessive.

Optical Guidance Systems (OGS). For the die-hard (and well-to-do) amateur astronomer, Optical Guidance Systems offers some impressive telescopes that are worth a look. OGS specializes in medium- to large-aperture Newtonians and Cassegrains that rank among the finest instruments in their respective classes.

Looking at its Newtonian line first, OGS offers several different instruments ranging in size from an 8-inch f/6 to a huge 24-inch f/4. Each may be purchased either as a tube assembly only or complete with a pier-supported, heavy-duty German equatorial mount. In either case, each telescope comes with an oversized focusing mount with adapters for 1.25- and 2-inch eyepieces, a straight-through 8 × 50 finderscope, and a choice of one high-quality eyepiece. Though far more expensive than more popular brands, these high-end telescopes are perfect for the amateur who demands the absolute best.

Optical Guidance Systems is also one of the few companies in the amateur market to offer both Classical and Ritchey-Chretien Cassegrain reflectors, with models ranging in aperture from 10 to 32 inches. The Classical design is available with focal ratios around f/16, while the Ritchey-Chretien variants are around f/9. Perhaps the most important benefit of the Ritchey-Chretien design is its freedom from coma, which is especially important if the instrument will be used for astrophotography.

One of the strongest selling points of Optical Guidance System's telescopes is their superb optics, supplied by Star Instruments of Flagstaff, Arizona. Unlike some manufacturers, who make all sorts of optimistic claims about optical quality, OGS simply states that its telescopes are guaranteed to have a *final* peak-to-valley wavefront of 1/4 wave or better. Furthermore, OGS certifies this claim by testing the optics with an interferometer, thus eliminating any guesswork.

Orion Optics (UK). Not to be confused with Orion Telescopes of California, Orion Optics of Great Britain sells several different Newtonian reflectors for amateur astronomers. All include aluminum tubes, 1.25-inch rack-and-pinion focusing mounts, traditional aluminum mirror mounts, and German equatorial mounts. Each telescope tube is secured to its mounting with a pair of well-made metal rings that are intended to allow the user to rotate the tube until the eyepiece is in a comfortable position. The primary and secondary mirrors are all made in-house by Orion, with respondents consistently commenting on the optics' high quality.

Least expensive of Orion Optics' telescopes are those in the Europa line, which are available in 4.25-inch f/8 (Europa 110), 6-inch f/5 (Europa 150), 8-inch f/6 (Europa 200), or 10-inch f/4.8 (Europa 250) apertures. In each case, the instrument's mirrors are manufactured by Orion Optics and then shipped to Taiwan, where they are mounted into the optical assembly. Each comes with a 6 × 30 straight-through finderscope and two 1.25-inch eyepieces and is mounted on a Taiwanese-made German equatorial mount and adjustable aluminum tripod. A 10 × 50 finder and 12-volt DC clock drive are available optionally on all telescopes, while 2-inch, rack-and-pinion focusing mounts are offered for an additional charge on the 6-, 8-, and 10-inch Europas. Overall, the mounting is adequate for the 4.25- and 6-inch instruments but is far too weak for the extra weight of the larger models.

Though more expensive, Orion Optics' SX Newtonian reflectors come mounted on sturdier, homegrown German equatorial mounts that are sufficient for visual observing but are also not solid enough for long-exposure astrophotography. Consumers may choose from three apertures—6-inch f/6 (SX150), 8-inch f/6 (SX200), and 10-inch f/4.5 (SX250). All optical and mechanical components are manufactured in Orion's shop in Crewe.

The three telescopes in Orion Optics' GX series are optically identical to the SX instruments; only the mountings are different. The GX telescopes are supported by a variation of Vixen's Great Polaris (GP) German equatorial mount on a steel pedestal. These offer a greater degree of stability while also permitting the addition of dual-axis stepper motors for accurate tracking during long-exposure photographic sessions. The Vixen mount also offers the capability of computer control.

Orion Optics' finest instrument is the 12-inch f/5.3 DX300 Newtonian reflector. Like the others, the DX300 includes an aluminum tube and rotating rings but comes with an upgraded 2-inch rack-and-pinion focusing mount for larger eyepieces and a 10 × 50 right-angle finderscope. The supplied Vixen DX German equatorial mount is a heavy-duty version of the GX mount supplied with the smaller telescopes. The DX mount includes manual slow-motion controls and is designed to accept an optional dual-axis, stepper-motor drive system. The DX300 is a heavy instrument that is best left in place rather than transported to distant observing sites.

Orion Telescopes (US). Now operating its catalog sales and showrooms under the name Telescope and Binocular Center, Orion Telescopes of Santa Cruz, California, offers several Newtonian reflectors. The most popular of these are the Orion Deep-Space Explorers, which are available in both standard and Premium models.

Standard Deep-Space Explorers are sold in three sizes: the 6-inch f/6, 8-inch f/6, and 10-inch f/4.5. All three are aimed at beginning amateurs or hobbyists who are operating on a budget and want a large aperture for a minimal investment. Like their chief competition, Celestron's Star Hopper telescopes and Meade's Starfinders, Orion's standard Deep-Space Explorer reflectors minimize on frills and standard accessories to keep prices low. All of the telescope tubes are made of spiral-wound cardboard. The Dobsonian mounts, made from pressboard covered with a black laminate, are sturdy and offer excellent vibration dampening.

So how do they compare against the similarly priced Celestron and Meade Dobsonians? One of the biggest pluses of the latter is that their optics are made of Pyrex glass, a material superior to the plate-glass mirrors found in standard Deep-Space Explorers. Pyrex has a lower coefficient of thermal expansion than plate glass, meaning that it will stabilize more quickly under varying temperatures. Another, often-overlooked point in favor of Pyrex is that it is stronger and less prone to cracking or fracturing than plate glass, a major consideration if many people will be handling the instrument.

The latest standard Deep-Space Explorers include rack-and-pinion focusing mounts and fully adjustable four-vane spider mounts to hold the diagonal mirror in place. Both are far superior to the older models that used helical focusers and single-vane diagonal mounts. A 25-mm Kellner eyepiece is supplied with each telescope. This eyepiece is roughly comparable to those that come with Meade and Celestron Dobsonians but inferior to the 26-mm Plössl that they used to supply. Standard Deep-Space Explorers also come with a 5 × 24 finder that you will want to replace almost immediately with either an 8 × 50 finder (the extra weight may require a counterweight) or a 1× aiming device, such as a Telrad (see chapter 6).

Orion Premium Deep-Space Explorers are available in three sizes: 10-inch f/5.6, 12.5-inch f/4.8, and 16-inch f/4.5. All feature high-quality mirrors of Pyrex glass coated with enhanced (94%) aluminizing. Owners commend the excellence of the images at both low and high power once the optics have acclimated to the temperature of the night air. Unfortunately, the primary mirror mounts still tend to trap heat near the mirror, slowing the equilibrium process. The diagonal mirrors are all held in place with traditional four-vane spider mounts.

Although none come outfitted with a finderscope, all three of the Premium Dobsonians come with a 26-mm Plössl eyepiece and are available with a choice of four different focusing mounts. As might be expected, the final price varies by up to $100, depending on which you choose. The least expensive, a low-profile 2-inch helical focuser, proves inconvenient to use with eyepieces that are not parfocal. The best and, naturally, most expensive is the low-profile JMI NGF-3 Crayford focuser. In between are two rack-and-pinion focusers, which also work well, but try to hold out for the JMI focuser.

The Dobsonian mounts used with the Premium Deep-Space Explorers, like their standard counterparts, are made from particle board but come covered with a white countertop-style laminate finish. Teflon pads riding on Ebony Star Formica produce smooth motions in azimuth. The altitude axis, consisting of plastic bearings riding on Teflon pads, also moves smoothly with little effort.

Like the standard models, the tubes of the Premium Deep-Space Explorers are spiral-wound cardboard Sonotubes. Because of its weight, a solid-tubed telescope can become a transportation nightmare when it comes time to lug it out into the country. Orion readily admits that the tube weight of the 12.5-inch "makes having a companion for setup desirable," while with the 16-inch "you'll definitely need a friend to help set up; this is a very big telescope." All three Premium instruments feature integral padded handles built into the altitude bearings but still require a fair amount of strength and room to move from home to observing site. If transportability by car is not a major concern and you are looking for the most aperture for the buck, the Orion Premium Deep-Space Explorers are great choices.

Several other reflectors bear the Orion name. The most recent addition is the R200SS-GP 7.9-inch f/4 Newtonian. Like several other instruments in Orion's collection, the R200SS is actually manufactured by Vixen Optical Industries of Japan. (Readers in other countries should note that this same instrument is also available directly through local Vixen dealers as the GP-R200S.) Such a short focal length, while terrific for transportation, will cause coma to play havoc with images. Orion sells a suitable coma corrector to help ease the problem, but this adds another $100 to an already expensive instrument. The R200SS includes a 1.25-inch rack-and-pinion focuser that also features a built-in adapter for attaching cameras directly to the telescope, a nice touch that eliminates the need for a separate camera mount (note that you will still need a T-ring for your camera—see chapter 6). The instrument's aluminum tube is held to a Great Polaris (GP) German equatorial mount with a pair of rotatable rings, allowing the user to shift the tube for balance and viewing comfort. The mounting is supported by an aluminum tripod that, while not as vibration resistant as wood, is adequate for the task. Orion packages the R200SS with a 26-mm Plössl eyepiece and a small 6 × 30 finder but no clock drive. For that, you will have to pay $200 for a single-axis drive; a dual-axis drive adds up to about $300 extra. Although the Orion incarnation of the R200SS is too new for an adequate evaluation here, I have received favorable reports from owners of the Vixen version in other countries. But even though this instrument may be a fine performer, bear in mind that you could well end up spending hundreds of dollars more than the current catalog price.

The Orion SkyView 4.5, a 4.5-inch f/8 Newtonian, is nearly a twin to Meade's 4500 Newtonian and Celestron's First 114 Premium. Like those others, the SkyView 4.5 accepts standard 1.25-inch eyepieces, giving the owner a much wider and better selection of eyepieces from which to choose. (More on this in the next chapter.) The Achilles' heel of these instruments is the German equatorial mounting; although superior to those supplied with the department-store versions of this imported scope, its weak equatorial head and aluminum legs are just not up to the task. The SkyView 4.5 retails for considerably less than the Celestron C4.5 but lacks the Celestron's solid mounting.

Orion's twin ShortTube 4.5-inch reflectors each have an effective focal length of f/8.8 but are folded into tubes about half as long as standard telescopes. How do they do this? The effective focal length is stretched by permanently installing a Barlow/corrector lens in the base of the focusing mount, allowing a compact package to house a system with greater focal length.

The ShortTube 4.5 EQ is housed in an 18-inch-long tube and includes a tiny 5 × 24 finder, two 1.25-inch Kellner eyepieces, and a rack-and-pinion focuser. The telescope comes with a German equatorial mount set atop a wooden tripod that is a clone of the department-store mountings. In this case, however, the mounting may not be that bad because the telescope weighs so little.

The ShortTube 4.5 EQ Ultra retails for more money but also includes a couple of extra features. Like the ShortTube 4.5, the Ultra comes with two 1.25-inch eyepieces but substitutes a 6 × 30 finder and a marginally more substantial equatorial mounting. The Ultra's aluminum tube assembly is held to the mounting in a pair of adjustable rings so that the rack-and-pinion focuser can be turned to a more convenient viewing angle if desired.

Parallax Instruments. This is one of few companies around to offer large-aperture, longer-focal-length Newtonian reflectors. Five high-quality telescopes highlight its lineup, ranging in size from the 8-inch f/7.5 PI200 to the 16-inch f/6 PI400. Each instrument boasts a diffraction-limited Pyrex primary mirror coated with enhanced aluminum. Overall manufacturing quality is excellent. In addition, all Parallax telescopes include 2.7-inch, low-profile eyepiece focusing mounts (with adapters for 1.25-inch and 2-inch eyepieces); aluminum tubes; and 8 × 50 finders. Their low-profile focusers combined with higher focal ratios mean that Parallax telescopes can use small diagonal mirrors, creating the smallest central obstruction of any reflector on the market. This is a *big* plus for those primarily interested in viewing the planets. The only extras needed are a couple of eyepieces and an adequate mounting, supplied either by Parallax or another company (see chapter 6).

Sky Designs. From Colleyville, Texas, comes a trio of large-but-portable Newtonian/Dobsonian instruments by Sky Designs—a 14.5-inch f/4.5, an 18-inch f/4.5, and a 20-inch f/4 that all follow the popular open-tube theme. Connecting the upper secondary-mirror cage to the lower primary-mirror box are eight aluminum poles, all secured in place by hex nuts and bolts. Unfortunately, as supplied, the Sky Designs system requires using a small wrench to tighten everything up, which is inconvenient. I would suggest that owners substitute either threaded knobs, or at least wing nuts, to make the job a little easier.

Fit and finish on both the wooden primary-mirror box and the secondary-mirror cage are excellent, with birch veneer plywood used throughout. All primary mirrors in Sky Designs instruments are made by the company rather than by a third-party supplier. Each is tested both on the bench as well as in the final optical tube assembly to ensure quality.

The biggest drawback to Sky Designs telescopes is that the secondary mirror is glued directly onto a wooden dowel. Adjustment is made by turning two sets of opposing screws into the dowel, which is more difficult than conventional spider mounts. In addition, Sky Designs does not include a provision for attaching wheels to move the instrument around once it is set up.

Sky Valley Scopes. Snohomish, Washington, is home base for another new name in the large-aperture Newtonian/Dobsonian market. By trade, Sky Valley owner Ken Ward has worked in the auto-body, custom-painting, and boat-building industries for more than twenty-five years. He has applied techniques and materials from all three fields to create some of the most colorful telescopes sold today. All are made from a combination of hardwood, honeycomb-core fiberglass, fiberglass cloth, and carbon fiber. Fiberglass is famous for its rigidity, but this very quality may make it more prone to transmitting vibrations than, say, wood, which has excellent damping properties. Ward's use of honeycomb-core fiberglass sandwiched with fiberglass cloth should help alleviate any such problems.

Sky Valley telescopes come with apertures from 12 to 18 inches in size. All are open-truss Dobsonians, with each available in two varieties: Ultra Lights, which are less expensive, and Rotating Tubes. The mirror box of the Ultra Lights is similar to those of other plywood-based Dobsonians, except that it uses honeycomb and fiberglass materials. All attachment points are reinforced to prevent cracking and other stress-related problems. The more advanced Rotating Tube models enable the observer to turn the tube assembly within the mounting to bring the eyepiece to a more comfortable position.

The upper assemblies that hold the diagonal mirrors and focusing mounts are also made from honeycomb-core fiberglass and fiberglass cloth. The eight aluminum truss poles that attach the upper assembly to the mirror box are secured with plastic thumbscrews, which is a plus. Unfortunately, the hardware is not held captive to the telescope, making it easy to drop one or more on the ground (or onto the mirror, if left uncovered) during assembly. As might be expected, the Teflon and Ebony Star Formica bearings offer very smooth motions.

Stargazer Steve. Many of us who became interested in astronomy as children back in the 1960s and 1970s might well have owned a 3-inch f/10 Newtonian reflector manufactured by Edmund Scientific Corporation. It seemed to be the quintessential first telescope of the era. Although the Edmund 3-inch f/10 is no longer made, Steve Dodson, also known as Stargazer Steve, an award-winning amateur telescope maker from Sudbury, Ontario, Canada, has resurrected and improved the design to produce the Sgr-3. The Sgr-3 is an excellent first telescope for young astronomers. Images of the Moon, planets, and brighter deep-sky objects are sharp and clear. To aid the view further, Dodson gives the primary mirror an enhanced aluminizing coating to produce 95% reflectivity. Both the primary and secondary mirrors are permanently mounted and collimated.

The telescope tube is 31 inches (787 mm) long, made from a vinyl-coated composite cardboardlike material, and painted flat black on the inside to minimize stray-light reflections. The tube assembly is held on a simple alt-azimuth mount of birch plywood and maple finished with clear varnish. The mounting and tripod are both lightweight yet sturdy. Motions in both altitude and azimuth are smooth. A wooden knob adjusts the tension on the altitude axis, which is important when using eyepieces of differing weights. The Sgr-3 comes with a 1.25-inch diameter 18-mm Kellner eyepiece and a peep-sight for aiming.

Stargazer Steve also offers a 4.25-inch f/10 Newtonian/Dobsonian kit for those who wish to assemble their own instrument from precut parts. The kit includes all parts and hardware as well as a manual and even a step-by-step VHS videotape. Again, optical and mechanical performance are very good.

StarMaster Telescopes. A new name in the world of telescopes, but one that is attracting a lot of attention, StarMaster Telescopes is owned and operated by

Rick Singmaster in Arcadia, Kansas. Its telescopes come in a wide variety of sizes to suit nearly anyone looking for a well-made, Dobsonian-mounted, Newtonian reflector.

Least expensive of the StarMasters is the Oak Classic, a 7-inch f/5.6 reflector. The optical assembly centers around a well-crafted primary mirror that the manufacturer touts as "superb, typically 1/8th wavefront or better." Testimonials from owners serve as evidence that this is not just an empty claim. Standard equipment on the Oak Classic includes a 2-inch low-profile helical focuser that also allows rapid coarse focusing when changing between eyepieces: superior Crayford focusers are available as for an additional charge. The diagonal mirror, held in place with a traditional four-vane spider mount, comes with 96% enhanced aluminizing, while the Pyrex primary is offered with standard aluminizing. Another nice touch is the all-steel, open-design primary-mirror cell that allows rapid cooling of the optics as well as easy collimation. The tube is a composite material coated with a pebbled acrylic/enamel finish, and the cradle, rocker box, and ground box are constructed from oak plywood and finished with a durable, clear polyurethane coat. Teflon-on-Formica bearings offer smooth motions in both altitude and azimuth. The StarMaster Oak Classic costs about one hundred and fifty dollars more than some of the better-known "Econo-Dobs" in this size class, such as those by Meade, Celestron, and Orion, but it is money well spent.

For the intermediate-to-advanced amateur, StarMaster also offers seven large-aperture telescopes from 10 to 22 inches. Each is based on the now-familiar, eight-tube truss design riding large, semicircular bearings and set on a low-riding Dobsonian base, first introduced by Obsession Telescopes. Mirrors, typically from Pegasus Optics, come with certification attesting to their high optical quality.

Each truss-mounted StarMaster comes with a 2-inch Crayford focuser (NGF focusers from JMI are available as an option), Telrad 1× aiming device, cloth light shroud, dust cover for the secondary mirror, primary-mirror cover, and a carrying case for the truss tubes. The tubes themselves are made from aluminum and can be painted black upon request. Unlike Obsessions and Starsplitters, however, the truss tubes are not covered with foam pipe-insulation sleeving. I'd urge buyers to retrofit the naked tubes with foam sleeves. You'll discover why the first time you set the telescope up on a winter's night: Those aluminum tubes can get mighty cold! The foam also tends to dampen vibrations better than bare tubes. You can find foam pipe insulation at plumbing, hardware, and home-improvement stores. Each of the larger truss models features oak-veneer plywood construction finished with a durable clear polyurethane. Motions in both altitude and azimuth are smooth thanks to the Teflon-on-Formica bearing surfaces.

A unique standard feature of 14.5-inch and larger StarMasters (offered as an option for an additional charge on the 12.5-inch) is a detachable mirror cell. By removing the mirror (the heaviest component of a large-aperture reflector), moving the telescope about becomes much easier. Once at the chosen sight, the mirror cell is then reattached to the telescope with little effect on collimation. Detachable wheelbarrow-style transport handles are also available but at extra cost.

Starsplitter Telescopes. This company has expanded greatly since (perhaps partially as a result of?) *Star Ware*'s first edition in 1994. Owner Jim Brunkella continues to offer the Starsplitter II large-aperture series but has added the Starsplitter Compact and Compact II lines as well. All Starsplitters are mounted on well-crafted Dobsonian mounts that feature excellent woodworking. Teflon pads riding on Ebony Star Formica let the telescopes move smoothly in both altitude and azimuth. In addition, each comes with a low-profile focusing mount (your choice of models manufactured by either JMI, AstroSystems, or JSL) and a Telrad aiming device.

Let's begin with the smallest Dobsonian telescopes in the line, the Starsplitter Compacts (Figure 4.5), available in four different apertures from 8 to 14.5 inches. All feature mirrors supplied by Nova Optics. Starsplitter Compacts look like no other commercially available telescope. Most reflectors use either a solid tube or an eight-pole truss design, but Starsplitter Compacts are based around a unique two-pole frame that was first conceived and used by Thane Bopp, an amateur telescope maker from Missouri.

The "Boppian" approach is the ultimate in portability. The open design also has the advantage of letting the optics acclimate to the outside temperature rapidly, although the process is slowed some by the closed bottom of the mirror box. Of course, with the optics so exposed to the air, mirror dewing is a

Figure 4.5 *The Starsplitter Compact, with its unique, two-pole "tube" design.*

problem on damp nights. The optional nylon light shield will help to slow dewing as well as reduce glare and stray light from ruining image contrast, but one respondent noted that the shroud occasionally sagged into the light path.

The frame tubes are held to the mirror box and the front secondary/focuser board with four J-shaped bolts and rest against aluminum channels. Although the end result is rigid, setting up can really test your agility. In his review of the 10-inch Starsplitter Compact in the May 1996 issue of *Astronomy* magazine, John Shibley noted that assembling the secondary board to the frame tubes was awkward but that it did seem to get a little easier with practice.

Collimation of the primary mirror is achieved by turning three thumb knobs on the bottom of the mirror box. A couple of correspondents felt that collimation was difficult with their early-model Compacts owing to the weakness of the mirror cell, but the manufacturer now uses newly designed aluminum mounts to eliminate this problem.

Recently, Starsplitter introduced its line of Compact II telescopes, available in three apertures from 10 to 14.5 inches and all with mirrors by Nova Optics. Compact IIs follow the more conventional eight-pole truss design used by many other manufacturers. The truss tubes are held to the mirror box with J-bolts as in the Compact models, but wooden clamps secure the tubes to the focuser/diagonal upper assembly, making assembly much easier.

Along with the conventional truss design of the Compact II comes a traditional four-vane spider mount for the diagonal mirror, which proves to be more stable than the curved-vane diagonal holder used on the Compacts. The cylindrical top end of the Compact II helps prevent the diagonal from dewing over and makes attaching a light shroud easier. Despite outward appearances, the Compact II telescopes are actually lighter in weight than those in the Compact series, an important consideration if you will be hauling your scope around a lot.

Starsplitter also makes *big* telescopes; the Starsplitter II series includes instruments from 12.5 to 30 inches in aperture size. The 12.5-, 14.5-, and 30-inch versions come with mirrors by Nova Optics, while the others use mirrors from Galaxy Optics. Starsplitter II telescopes are also designed around open-truss tubes that allow for comparatively easy storage. All come with removable wheelbarrow handles for moving the instrument around prior to setup. Like the smaller Starsplitters, their design employs captive hardware, making separate tools such as wrenches and screwdrivers unnecessary for setting up and tearing down the instrument. The truss tubes are held to both the mirror box and the secondary-mirror assembly with thumbscrews and mating wooden blocks—a simple, useable design. If the overall width of the telescope is a concern, you should know that the truss tubes ride inside the mirror box, rather than outboard as on Obsession telescopes, adding another three or four inches to the width of the mirror box.

From my research for the first edition of this book, I decided to purchase an 18-inch Starsplitter II, reasoning that it was the largest telescope that could fit into my car at the time. So how has it held up? Overall, quite well. The telescope takes about ten minutes to set up and another few minutes to collimate by turning one of three large knobs behind the mirror. The Galaxy optics in my 18-inch are quite good, although the main mirror has a slightly turned edge. Though noticeable when performing the star test, most of the time my observing limits are set by atmospheric rather than optical problems.

Here's the bottom line: Starsplitter offers some intelligently designed, well-made instruments at some very attractive prices. If I were in the market for an 8- to 10-inch telescope, however, I would probably lean more toward the Compact II line rather than the Compact models.

Recently, Starsplitter began to import a pair of Taiwanese-made Newtonian reflectors intended for novice stargazers. Starsplitter Gems (an acronym for German equatorial mount) come in two apertures: a 4.5-inch f/8 and a 6-inch f/5. The former is virtually a twin to similar models sold by Orion Telescopes, Meade, and Celestron. Both Gems feature aluminum-tubed optical tube assemblies, 1.25-inch rack-and-pinion focusing mounts, and too-small 6 × 30 finderscopes and come with two Plössl eyepieces.

Image quality is acceptable, producing good views of the Moon, planets, and brighter deep-sky objects. Both instruments feature plate-glass mirrors held in adjustable, albeit closed, mirror cells; these, along with the telescopes' aluminum tubes, may retard temperature adjustment when the instruments are first brought outside from a warm room. Note that the 6-inch Gem is also available in a premium model, in which the Taiwanese plate-glass optics are replaced with Pyrex mirrors by Stellar Optical of Utah, though at a hefty additional cost of $250.

The mounting used with both the 4.5- and 6-inch Gems is a Taiwanese replica of Vixen's Polaris mount on an aluminum tripod. The telescopes are held in place by two adjustable rings that let users rotate the tubes for easier viewing. The mounts include built-in polar alignment sighting devices that make alignment with the celestial pole easier than it would be otherwise. Although neither comes with a motorized clock drive, both mountings include manual slow-motion controls; a clock drive may be retrofitted.

Price may be a stumbling block with the Gems, however. For the price of the 6-inch Premium model, you can purchase an 8-inch Meade Starfinder Equatorial, which also comes with Pyrex optics and a clock drive. Some may consider the Meade's cardboard tube a drawback, although with proper care, it should easily last for years.

Takahashi. In addition to its outstanding line of refractors, Takahashi offers several Newtonian reflectors: the 5.2-inch f/6 MT-130, the 6.3-inch f/6 MT-160, the 7.9-inch f/6 MT-200, and the 9.8-inch f/6 MT-250. Optical performance of all three is typical of Takahashi: exceptional! Images are sharp and clear with good contrast and little evidence of aberrations. Each comes with a smooth rack-and-pinion focuser designed to accept 1.25-inch eyepieces. All are supplied with 18-mm and 7-mm orthoscopic eyepieces and small finderscopes.

Mechanically, all of the MT reflectors ride in rotating tube rings atop nicely executed German equatorial mounts. The MT-130 comes with the EM-2 mount, complete with a motorized clock drive on the right-ascension axis and manual slow-motion control on the declination axis. Takahashi describes the mount as designed for visual observation, although some astrophotography is certainly possible. The MT-160 features the larger EM-200 equatorial mount, which includes dual motor-driven axes and is suitable for long-exposure astrophotography. Finally, the MT-200 and MT-250 come mounted on the impressive NJP German equatorial mount, which has dual motor-driven axes and multiple counterweights for fine balancing. It sits atop a steel pedestal rather than a wooden tripod like the others.

Takahashi manufactures several Mewlon Cassegrain reflectors based on the Dall-Kirkham version of the Cassegrain optical system. You'll recall from the brief discussion in chapter 2 that Dall-Kirkham Cassegrains use mirrors with simpler curves than either classical or Ritchey-Chretien Cassegrains yet still produce excellent results. Takahashi Mewlon instruments currently come in three aperture sizes, the 8.3-inch f/11.5 M-210, the 9.8-inch f/12 M-250, and the 11.8-inch f/11.9 M-300. The M-210 sits astride the EM-2 or, optionally, the EM-10 German equatorial mount, while the M-250 comes on the EM-200 mount standard. The M-300 is equipped with the EM-500 mounting.

Takahashi recently introduced its 8.3-inch CN-212 convertible Newtonian/Cassegrain reflector. Such a two-way telescope is possible because both the classical Cassegrain and Newtonian optical designs are based around parabolic primary mirrors. By swapping secondary mirrors, the CN-212 can change between an f/12.4 classical Cassegrain and an f/3.9 Newtonian. A keyed secondary-mirror mount helps to keep the secondary mirror in line, while the mating four-vane spider helps to maintain optical collimation after conversion. The CN-212 is available on either the EM-2 or EM-200 mount, the latter being recommended for long-exposure, through-the-telescope photography.

Lastly, Takahashi sells a line of specialized hyperbolic photo/visual astrographic reflectors that use hyperbolic concave primary mirrors, instead of the usual parabolic mirrors found in standard Newtonians, and four-element field corrector/flattener lens sets to yield sharp, flat photos of wide slices of the sky. The 6.3-inch f/3.3 Epsilon 160, 8.3-inch f/3.0 Epsilon 210, and 9.8-inch f/3.4 Epsilon 250 are optimally designed for astrophotography, but all may also be used with standard 1.25- and 2-inch eyepieces.

Tectron Telescopes. Some of the best large-aperture telescopes sold today bear the Tectron name, although they are facing ever-increasing competition from a growing field of manufacturers. Tectrons have changed little since the first edition of *Star Ware* was published in 1994. Back then, I summed up these telescopes in one word: superb. They still are.

Tectrons, like most other "higher-order" large-aperture Newtonians, feature open-truss optical tube assemblies set on finely crafted, Dobsonian-style, alt-azimuth mounts. Assembly of the 15-, 18-, and 20-inchers can be performed by one person, while the 24- and 32-inchers really require two. Once erected on their oak-plywood mounts, the telescopes are wonderfully stable and a joy to use. Tectrons are designed to be compact when disassembled and stored; even their largest 32-inch telescope stores in a 41" × 41" × 26" cube.

Tectron uses optics from a variety of sources, such as Galaxy Optics, Nova Optics, and Pegasus Optics. I recall an outstanding view of the Orion Nebula I had through a Tectron a few years ago at the Winter Star Party in the Florida Keys. Even planetary images are sharp and clear at full aperture (a quality that some "fast" Newtonians are not known for). All secondary mirrors used in Tectron telescopes have enhanced aluminum coatings to increase their reflectivity to 96%; the primaries come with standard aluminizing only. The 18-, 24-, and 32-inch models come with a black-cloth light shroud; a shroud is available for the 15- and 20-inch Tectrons at extra expense.

When not making telescopes, Tectron owner Tom Clark runs a computer-numerical-controlled (CNC) machine shop that primarily turns out components for both military and commercial aircraft, industries that require a high level of quality control. That same degree of precision is built into Tectron telescopes. All of their features come together to produce excellent instruments that are well regarded by their owners.

As I see it, Tectron telescopes have three shortcomings. One involves the closed mirror box, which hinders cooling of the optics. Large mirrors can take hours to acclimate to nighttime temperatures even in an open-ended box, let alone one that is closed. Another problem surfaces when moving Tectron telescopes around. The 20-inch and larger Tectrons offer snap-on casters as an option, but they do not work as well as the wheelbarrow-handle casters used by some others. (Clark doesn't recommend them, as he feels that 20-inch and larger telescopes should really be assembled by two people anyway.) Finally, Tectron's trusses are assembled using large knurled knobs. Although no tools are required, the knobs are not permanently tied to the telescope and can get misplaced easily at night.

Despite these weaknesses, Tectron telescopes represent a very good buy because they cost hundreds of dollars less than their chief competition, Obsession, Starsplitter, and Starmaster telescopes. As one owner put it, "Although the telescope doesn't look as 'sexy' as a Starsplitter or an Obsession, you can tell that a lot of thought went into its design and construction."

Catadioptric Telescopes

In this edition, I have decided to divide this section into two separate parts: one for Schmidt-Cassegrains and one for Maksutovs. In this way, readers can directly compare apples with apples and oranges with oranges, so to speak.

Schmidt-Cassegrain Telescopes

Celestron International. This company is renowned as the first to introduce the popular 8-inch Schmidt-Cassegrain telescope back in 1970. While only one basic model—the Celestron 8 (C8)—was sold back then, there are now many variations of the original design from which to choose, ranging from bare-bones telescopes to extravagant, computerized instruments. Celestron tells us, however, that all of its 8-inch SCTs share similar optics with the same optical quality, regardless of model and price. All feature Starbright enhanced coatings for improved light transmission and image contrast.

Images through Celestrons are usually good—sometimes excellent but sometimes less than excellent. Maintaining quality in a mass-produced instrument as sophisticated as a Schmidt-Cassegrain while also trying to maintain an attractive price is a difficult chore. Because of the comparatively large central obstruction caused by their large secondary mirrors, Schmidt-Cassegrains as a breed generally lack the high degree of image contrast seen in refractors and many Newtonians. It is also important to note up front that both Celestron and Meade (discussed later) Schmidt-Cassegrain telescopes suffer from something called *mirror shift*. To focus the image, both manufacturers chose to move the primary mirror back and forth rather than the eyepiece, which is more common with other types of telescopes. Unfortunately, as the mirror slides in its track, it tends to shift, causing images to jump. The current telescopes produced by both companies have less mirror shift than earlier models, but this problem is still evident to some degree.

Celestron offers several variations of the C8, with equipment levels varying dramatically from one model to the next. At the low end of the scale are the Celestar 8 and the Great Polaris–Celestron 8, or GP-C8.* The GP-C8 comes equipped with a high-quality 26-mm Plössl eyepiece, while only a so-so 25-mm SMA (Kellner) eyepiece is included with the Celestar. Both only offer small 6 × 30 finderscopes. Owners quickly discover they need to buy new finderscopes (an 8 × 50 would be nice) and a couple of better-quality eyepieces to make the packages complete.

The GP-C8, the next generation of the popular Super Polaris–C8 (SP-C8), comes with a tripod-supported Great Polaris German equatorial mount. The GP features an improved latitude adjustment system but is otherwise comparable to the older SP mounting. A motor drive is not supplied as standard, although a DC drive is available as an option. The GP mount and its mated, wooden tripod are surprisingly steady and prove quite adequate for supporting a telescope of this size.

The Celestar 8 is supplied on a fork equatorial mount, a DC spur-gear clock drive, and a "wedgepod," which incorporates in one unit both a fixed-height tripod and the wedge needed for tilting the telescope at the proper angle so that it can track the sky. Respondents seemed divided about the wedgepod concept,

*See note on page 110.

with many noting that the mounting was not as stable as they had hoped. Others were fond of the wedgepod, saying "I'm not very gadget oriented, but this telescope is a peach to set up; even I can screw in a couple of bolts." The self-contained clock drive is powered by a single 9-volt alkaline battery, a great feature for observers who spend most of their time under the stars far from civilization. Although fine for the visual observer who just wants to keep a target somewhere in the field for an extended period, the drive is not as accurate as a worm-gear drive. For that, you need the Celestar 8 Deluxe, which I'll describe in a bit.

For those who want their telescope to tell them where to go, the Celestar 8 Computerized comes with factory-installed Advanced Astro Master digital setting circles. Once set up, the Astro Master lets the observer select from more than ten thousand sky objects. Choose a target, and the computer will steer you to its location in the sky. Pointing is manual; for motorized "go-to" slewing, see the following discussion of the Celestron Ultima 2000. Polar alignment is not necessary in order to use the Advanced Astro Master, although it is still required for tracking.

Both Celestar instruments are also available in deluxe editions. The Celestar Deluxe and Celestar Deluxe Computerized offer several improvements, including a 26-mm Plössl eyepiece, a 9 × 50 straight-through finder, drive motors on both the right-ascension and declination axes, a standard equatorial wedge and adjustable-height tripod, and a worm-gear drive system with a built-in Periodic Error Correction, or PEC, circuit for greater tracking accuracy. Theoretically, a worm-gear clock drive should track the stars perfectly if it is constructed and accurately aligned to the celestial pole, but this is not the case in practice. No matter how well machined a clock drive's gear system is or how well aligned an equatorial mount is to the celestial pole, the drive mechanism is bound to experience slight tracking errors that are inherent in its very nature. These errors occur with precise regularity, usually keeping time with the rotation of the drive's worm. The PEC eliminates the need for the telescope user to continually correct for these periodic wobbles. After the observer initializes the PEC's memory circuit by switching to the record mode and guiding the telescope normally with the hand controller (typically a five- to ten-minute process), the circuit plays back the corrections to compensate automatically for any worm-gear periodicity. Be aware, however, that the PEC will *not* perform its function without being retaught every time the clock drive is switched on, and neither will it make up for sloppy polar alignment. (See further discussion on this under "Meade Instruments" in this section.)

Riding the growing wave of CCD astronomy, Celestron offers the Fastar 8, a Schmidt-Cassegrain instrument designed specifically for use with the PixCel 255 CCD imaging system (sold separately), manufactured by Santa Barbara Instrument Group (SBIG) and described in chapter 6. This unique instrument is designed to hold the camera either conventionally behind the primary mirror or in front of the telescope in place of the secondary mirror. The former configuration, with the conventional secondary mirror in place, operates at f/10 like other Celestron Schmidt-Cassegrains. But by replacing the specially outfitted secondary-mirror holder with the CCD imager, the Fastar records images at an effective focal ratio of f/1.95. This, coupled with the tremendous light-gathering ability of a CCD imager, creates an optical system that can record faint objects in a matter of seconds rather than minutes or even hours as with conventional astrophotographic methods. And although many CCD imagers are plagued with small fields of view, the PixCel covers a 0.33° × 0.5° area when operated at the Fastar's f/1.95 focus. The only drawback to this unconventional setup is the large central obstruction caused by the CCD imager (41% by diameter), which lowers image contrast even more than usual. Image processing should be able to compensate for much of this effect, but it is still a consideration.

This is more than just a CCD imaging system, however; it can also be used for visual observing and conventional astrophotography just like all other Celestron 8-inch instruments. The Fastar comes with the same standard accessories listed above for the Celestar 8 Deluxe as well as a digital counter linked to the focuser so that precise settings may be easily duplicated during future sessions. Make no mistake, the Fastar 8/PixCel 255 CCD imager is an expensive investment, but it holds great promise for anyone who wants to surf the CCD-imaging wave.

One of the most sophisticated telescopes on the amateur market is Celestron's long-awaited flagship, the Ultima 2000 (Figure 4.6), which features a sophisticated, self-contained, computerized, worm-gear clock drive with a PEC circuit for improved tracking accuracy. The drive is rated to run on eight AA alkaline batteries for 24 hours under normal usage, with a built-in voltmeter to constantly monitor the batteries' condition.

For visual observing, the telescope is mounted with its fork mount in alt-azimuth mode. After being initialized (accomplished by aiming the telescope

Figure 4.6 *The Ultima 2000, Celestron's flagship 8-inch Schmidt-Cassegrain instrument.*

at two known bright stars and pressing a button), the onboard computer will track the instrument across the sky. The Ultima 2000 also features automatic go-to operation; that is, once an object is selected using the hand controller, the telescope will quickly and quietly slew to its location automatically. Users note that in practice, pointing accuracy is within about 15 arc-minutes, with some problems experienced if the telescope is aimed straight at the zenith (in the alt-azimuth mode).

All this assumes that the instrument is balanced properly; the Ultima 2000 simply will not work well if the counterweighting isn't nearly perfect. If you switch from, say, a lightweight Plössl eyepiece to a heavy 2-inch wide-field eyepiece, you will probably have to adjust the counterweighting, or the Ultima may well go into an endless slew, never reaching its destination. Owners note that this is a mildly frustrating experience, but it becomes easier to deal with over time. The good news is that the telescope will not lose its directions while doing the counterbalancing, because it can be moved manually without forgetting its place in the sky. Of course, if you want to find objects the old-fashioned way (i.e., by eye) the Ultima 2000 comes with a 7×50 straight-through finder that is adequate, but can be difficult to look through when aimed near the zenith.

Amateurs who need to consider portability over aperture size should consider Celestron's C5, a 5-inch f/10 Schmidt-Cassegrain instrument. The C5 is available in two variants: the C5 Spotting Scope and the C5+. Because both feature identical optical components enhanced with Starbright coatings, the only real difference between the models is in their mountings. The spotting scope is sold unmounted but comes with a built-in tripod socket for attaching it to a camera tripod. The C5+ comes mounted on a single-arm fork mount featuring a built-in clock drive powered from a single 9-volt battery, which Celestron claims will last about fifty hours. The C5+ optical tube assembly is mounted on a dovetail plate that permits easy removal for transport, a nice touch for jet-setting astronomers.

For those who want a little more aperture, Celestron makes the CG-9¼ and Ultima 9¼. Both gather 34% more light than their 8-inch relatives and are excellent performers. Their optical tube assemblies are identical, the only difference being in the mounting. Quality of construction is high, with the aluminum tube painted with Celestron's trademark smooth, gloss-black finish. A digital readout counter, similar to that found on the Fastar 8, is incorporated into each of the Celestron 9.25-inch f/10 twins.

The Ultima 9¼ is fork mounted and comes with a wedge and adjustable tripod. The electronics are powered for up to 24 hours by a single 9-volt battery for totally cordless operation; an optional power cord can connect the telescope to an external power source. The accuracy of the drive without the PEC circuit is fine for visual observing, while using it maintains the very precise tracking rate needed for long-exposure astrophotography. Celestron packages the Ultima 9¼ with a hand controller, a 26-mm Plössl eyepiece, 1.25-inch star diagonal, and an inadequate 6 × 30 finderscope; a declination drive motor is available at extra cost.

The CG-9¼ is mounted on a high-quality German equatorial mount manufactured for Celestron by Scott Losmandy of Hollywood General Machining. The optical tube assembly is attached to the equatorial mount's head by means of a 17-inch-long dovetail bar that spans the length of the tube. This freedom lets the user set telescope balance precisely—a big plus when adding cameras or other accessories. Workmanship is outstanding, with each component of the mounting machined from either stainless steel or aluminum. Everything comes together to produce one of the smoothest-moving mounts in the size class. But as good as the equatorial mount is, the aluminum tripod used to support it is weak. Even with the legs fully collapsed, the tripod is barely able to hold the telescope steady enough for visual observations.

Still hungry for even more aperture? The Celestron Ultima 11, an 11-inch f/10 SCT, is basically an overgrown version of the Ultima 9¼, sharing all of its strengths (portability, extensive choice of accessories, etc.) and weaknesses (mirror shift, relatively low image contrast, etc.). Housed in the base of the Ultima 11 is the same accurate drive system that is packaged with the Ultima 9¼, featuring a stainless-steel worm gear and matching worm. The mounting was redesigned several years ago to improve the rigidity of the fork tines, wedge, and tripod, giving superior performance compared to older models. The instrument also comes with a 7 × 50 finder, a 1.25-inch prism diagonal (it used to come with a 2-inch diagonal), a 26-mm Plössl eyepiece, and a clock-drive hand controller as standard equipment, but again, a declination motor will cost you a little more.

Although the optical assembly and accessories are the same as the Ultima 11's, the Celestron CG-11 comes mounted on a rock-steady German equatorial mounting made by Losmandy. Telescope and mounting slide together using a dovetail bar, like the CG-9¼, for great adaptability when adding accessories.

The G-11 mount is one of the few commercial units that successfully combines portability with stability in one neat parcel. The dual-axis drive system marries two 5.6-inch worm-gear matching stainless-steel worms to a crystal-controlled circuit that can run on either 115 volts AC or 12 volts DC. A PEC circuit is also included to help the user correct for any periodic gear error in right ascension, while the declination drive features a unique Time Variable Constant (TVC) setting to eliminate backlash. One of the neatest features of all is the mount's optional polar-alignment telescope. Through the illuminated reticle, the observer sees patterns of the Big Dipper and Cassiopeia etched into it. When the patterns line up with the stars in the sky, the mounting is aligned. All of these features, and more, combine to make the Celestron CG-11 the most versatile large-aperture SCT available today.

By adding extra counterweights to help balance the additional weight, Celestron also uses the G-11 mounting to support its CG-14, a 14-inch f/11 Schmidt-Cassegrain telescope. The CG-14 is truly an observatory-class instrument that can also be made transportable with two or more people. Unfortunately, the entire package taxes the upper capacity limit of the G-11 mount and, in fact, can prove a little wobbly depending on whether any accessories, such as a CCD camera, are attached.

Meade Instruments.   Meade opened its doors in 1972 as a mail-order supplier of small, imported refractors. Its first homegrown instruments were 6- and 8-inch Newtonian reflectors, but it was to make its mark with the introduction of the Model 2080 8-inch Schmidt-Cassegrain telescope in 1980. Since then, Meade has grown to become the world's largest manufacturer of telescopes for the serious amateur.

Since Meade's beginnings, its bread and butter have been its line of 8-inch Schmidt-Cassegrain telescopes. Today, it offers three variations: the LX10, the LX50, and the LX200, in order of increasing price and sophistication. Each includes the same f/10 optical tube assemblies, although the LX200 is also available as an f/6.3 instrument, and all include fully "super multi-coated" optics. Meade's light transmission and image contrast are comparable to Celestron's, with most owners rating the optical quality as good to excellent.

The Meade 8-inch f/10 LX10 is designed for those who need the portability of a fork-mounted Schmidt-Cassegrain but may not have the money (or desire) for an electronics-laden instrument. Unfortunately, its fork arms are not as substantial as those of the LX50 and LX200 and therefore do not dampen vibrations as well. Built into the base of the LX10 mount is a DC-powered, worm-gear, right-ascension clock drive that is powered by four AA batteries; a declination motor drive is available optionally. Included with the LX10 is a much-too-small 6 × 30 straight-through finderscope and a 25-mm MA (modified achromat, similar to a Kellner) eyepiece. Although the manufacturer also includes an equatorial wedge with the LX10 base price, the mating tripod is sold separately. Many purchase the fixed-height LX10 Field Tripod, but the adjustable-height Field Tripod used with the LX50 and LX200 is much sturdier and well worth the extra cost.

Meade's moderately priced 8- and 10-inch f/10 LX50 series of Schmidt-Cassegrains are mounted on the same fork mounts as the LX200s, although they do not feature the same sophisticated electronics. As such, they must be polar aligned before the instruments will track the sky, as with any telescope that relies on traditional equatorial mounting. What they do offer, however, is an accurate worm-gear, DC clock-drive tracking system powered by either six AA batteries or directly from a car's cigarette lighter, the latter using a 25-foot cord that comes standard. Connections for adding digital setting circles, such as Meade's own Magellan II system, are built into the base and one fork arm. Again, this will not convert an LX50 to an LX200; the telescope must still be pointed manually using cues from the Magellan's hand controller.

Meade did it again by introducing the LX200 series of 8- and 10-inch Schmidt-Cassegrain telescopes, both available with either f/6.3 or f/10 focal ratios. All LX200 series scopes are mounted on computer-controlled fork mounts that automatically slew them across the sky from object to object simply by entering the desired target into the handheld controller. After the telescope is set up, its computerized drive system is initialized by aiming toward 2 of the 351 stars in the LX200 database (only 1 star is needed if the telescope is level and the observing site's location is known). Unlike the LX50, the LX200 will automatically slew to each of the more than sixty-four thousand objects stored in its memory at rates up to 8° per second. The LX200 is not as quiet or as fast as Celestron's Ultima 2000, which slews at up to 15° per second, but is much more tolerant of weight imbalances thanks to the use of higher-torque motors. But the LX200 requires 18 volts DC and so cannot be plugged directly into a 12-volt cigarette lighter. Instead, the LX200 comes with a 115-volt AC adapter; an optional (and expensive) 12- to 18-volt adapter is available. Some say that they have powered their LX200 with just 12 volts DC, but Meade does not recommend this.

The LX200 is normally set up in alt-azimuth configuration, which proves much steadier than tilting the fork mount by an equatorial wedge. The only shortcoming is that for long-exposure photography, alt-azimuth-induced field rotation can only be eliminated by using an equatorial wedge or the newly introduced Meade #1220 Field De-rotater. Although Meade does not supply a wedge with any of the LX200s, the base is designed to fit on Meade's standard-design wedge or its beefier Superwedge.

Like Celestron's Ultima 2000, the LX200's Smart Drive features a PEC circuit that lets the user compensate for minor periodic worm-gear inaccuracies (see the description under "Celestron" earlier in this section). The Celestron must be reprogrammed each time the drive is switched on, but the Meade Permanent PEC remembers the steps needed to compensate for the inaccuracies forever.

As for choosing between focal ratios, I would almost invariably advise against the f/6.3. The reason is simply a matter of image contrast. Because of the large central obstructions from their secondary mirrors, all SCTs suffer from lower image contrast than refractors and most Newtonian reflectors. Therefore, it makes sense to make the secondary mirror as small as possible. To achieve the faster focal ratio, however, the Meade f/6.3 scopes must use a larger secondary mirror than the equivalent f/10s, thereby decreasing contrast. The 8-inch f/6.3 has a 3.45-inch central obstruction (18.6% by area, 43.1% by diameter), compared to 3.0 inches for the f/10 (14.1% by area, 37.5% by diameter); the 10-inch f/6.3's central obstruction measures 4.0 inches (16.0% by area, 40% by diameter), while the f/10's measures 3.7 inches (13.7% by area, 37% by diameter). If focal ratio is of concern to you (attention, photographers), consider purchasing an f/10 SCT and a Meade or Celestron Reducer/Corrector attachment that cuts the focal ratio to f/6.3 (reviewed in chapter 5).

Meade also produces 12- and 16-inch f/10 Schmidt-Cassegrain telescopes on overgrown LX200 mountings. Impressive in both their aperture and their weight, these observatory-class instruments are also transportable with difficulty. The 12-inch LX200 sits atop the Giant Field Tripod, while the 16-inch is available on either the Supergiant Field Tripod, the Permanent Alt-azimuth Pier, or the Permanent Equatorial Pier, the latter two designed for installation in an observatory. Operation is impressive, and although the optics are usually quite good, their performance still does not compare with a well-engineered Newtonian reflector of similar aperture. Of course, most large-aperture Newtonians are supported by Dobsonian mounts, making long-exposure astrophotography much less practical than with an LX200.

So which brand is better: Celestron or Meade? That question has been pondered by amateurs now for more than two decades. Both companies have been criticized openly in the past for poor quality, but they seem to have cleared up most of the deficiencies. Now both companies produce optics of about equal quality, although most observers agree that Schmidt-Cassegrains universally produce inferior images compared to those obtained through high-quality refractors and Newtonians. The Meade 8-inch LX10 and Celestar 8 are both competent, no-frills telescopes, although at a price in excess of one thousand dollars, I am hesitant to label them "beginners' telescopes." Personally, assuming the optics are comparable, I like the LX10 better for its improved mounting. Meade's LX50 and Celestron's Celestar 8 Deluxe are nice midrange instruments, the better choice depending on price and availability. Celestron's 8-inch Ultima 2000 is faster and quieter than Meade's 8-inch LX200, but the Meade is less sensitive to exact balancing. For larger mountings, I still like the CG-11. The CG-14 is too heavy for its mounting, so in that case the mammoth Meade 16-inch LX200 gets the nod. But let's not forget a new offering (at least to us here in the United States) imported from Vixen by . . .

Orion Telescopes (US). This firm recently entered to the Schmidt-Cassegrain fray by introducing the Vixen VC200L 8-inch f/9 telescope. After just one look, you can see that the VC200L is something completely different. Vixen calls its optical system VISAC, which stands for Vixen Sixth-Order Aspherical Catadioptric. Unlike conventional SCTs, the VC200L does not use a corrector plate. Instead, it relies on a compound-curve, parabolic primary mirror; a convex secondary mirror; and a three-element corrector lens built into the focuser to bring an image into focus. The secondary mirror is positioned in the front of the open tube with a four-vane spider mount that is similar to those used in traditional Cassegrain reflectors.

Other notable features on the VC200L include a rack-and-pinion focusing mount, which eliminates image shift common with other SCTs, and a slide bar that allows the user to balance the telescope assembly on its mounting. The venerable Great Polaris German equatorial mount and an aluminum tripod are used to support the 15-pound VC200L. While adequate, the aluminum tripod is apt to transmit more vibration than the wooden legs used with the Celestron GP-C8. Also included are a small 6 × 30 finder, a 26-mm Plössl eyepiece, and a 1.25-inch star diagonal. An optional adapter ring is available for 2-inch star diagonals and eyepieces, as is a motorized clock-drive system. Also optional is the SkySensor 2000 motor-driven pointing system (designed to fit any Great Polaris or older Super Polaris mount, not just the one used with the VC200L), which will automatically slew the telescope across the sky after selecting one of the 6,500 objects in its built-in database.

Although the VC200L is a newcomer to the United States, it has a long track record in other countries. One respondent from Australia notes that his produces "perfect images both visually and photographically. The VC200L has superior overall design; no other Schmidt-Cassegrain comes close to its superlative optics. Images are razor sharp even at 400× when atmospheric conditions permit." On the other hand, one new owner here in the United States found that he could never focus star images, but as another owner explains, "The VC200L comes with a 100-mm-long extension tube for direct viewing or for attaching a camera. Users must take this off, screw in the 1.25" eyepiece adapter, then slip in the star diagonal; works every time!" The only drawback is the price, which is higher than comparably sized Meade and Celestron instruments.

Maksutov Telescopes

Ceravolo Optical Systems. A pair of outstanding Maksutov-Newtonian telescopes, the 5.7-inch f/6 HD145 and 8.5-inch f/6 HD216, represents this company's current offerings. Both of these fine instruments feature all-aluminum construction and low-profile helical focusing mounts for both 1.25- and 2-inch eyepieces. Unlike inexpensive helical mounts, these custom focusers allow rapid, coarse focusing as well as smooth, fine focusing. The HD145 and HD216 Mak-Newt are unhindered by coma, astigmatism, and other aberrations that plague lesser instruments. Contrast is also very good, thanks to small secondary mirrors, a 1-inch minor axis on the HD145 (central obstruction only), and a 1.5-inch minor axis on the HD216. When mated with a Tele Vue 22-mm Panoptic eyepiece, for instance, the HD145's view encompasses 1.8°, nearly four times the diameter of the full Moon, with pinpoint star images seen across the entire field. At higher powers, fine planetary detail comes through clearly, rivaling that seen with apochromatic refractors of similar aperture and focal ratio but costing double or triple the price. Few telescopes are able to compete with such impressive performance.

A caveat: Neither the HD145 nor the HD216 is available from stock but rather is constructed on a per-order basis only. As a result, delivery time ranges between six months to more than a year. The telescope as delivered does not include an eyepiece, finderscope, or mounting, although various Astro-Physics and Losmandy equatorials customized for the instruments are available, as are tube rings to retrofit an existing mounting. For those looking for the performance of a short-focus apochromatic refractor but with a little different spin, a Ceravolo Mak-Newt may be for you.

Intes. These Russian/Ukrainian-made instruments have gone through a number of changes since they were first introduced a few years ago. At first, Intes telescopes suffered from inconsistent optical quality, but since then they have certainly improved. Some inconsistency is still evident, but overall they deliver what they promise.

The smallest members of the Intes line of Maksutov telescopes are the 6-inch f/10 MK 67 and 6-inch f/6 MK 69. Both include enhanced and multi-coated optics that promise high-contrast views of planets and deep-sky objects alike. Image shift when focusing is eliminated because Intes uses a 2-inch Crayford-style focuser to bring images into focus instead of shifting the primary mirror as in most other catadioptric telescopes. Both include 7 × 50 finder-scopes and carrying handles but are sold mountless. Although the MK 67 is available only as a Maksutov-Cassegrain, the MK 69 is available in either a Maksutov-Cassegrain or more expensive Maksutov-Newtonian configuration. Intes also offers the MK 91 9-inch f/13.5, which is, in effect, a scaled-up version of the MK 67. The heart of this instrument are primary and secondary mirrors made from Zerodur, a special glass with excellent thermal characteristics. Both mirrors are coated with enhanced (96% reflectivity) aluminizing for bright views.

Intes-Micro, a subdivision of Intes, manufactures three interesting f/10 Maksutov-Cassegrains from 6 to 10 inches in aperture size. All are closer in form to other telescopes of this genre, with internal image focusing accomplished by moving the mirror along a ball-bearing–supported track. Although this will inevitably lead to some degree of image shift, it also allows owners to use standard accessories from Meade and Celestron. A mirror-style star diagonal, 30-mm Kellner eyepiece, 2.4× Barlow, and 12 × 55 straight-through finder are included as standard equipment. You will probably want to upgrade the eyepiece and finder in fairly short order.

Meade Instruments. Of all the new telescopes introduced recently, none has attracted more attention than the little Meade ETX, a 90-mm f/13.8 Maksutov-Cassegrain designed for maximum portability while also delivering outstanding images. It certainly succeeds on both counts, and at a terrific price. Images are absolutely textbook perfect, with very good contrast and clear diffraction rings. Focusing is precise with no mirror shift detected, giving some wonderful views of brighter sky objects such as the Moon and the planets. To quote one owner, "The Meade ETX delivers optical performance well beyond its price class." The optics are housed in a deep-purple aluminum tube that is smooth and nicely finished.

The ETX is available in two versions, the equatorially mounted Astro (see Figure 6.8) and the unmounted Spotting Scope. Both feature the same optical-tube assembly. A metal plate on the bottom of the tube can be used to attach the ETX Spotter to any standard photographic tripod. The Spotting Scope also includes a 26-mm Meade Super Plössl eyepiece, a tiny 8 × 21 finderscope, and a very nice screw-on dust cap as well as a 45° erecting diagonal prism for upright terrestrial views. Unfortunately, the finder will likely prove unusable for most users simply because it is mounted so close to the tube. I found it difficult, if not impossible, to look through the telescope as it raised in altitude, causing my nose to frequently scrunch up against the eyepiece. It might be best to replace the finder with one of the smaller 1× aiming devices described in chapter 6 or make your own using the plans in chapter 7.

The ETX Astro includes the dust cap, eyepiece, and finder that are supplied with the Spotter and comes mounted on a miniaturized, clock-driven, fork equatorial mount made mostly from molded plastic. The DC clock drive runs for more than fifty hours on three common AA batteries. Some respondents commented on the amount of backlash in the drive gears, but once it works itself out, the drive tracks well, keeping objects in view for half an hour or more. The model I tested had little problem with the clock drive.

Meade introduced its 7-inch f/15 Maksutov-Cassegrain telescope (Figure 4.7) in 1995, dressed as either an LX50 or LX200. As with Meade's Schmidt-Cassegrains, the only difference is the mountings; the Maksutov optical tube assemblies are identical. And those optics are excellent. The views through Meade 7-inch Maks impress me as sharper and clearer than those obtained through their Schmidt-Cassegrain counterparts. Image contrast is much higher and their focus snaps in, with no appreciable mirror shift evident in the several models that I tested. Each Mak includes a built-in fan to help acclimate the optics to the ambient outdoor temperature, speeding up the time between setup and optimal optical performance.

Perhaps the Mak's biggest drawback is its tube length. The Maksutov uses the same fork mount as the Meade 8-inch SCTs, but its tube is 3 inches longer (19 inches versus 16 inches), making it impossible to stow between the fork tines during transport and storage. Photographers should also note that the Mak's slow focal ratio of f/15 yields narrower fields and requires longer photographic

Figure 4.7 *Meade's 7-inch LX200 Maksutov cata-dioptric telescope.*

exposures than 8-inch Schmidt-Cassegrains, but if I were in the market for an 8-inch SCT, I would give the Meade 7-inch Maksutov *very* strong consideration.

Orion Optics (UK). In addition to selling several different Newtonian reflectors, Orion Optics of Great Britain is the only company to manufacture Schmidt-Newtonian catadioptric telescopes. The XSN200 and GPX200 are both 8-inch f/4 instruments that come packed in an aluminum tube measuring only 27 inches in length. Both feature identical optical assemblies wrapped in painted aluminum tubes and include 1.25-inch rack-and-pinion focusing mounts and 6 × 30 finderscopes. The XSN200 is mounted on Orion's own German equatorial mount, while the GPX200, on Vixen's Great Polaris (GP) mount, is more suitable for long-exposure photography or imaging using a CCD camera.

Orion Telescopes (US). This firm has just recently thrown its hat into the Maksutov ring by introducing the Argonaut 150, a 6-inch f/12 Maksutov-Cassegrain instrument. The Argonaut is manufactured in Russia and bears more than a passing resemblance to Intes instruments. Although I have not heard from anyone who has had the opportunity to use an Argonaut, I would expect it to perform comparably to Intes instruments of similar aperture. The Argonaut comes with a 2-inch Crayford-style focusing mount, a padded carrying case, and a small 7 × 35 straight-through finder as well as Orion's 30-day money-back guarantee. Drawbacks? Trying to get your eye to the finder may well prove difficult, as it so often does with this type of instrument. Consider augmenting the finder with a lightweight 1× aiming device, such as the Orion EZFinder or one of the others detailed in chapter 6. Consumers should also note that the Argonaut does not come with an eyepiece, star diagonal, or mounting, all of which are essential, but it does feature a mounting plate for attaching the telescope to a camera tripod. A built-in handle is also handy for carrying the instrument into the field.

Questar Corporation. Just as it has ever since the mid-1950s, Questar continues to sell exceptional quality Maksutov-Cassegrain telescopes. A couple of years ago, it looked as though Questars were going to become things of the past as the company declared Chapter 11 bankruptcy. But with the business end of things now corrected, Questar is heading into the twenty-first century on an even keel.

Although its business focus now includes such diverse instruments as long-distance microscopes and surveillance systems, Questar still manufactures its famous 3.5-inch f/15 Maksutov, but its exceptional quality does not come cheaply! For instance, the Standard 3.5-inch Maksutov from Questar retails for about twice the price of a Meade or Celestron 8-inch Schmidt-Cassegrain!

Questar telescopes come outfitted with many little niceties that add to the user's pleasure. One of the best is actually very simple: a built-in telescoping dew cap that effectively combats fogging of the front corrector plate. Another welcome plus is the screw-on solar filter that allows safe viewing of our nearest star; no other telescope comes supplied with one as standard equipment. The tabletop fork equatorial mounting is both smooth and rigid, while the built-in clock drive accurately tracks the sky. And the quality of the all-metal assembly is without parallel.

The Questar Standard 3.5 is permanently mounted on the fork mount, while the more expensive Duplex version allows the optical tube assembly to be removed and attached to a standard photographic tripod. Two mountless spotting-scope versions of the Questar 3.5 are also available. Traveling astronomers should consider the Powerguide II option, which switches the clock drive from 120-volt (220-volt optional) AC to 9-volt DC. The manufacturer states that a common 9-volt alkaline battery will operate the drive for up to fifty hours.

As nice as Questars are, a couple of idiosyncrasies plague both models. For one, only Questar's own custom-made Brandon eyepieces fit into their eyepiece holders. Owners should note that Tele Vue offers an adapter that permits use of standard 1.25-inch eyepieces (they hope Tele Vue's) in Questars. The other weakness, in my opinion, is actually looked upon as a plus by many people: Instead of equipping the telescope with a separate finder, Questar builds one right into the instrument. By flipping a lever, the observer can switch back and forth between the finder (4× or 6×, depending on whether or not the built-in Barlow lens is engaged) and the main telescope without ever leaving the instrument's eyepiece. A lot of practice is required before the telescope can be accurately aimed toward a target, but many people believe this to be a great convenience, so I guess it is largely a matter of personal preference. And as the Meade ETX proves, a separate finder would be difficult to incorporate on such a small instrument.

Naturally, the small size of the Questar 3.5 dramatically limits what can be seen, but the images of what is visible are exquisite. This appeals to the many amateurs who prefer image quality over aperture quantity. There is certainly something to be said for that philosophy. Think of it in terms of comparing a fine painting to a snapshot photograph. For most people, the snapshot adequately shows the scene, although it may miss some of the finer nuances. The painting, however, reaches more deeply to touch the soul of those who appreciate such things. After all, a Questar is as much a work of art as it is a scientific instrument.

The Scorecard

With so many telescopes and so many companies from which to choose, how can the consumer possibly keep track of everything? Admittedly, it can be difficult, but hopefully appendix A, which lists and sorts by price range all of the telescopes and binoculars mentioned in this chapter, will help a little. Placement within each range is based on "street" prices, not necessarily the manufacturer's suggested retail prices. Frequently, these are artificially inflated, perhaps in an effort to make consumers believe that they are getting deals.

There are many things to look for when telescope shopping. If you are thinking about buying binoculars or a refractor, make certain that all the optics are fully coated with a thin layer of magnesium fluoride to help reduce lens flare and increase contrast. As mentioned before, multiple coatings are the best. For reflectors and catadioptrics, check to see if their mirrors have enhanced aluminum coatings to increase reflectivity. Find out if the telescope comes with more than one eyepiece. Is a finderscope supplied? If so, how big is it? Though a 6 × 30 finder might be all right to start with, most observers prefer at least an 8 × 50 finder; anything smaller than 6 × 30 is worthless. If the telescope does not come with a finder, then one must be purchased separately before the instrument can be used to its fullest potential. Next, take a long, hard look at the mounting. Does it appear substantial enough to support the telescope securely, or does it look too small for the task? Hit the side of the telescope tube lightly with the ball of your hand while peering through the eyepiece at a target. If the vibrations disappear in less than three seconds, the mounting has excellent damping properties; three to five seconds is good; five to ten seconds is only fair; and greater than ten seconds is poor. Remember all that we have gone over in this chapter up to now, and above all, be discriminating.

Without a doubt, the best way of getting to know many kinds of telescopes firsthand is by joining a local astronomical society. Chances are good that at least one member already owns the telescope that you are considering and will happily share personal experiences, both good and bad. Plan on attending a club observing session, or star party as it may be called. Here, members bring along their telescopes and set them up side by side to share with each other the excitement of sky watching. To find the club nearest you, contact the Astronomical League, or if you have access to the Internet, check out the World Wide Web. Information and addresses are found in appendix E. You can also contact a local museum or planetarium to find out if there is an astronomy club in your area.

Look through every instrument at the star party. Bypass none, even if you are not considering that particular kind of telescope. When you find one on your list, speak to its owner. If the telescope is good, he or she will brag just like a proud parent. If it is poor, he or she will be equally anxious to steer you away from making the same mistake. Listen to the wisdom of the owner and compare his or her comments with the advice given in this chapter.

Next, ask permission to take the telescope for a test drive so that you may judge for yourself its hits and misses. Begin by examining the mechanical integrity of the mounting. Tap (gingerly, please) the mounting. Does it vibrate? Do the vibrations dampen out quickly, or do they continue to reverberate? Try the same test by rapping the mounting and tripod or pedestal. How rapidly does the telescope settle down?

Working your way up, check the mechanical components of the telescope itself. Does the eyepiece focusing knob(s) move smoothly across the entire length of travel? If you are looking at a telescope with a rack-and-pinion or Crayford-style focusing mount, does the eyepiece tube stop when the knob is turned all the way, or does it separate and fall out? Is the side-mounted finderscope easily accessible?

When you are satisfied that the telescope performs well mechanically, examine its optical quality. By this time, no doubt the owner has already shown you a few showpiece objects through the telescope, but now it is time to take a more critical look. One of the most telling ways to evaluate a telescope's optical quality is to perform the "star test" outlined in chapter 8. It will quickly reveal if the optics are good, bad, or indifferent.

How should you buy a telescope? Some manufacturers sell only factory-direct to the consumer, while others have networks of retailers and distributors. When it comes time to purchase a telescope, shop around for the best deal but do not base your choice on price alone; be sure to compare delivery times and shipping charges as well! Some of the more popular telescopes, such as those from Celestron, Tele Vue, and Meade, are available from dealer stock for immediate delivery. At the opposite end of the telescope spectrum are other companies whose delivery times can stretch out to weeks, months, or even more than a year! Consult appendix C for a list of distributors, or contact the manufacturer for your nearest dealer.

Once you decide on a telescope model, it is best, if possible, to purchase the telescope in person. Not only will you save money in crating and shipping charges, but you will also be able to inspect the telescope beforehand to make sure all is in order and as described. Though most manufacturers and distributors strive for customer satisfaction, there is always the possibility of trouble when merchandise is mail-ordered.

"Don't Worry ... The Check Is in the Mail"

A problem that I had with a well-known source of astronomical equipment led to this section of the book. Briefly, an eyepiece that I received from this supplier had a small chip in one of the inner lens elements. I called them and was told to return it for replacement or refund. I opted for replacement, so I packaged it up and sent it off. Two weeks went by, but nothing happened. I called again and was assured that the faulty eyepiece had been received and that a new eyepiece would be sent to me by week's end. Fourteen more days went by, and still nothing. I called them again, at which time the owner assured me the eyepiece had been sent out a few days earlier and that it should reach me any day. Two more weeks elapsed, and with no eyepiece in hand, I wrote to the owner demanding an explanation. Within a week, I received a refund check and a surprisingly nasty letter stating that the package containing my returned eyepiece had arrived damaged because of my negligent packing (earlier, I had been told it had arrived in fine shape). Because of my "attitude," they did *not* want me as a customer again! The company in question? Sorry, I cannot mention it by name, but I can say that it is *NOT* listed anywhere in this book. It is hoped that it will be made conspicuous by its absence! I heard from many readers of the first edition who guessed correctly; can you?

Happily, this unfortunate experience is the exception, not the rule. The vast majority of astronomical companies are owned and operated by competent, friendly people. They want happy customers (remember, a happy customer is a repeat customer) and guard their good reputations jealously. Most are willing to bend over backwards to see that problems are resolved to their customers' satisfaction. But what can the consumer do if he or she is dissatisfied with a manufacturer or distributor?

Begin on the right foot. Before returning a defective piece of merchandise, always speak to the manufacturer first about the problem. Request instructions for the most expeditious way to return the item for replacement or refund. Conform to the directions precisely, but to protect yourself always follow up the conversation with a letter in which you repeat the nature of the problem as well as the desired outcome. Send the letter by certified mail, return receipt requested, and keep a copy for your records.

Allow the company a reasonable length of time to respond to your complaint, typically two to four weeks. If after that time a satisfactory resolution has not been reached, write to the company again and inquire as to the delay. State that you expect a response within a given period of time, say, ten business days. Once again, send the letter by certified mail, return receipt requested, and keep a copy for your records. If there is still no response, call the company and find out the owner's name. Write to him or her directly, recounting all that has happened since the item was ordered.

By now, the predicament should have been resolved, but if it has not, then it's time to take action. The major astronomical periodicals do not have on-staff consumer advocates, but they do take an active interest in consumer satisfaction with all who advertise in their magazines. Write to them with your complaint, being certain to send a copy to the president/owner of the offending company. In addition, send a copy of the letter to the Astronomical League (see appendix E for the address). The league is also interested in customer satisfaction and may offer assistance. If you suspect mail fraud, also contact your local postmaster or postal inspector or write Consumer Advocate, Customer Services Department, Postal Service, 475 Longhand Plaza West S.W., Room 5910, Washington, D.C. 20260-6320. I would also like to hear about your problems for future reference. I received several complaints about a particular dealer after the first edition went to press and subsequently dropped its listing in this edition.

Most consumer advocates recommend charging all mail-order purchases to a major credit card; do not use a check or money order if possible. Using a credit card gives you certain powers that are not available any other way. On the back of every credit card's monthly statement, you will find instructions that clearly describe steps to be taken in the event of a consumer problem. Usually, the card requires that the consumer describe in specific detail the exact nature of the problem and provide copies of all receipts and documentation. The charge will then be put in contest until the problem is resolved. If a charge is contested, the consumer is not responsible for any interest that may accrue as a result. When a final determination is made, either a credit will be issued to the charge account, or the balance plus interest will be due.

Contesting a charge should be viewed as a measure of last resort. Only put a charge in contest when a bona fide problem exists and the vendor refuses to cooperate. For instance, just because you decided that you don't like an item anymore is not reason enough to contest a charge, but poor quality or workmanship is. See the difference?

Honest Phil's Used Telescopes

What if you took the pop quiz in the last chapter and found out that the best telescope for your needs was, say, an 8-inch Schmidt-Cassegrain, but you cannot afford to spend $1,500 to get the instrument you want? What can you do? If you needed to buy a car but could not afford to buy a new one, the odds are that you would check the classified advertisements for used cars, right? If it works for cars, then why not for telescopes? All other things being equal, an old telescope, such as my 30-year-old 8-inch Newtonian seen in Figure 4.8, will work just as well as a new telescope as long as it was treated kindly. The best source of used equipment is *The Starry Messenger*, a monthly periodical devoted solely to used telescopes, binoculars, and accessories. *Astronomy* magazine also has a small listing of readers' classified advertisements in each issue. For Internet Web surfers, check the used-equipment advertisements listed in appendix E.

Look through the classified advertisements to see if anything strikes your fancy. If possible, restrict your search to an area that is within a day's drive so that you can check out the telescope firsthand instead of relying on a stranger's word. One person's treasure is another person's junk (and vice versa)!

Figure 4.8 *A blast from the past: the author's Criterion Dynascope RV8 (purchased new in 1971) piloted by his daughter, Helen (also purchased new, but in 1984).*

Table 4.1   **The Best Telescopes of Yesteryear**

1. Astro-Physics 6-inch f/12 "Super Planetary" apochromatic refractor
2. Brandon 3.7-inch f/7 apochromatic refractor
3. Cave 6-inch (or larger) f/8 Newtonian reflector
4. Celestron SP-C102F, a 4-inch f/9 fluorite apochromatic refractor
5. Celestron Super C8+, an 8-inch f/10 SCT
6. Criterion "Dynascope" RV-6 (6-inch f/8) or RV-8 (8-inch f/7) Newtonians
7. Fecker Celestar 6-inch Newtonian reflector
8. Juno-14, a 14-inch Newtonian reflector, Dobsonian mounted on an equatorial table, by Jupiter Telescope Company
9. Optical Craftsmen "Connoisseur" 6-inch f/8 Newtonian reflector or larger
10. Quantum 4, a 4-inch Maksutov-Cassegrain

What should you look for in a used telescope or binoculars? Essentially the same things as in a new one. You want to check the instrument both optically and mechanically. Inspect the instrument for any damage or mishandling. Are the optics clean? Did the owner store the telescope properly?

Table 4.1 might help you find a prince among the frogs by listing an inventory of the ten best telescopes of yesteryear, as voted by survey respondents, listed in alphabetical order. Bear in mind that prices may vary greatly for the same instrument depending on its condition.

Congratulations, It's a Telescope!

Be the telescope new or be it used, read its instruction manual from cover to cover. Absorb all the information it has to offer. Remember, the universe has been around for billions of years; it will still be there when you get your telescope together! If you have any questions, don't hesitate to call the dealer or manufacturer from whom you purchased the instrument. By following all of the steps here as well as the other suggestions found throughout the chapters yet to come, you will be well on your way toward a fantastic voyage that will last a lifetime.

Note: *Stop the presses!* Late word as this edition goes to press is that Celestron has eliminated the GP-C102 and GP-C102ED refractors, the C4.5 Newtonian reflector, and the GP-C8 Schmidt-Cassegrain. In their places are the 4-inch f/10 C102-AZ achromatic refractor, sold on an alt-azimuth mounting; 3.1-inch f/11 C80-HD and 4-inch f/10 C102-HD refractors, each on the new CG-4 German equatorial mount (an apparent clone of the Great Polaris mount); 4.5-inch f/8 C114-HD and 6-inch f/7 C150-HD Newtonians, both on the CG-4 mounting; and 8-inch f/10 G-8 Schmidt-Cassegrain, on a slightly enlarged CG-5 German equatorial mount. Although all are too new to have been fully reviewed here, they look promising.

5

The Eyes Have It

Have you ever tried to look through a telescope without an eyepiece? It doesn't work very well, does it? Sure, you can stand back from the empty focusing mount and see an image at the telescope's focal plane. But without an eyepiece in place, the telescope's usefulness as an astronomical tool is greatly limited, to say the least.

Until recently, eyepieces (Figure 5.1) were thought of as almost second-class citizens whose importance was considered minor compared to a telescope's prime optic. With few exceptions, many eyepieces of yore suffered from tunnel vision as well as an assortment of aberrations. The 1980s, however, saw a revolution in eyepiece design. Suddenly advanced optical designs brought resolution and image quality to new heights and pushed their lackluster cousins to the side. With the possible exception of selecting the telescope itself, picking the proper eyepiece(s) is probably the most difficult choice facing today's amateur astronomers.

Although eyepieces (or oculars, if you prefer) are available in all different shapes and sizes, let's begin the discussion here with a few generalizations. Figure 5.2 shows a generic eyepiece with its components labelled. Regardless of the internal optical design, the lens element(s) closest to the observer's eye is always referred to as the *eye lens*, while the lens element(s) farthest from the observer's eye (that is, the one facing inward toward the telescope) is called the *field lens*. A *field stop* is usually mounted just beyond the field lens at the focus of the eyepiece, giving a sharp edge to the field of view as well as preventing peripheral images of poor quality from being seen.

Although the eyepiece must be sized according to the diameter of the eyepiece optics, the barrel (the part that slips into the telescope's focusing mount)

Figure 5.1 *A selection of premium Plössl eyepieces. Photo courtesy of Tele Vue Optics, Inc.*

is always one of three sizes. Most amateur telescopes use 1.25-inch diameter eyepieces, a standard that has been around for years. At the same time, many less expensive, department-store telescopes are outfitted with 0.965-inch oculars. Finally, the astronomical community has received a boon in a whole new breed of giant eyepieces with 2-inch barrels.

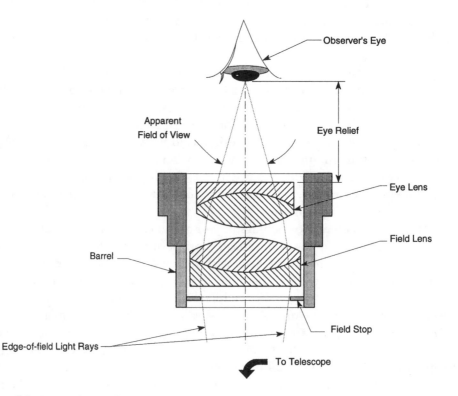

Figure 5.2 *A generic eyepiece showing components.*

Before looking at specific eyepiece designs, we must first become fluent in "eyepiece-ese," terms that describe an eyepiece's characteristics and performance. There are surprisingly few. Perhaps most important of all is *magnification* (or maybe I should say *lack of magnification*). As previously outlined in chapter 1, magnification is equal to the focal length of the telescope divided by the focal length of the eyepiece. Therefore, the longer the telescope's focal length, the greater the magnification from a given eyepiece.

Wouldn't it be nice if one specific magnification value would work well in every telescope for every object in the sky? Sadly, this is simply not the case. For certain targets, such as widely scattered star clusters or nebulae, lower powers are called for. To get a good look at the planets or smaller deep-sky objects (e.g., planetary nebulae and smaller galaxies), higher powers are required. If the magnification is too high for a given telescope, then image integrity will be sacrificed.

Just how much magnification is too much, and how much is just right? A good rule of thumb is to use only as much magnification as needed to see your target object. If you own a smaller telescope (that is, 8 inches or smaller in aperture), the oft-repeated rule of not exceeding 60× per inch of aperture is suggested. This means that an 8-inch telescope can operate at a maximum of 480×, but remember, this value is *not* cast in stone. It all depends on your local atmospheric conditions and the instrument. Given excellent optics, you may be able to go as high as 90× or even 100× per inch on some nights, while on others, 30× per inch may cause the view to crumble.

On the other hand, larger telescopes (e.g., instruments greater than 8 inches in aperture), especially those with fast focal ratios, can seldom meet or exceed the 60×-per-inch rule. Instead, they can handle a maximum of only 30× or 40× per inch.

The choice of the right magnification must be based largely on past experience. If you are lacking that experience, do not get discouraged; it will come with time. For now, use Table 5.1 as a guide for selecting the maximum usable magnification for your telescope.

Notice how the table stops at 300× for all telescopes beyond 8 inches in aperture? Experience shows (and I'm sure that some reading this now will disagree vehemently) that little is gained by using more than 300× to view an object, regardless of aperture. Only on those rare nights when the atmosphere is at its steadiest and the stars do not appear to twinkle can this value be bettered.

Another important consideration when selecting an eyepiece is the size of its *exit pupil*, which is the diameter of the beam of light leaving the eyepiece and traveling to the observer's eye, where it enters the pupil. You can see the exit pupil of a telescope or binocular by aiming the instrument at a bright surface, such as a wall or the daytime sky. Back away and look for the little disk of light that appears to float just inside the eye lens.

Table 5.1   *Telescope Aperture versus Maximum Magnification*

Telescope (in.)	Aperture (cm)	Magnification	
		Theoretical (60×/in.)	Practical
2.4	6	144	100
3.1	8	186	125
4	10	255	170
6	15	360	240
8	20	480	300
10	25	600	300
12.5	32	750	300
14	36	840	300
16	41	960	300
18	46	1080	300
20	5	1200	300
25	64	1500	300
30	76	1800	300

How can you find out the size of the exit pupils produced by your eye-pieces in your telescope? Easily, using either of the two formulas shown below:

$$\text{Exit pupil} = D/M$$

where

D = the diameter of the telescope's objective lens or primary mirror in millimeters

M = magnification

or

$$\text{Exit pupil} = F_e/f$$

where

F_e = the focal length of the eyepiece in millimeters

f = the telescope's focal ratio (its f-number)

Knowing the diameter of the exit pupil is a must, for if it is too large or too small, the resulting image may prove unsatisfactory. Why? The pupil of the human eye dilates to about 7 mm when acclimated to dark conditions (though this varies from one person to the next and shrinks as you age). If an eyepiece's exit pupil exceeds 7 mm, then the observer's eye will be incapable of taking in all the light that the eyepiece has to offer. Many optical authorities would be quick to point out that an excessively large exit pupil wastes light and resolution. This is not necessarily the case for the owners of refractors. Say you own a 4-inch f/5 refractor and wish to use a 50-mm eyepiece with it. The exit pupil resulting from this combination is 10 mm. Though the exit pupil is technically too large, this telescope-cum-eyepiece would no doubt provide a wonderful low-power, wide-field view of rich Milky Way starfields when used under dark skies.

The key phrase in that last sentence is *when used under dark skies*. If the same pairing was used under mediocre suburban or urban sky conditions, then the contrast between your target and the surrounding sky would suffer greatly. This is because the eyepiece is transmitting not only starlight but also skyglow (light pollution)—*too much* skyglow.

What about using a 50-mm eyepiece with a 4-inch f/5 *reflector*? Sorry, not a good idea. With conventional reflectors as well as catadioptric instruments, obstruction from the secondary mirror will create a noticeable black blob in the center of view when eyepieces yielding exit pupils greater than about 8 mm are used. With these telescopes, it is best to stick with eyepieces of shorter focal lengths.

On the other hand, if the exit pupil is too small, then the image will be so highly magnified that the target may be nearly impossible to see and focus. Just as there is no single all-around best magnification for looking at everything, neither is there one exit pupil that is best for all objects under all sky conditions. It depends on what you are trying to look at. Table 5.2 summarizes (rather subjectively) my personal preferences.

Although magnification gives some feel for how large a swath of sky will fit within an eyepiece's view, it can only be precisely figured by adding another ingredient: the ocular's *apparent field of view*. Nowadays, most manufacturers will proudly tout their eyepieces' huge apparent fields of view because they know that big numbers attract attention. Unfortunately, few take the time to explain what these impressive figures actually mean to the observer.

The apparent field of view refers to the eyepiece field's edge-to-edge angular diameter as seen by the observer's eye. Perhaps that statement will make more sense after this example. Take a look at Figure 5.3a. Peering through a long, thin tube (such as an empty roll of paper towels), the observer sees a very narrow view of the world—an effect commonly known as *tunnel vision*. This perceived angle of coverage is also known as the apparent field of view. To increase the apparent field of view in this example, simply cut off part of the cardboard tube. Slicing it in half (Figure 5.3b), for instance, will approximately double the apparent field, resulting in a more panoramic display.

Table 5.2 **Suggested Exit Pupils for Selected Sky Targets**

Target	Exit Pupil mm
Wide star fields under the best dark-sky conditions (e.g., large star clusters, diffuse nebulae, and galaxies)	6 to 7
Smaller deep-sky objects; complete lunar disk	4 to 6
Small, faint, deep-sky objects (especially planetary nebulae and smaller galaxies); double stars, lunar detail, and planets on nights of poor seeing	2 to 4
Double stars, lunar detail, and planets on exceptional nights	0.5 to 2

Figure 5.3 *Simulated view through eyepieces with (a) a narrow apparent field of view and (b) a wide apparent field of view. Photograph of M33 by George Viscome (14.5-inch f/6 Newtonian, Tri-X film in a cold camera, 30-minute exposure).*

In the world of eyepieces, the apparent field of view typically ranges from a cramped 25° to a cavernous 80° or so. Generally, it is best to select eyepieces with at least a 40° apparent field because of the exaggerated tunnel-vision effect obtained through anything smaller. An apparent field in excess of 60° gives the illusion of staring out the porthole of an imaginary spaceship. The effect can be really quite impressive.

Naturally, eyepieces with the largest apparent fields of view do not come cheaply! Some, especially the long-focal-length models, are quite massive both in terms of weight and cost. Typically, they must be made from large-diameter lens elements and may be available in 2-inch barrels only, restricting their use to medium- and large-aperture telescopes only. Some are so heavy that you may actually have to rebalance the telescope whenever they are used. More about this when specific eyepiece designs are discussed later in this chapter.

By knowing both the eyepiece's apparent field (typically specified by the manufacturer) and magnification, we can calculate just how much sky can squeeze into the ocular at any one time. This is known as the *true* or *real field of view* and can be approximated from the following formula:

$$\text{Real field} = F/M$$

where
 F = the apparent field of view
 M = magnification

To illustrate this, let's look at an 8-inch f/10 telescope and a 40-mm eyepiece. This combination produces 50× and a 4-mm exit pupil. Suppose this particular eyepiece is advertised as having a 45° apparent field. Dividing 45 by 50 shows that this eyepiece produces a real field of 0.9°, almost large enough to fit two full Moons edge to edge.

Another term that is frequently encountered in eyepiece literature but rarely defined is *parfocal*. This simply means that the telescope will not require refocusing when one eyepiece is switched for another in the same set. Without parfocal eyepieces, the observer may lose a faint object during the second focusing should the telescope be moved accidentally during the change. Please keep in mind that even when an eyepiece is claimed to be parfocal, that does not mean that it is *universally* parfocal. Two eyepieces of the same optical design, say a 26 mm from Brand X and a 12 mm from Brand Y, that claim to be parfocal are most likely not parfocal with each other. Similarly, two eyepieces of different optical designs may be manufactured by the same company and declared parfocal, but they are not likely to be parfocal with each other.

Finally, a well-designed eyepiece will have good *eye relief*, which is the distance from the eye lens to the observer's eye when the entire field of view is seen at once. Less expensive eyepieces may offer eye relief of only one-quarter times the ocular's focal length. This is much too close to view comfortably. Some modern designs maintain an eye relief of 20 mm regardless of the eyepiece's focal length, making observing more enjoyable, especially for those who must wear eyeglasses. Of course, there can be too much of a good thing. Excessive eye relief can make it difficult for the observer to hold his or her head steadily while hovering above the eyepiece.

Image Acrobatics

Because eyepieces can suffer from the same aberrations and optical faults as telescopes, it might be wise to list and define a few of their more common problems. Some have already been defined in earlier chapters, but now we will concentrate on their impact on eyepiece performance. (Of course, if any of the following conditions exist all of the time regardless of eyepiece used, then the problem likely lies with the telescope, not the eyepiece.)

If star images near the field's center are in focus when those near the edge are not, then the eyepiece suffers from *curvature of field*. The overall effect is an annoying unevenness across the entire field of view. Another flaw found in lesser eyepieces is *distortion*, which is most readily detectable when viewing either terrestrial sights or large, bright celestial objects such as the Moon or Sun. This condition is usually characterized by a warping of the scene in a way similar to the effect seen through a "fish-eye" camera lens.

Chromatic aberration, previously defined in earlier chapters, is nearly extinct in today's eyepieces thanks to the use of one or more achromatic lenses.

Still, some less-sophisticated eyepieces, often sold as standard equipment, suffer from this ailment. If an ocular transmits chromatic aberration, the problem will be immediately detectable as a series of colorful halos surrounding all of the brighter objects found toward the edge of the field of view; the center of view is usually color-free.

Spherical aberration has also been all but eliminated in most (but not necessarily all) eyepieces of modern design. If an eyepiece is free of spherical aberration, then a star should look the same on either side of its precise focus point. When spherical aberration is present, however, the star will change its appearance from when it is just inside of focus compared to when it is just outside of focus. This predicament is the result of uneven distribution of light rays at the eyepiece's focal point. These days, if spherical aberration is present, chances are it is being introduced by the telescope's prime optic (main mirror or objective lens) and not the eyepiece. At low powers (large exit pupils), it can also be introduced by the observer's own eye.

Just as with objective lenses and corrector plates, most eyepiece lenses are now coated with an extremely thin layer of magnesium fluoride. Coatings reduce *flare* and improve *light transmission,* two desirable characteristics for telescope oculars. A bare lens surface can reflect as much as 4% of the light striking it. By comparison, a single-coated lens reduces reflection to about 1.5% per surface. As already mentioned in chapter 3's discussion of binoculars, a lens coated with the proper thickness of magnesium fluoride exhibits a purplish hue when held at a narrow angle toward a light. Top-of-the-line eyepieces receive multiple coatings, reducing reflection to less than 0.5%. These show a greenish reflection when turned toward a light.

Eyepiece Evaluation

Galileo, Kepler, and Newton had it pretty easy when it came to selecting eyepieces. Look at their choices! Galileo used a single concave lens placed before the objective's focus. It produced an upright image, but the field of view was incredibly small and severely hampered by aberrations. Kepler improved on the idea by selecting a convex lens as his eyepiece. It gave a wider, albeit inverted view but still suffered from aberrations galore. Progress in eyepiece design was slow in the early years.

Here we will see just how far the art of eyepieces has progressed while evaluating which designs are best suited for the telescope you chose earlier.

Huygens. The first compound eyepiece was concocted by Christiaan Huygens in the late 1660s (just as with new telescope designs, an eyepiece usually bears the name of its inventor). As can be seen in Figure 5.4a, Huygens eyepieces contain a pair of plano-convex elements. Typically, the field lens has a focal length three times that of the eye lens.

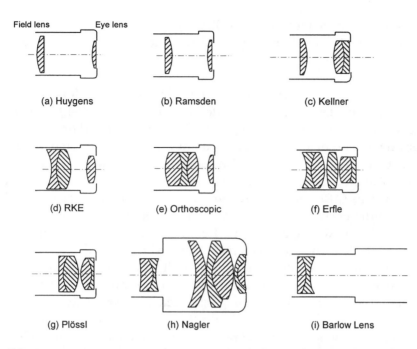

Figure 5.4 *The inner workings of several eyepiece designs from both yesterday and today. Eyepieces shown are (a) Huygens, (b) Ramsden, (c) Kellner, (d) Edmund's RKE, (e) Orthoscopic, (f) Erfle, (g) Plössl, and (h) Nagler. A typical Barlow Lens is shown in diagram (i).*

In the past, Huygens eyepieces were supplied as standard equipment with telescopes of f/15 and greater focal ratios. At longer focal lengths, these oculars can perform marginally well, although their apparent fields of view are narrow. In telescopes with lower focal ratios, however, image quality suffers from spherical and chromatic aberrations, image curvature, and an overall lack of sharpness. Do you own a Huygens eyepiece? If so, it will probably have an **H** on the barrel; for instance, **H25mm** indicates a Huygens eyepiece with a 25-mm focal length.

Ramsden. Devised in 1782, this design was the brainchild of Jesse Ramsden, the son-in-law of John Dollond (you may recall him from chapter 2 as the father of the achromatic refractor—small world, isn't it?). As with the Huygens, a Ramsden eyepiece (Figure 5.4b) consists of two plano-convex lenses. Unlike the earlier ocular, however, the Ramsden elements have identical focal lengths and are flipped so that both convex surfaces face each other.

In most cases, the lenses are separated by about two-thirds to three-quarters of their common focal length. This is, at best, a compromise. Setting the elements closer together improves eye relief but dramatically increases aberrations. Going the other way will decrease the design's inherent faults, but

eye relief quickly drops toward zero. Therefore, like the Huygens, it is probably best to remember the Ramsden for its historical significance and pass it by
in favor of other designs.

Kellner. Over six decades of experimentation took place before an improvement to the Ramsden eyepiece was developed. Then, in 1849, Carl Kellner
introduced the first achromatic eyepiece (Figure 5.4c). Based on the Ramsden,
Kellner eyepieces replace the single-element eye lens with a cemented achromat, which greatly reduces most of the aberrations common to Ramsden and
Huygens eyepieces. Sometimes called Modified Achromats (MAs) or Achromatic Ramsdens (ARs), Kellners feature fairly good color correction and edge
sharpness, little curvature, and apparent fields of view ranging between 40°
and 50°. In low-power applications, Kellners offer good eye relief, but this
tends to diminish as the eyepiece's focal length shrinks.

Perhaps the greatest drawback of the Kellner design is its propensity for
internal reflections. I have often heard Kellners referred to as "haunted eyepieces" because of their ghost images, which are especially noticeable on
bright objects. But thanks to antireflection optical coatings, this effect is
almost eliminated when Kellners are used with 8-inch and smaller telescopes.

Given all of their pluses and minuses, Kellners represent a good buy for
budget-conscious owners of small- to medium-aperture telescopes who are in
the market for a low- to medium-power eyepiece.

RKE. From Edmund Scientific Company comes this fresh twist on the Kellner
eyepiece. Instead of using an achromatic eye lens and a single-element field
lens, the RKE (short for Rank-modified Kellner Eyepiece after its inventor,
Dr. David Rank; Figure 5.4d) does just the opposite. The computer-optimized
achromatic field lens and single-element eye lens combine to outshine the
Kellner in just about every respect. Actually, performance of the three-element
RKE is most comparable to that of the four-element orthoscopic, with a moderate apparent field of view (45° for the RKE) as well as good color correction
and image clarity. Both work well at all focal lengths in all telescopes.

Orthoscopic (Abbe). Introduced in 1880 by Ernst Abbe, the orthoscopic eyepiece has become a perennial favorite of amateur astronomers. As shown in
Figure 5.4e, it consists of a cemented triplet field lens matched to a single
plano-convex eye lens. What results is close to a perfect eyepiece, marred by
neither chromatic nor spherical aberration. There is also little evidence of
ghosting or curvature of field. "Orthos" offer flat views with apparent fields
between 40° and 50° and excellent eye relief. Color transmission and contrast
are superb, especially when combined with today's optical coatings. With
either a long focal length for low power, a short focal length for high power, or
anywhere in between, orthoscopic eyepieces remain one of the best eyepieces

for nearly all amateur telescopes. To my eye, they yield equal or higher image contrast than many of the new-generation "super-deluxe" eyepieces, especially when aimed at the planets.

Erfle. The Erfle, the granddaddy of all wide-field eyepieces, was originally developed in 1917 for military applications. With apparent fields of view ranging between 60° and 75°, it was quickly embraced by the astronomical community as well. Internally, Erfles consist of either five or six elements; one variety uses two achromats with a double convex lens in between, while a second has three achromats, as shown in Figure 5.4f.

Erfles give observers an outstanding panoramic view of the deep sky. The spacious view takes its toll on image sharpness, however, which suffers from astigmatism toward the field's edge. For this reason, Erfles are inappropriate for lunar and planetary observations or any occasion that calls for higher magnification, but in low-power, wide-angle applications, they are very impressive.

Zoom. These eyepieces combine a wide variety of focal lengths into one package, typically ranging from about seven to twenty-one millimeters. Sounds too good to be true? Unfortunately, it usually is! Most zoom eyepieces are compromises at best. For one thing, aberrations are frequently intensified in zoom eyepieces, perhaps owing to poor optical design or because the lenses are constantly sliding up and down in the barrel. Another problem is that their apparent fields of view are not constant over the entire range; the widest ones occur at high power but shrink rapidly as magnification drops. Finally, many are not parfocal across their entire range, requiring you to refocus whenever the zoom eyepiece is zoomed.

Plössl. One of the most highly regarded eyepieces around today, the Plössl (Figure 5.4g) features twin close-set pairs of doublets for the eye lens and the field lens. The final product is an excellent ocular that is comparable to the orthoscopic in terms of color correction and definition but with better eye relief and a larger apparent field of view. Ghost images and most aberrations are sufficiently suppressed to create remarkable image quality.

Though it was developed in 1860 by G. S. Plössl, an optician living in Vienna, Austria, the Plössl eyepiece took more than a century to catch on among amateur astronomers. Back in the 1970s, when I first became seriously engrossed in this hobby, I heard about a mysterious eyepiece called the Plössl. Try as I might, I could not find much published information on it, other than that I could not afford one! Perhaps the enigma of the Plössl was heightened by the fact that back then, the only company to offer them was Clavé in France, and their distribution was limited. Then, in 1980, the Plössl hit the big time. That year, Tele Vue Optics introduced a line of Plössls that was to start an eyepiece revolution among amateur astronomers. They were an instant success.

Many companies now offer Plössls. All are quite good, but some are perceptibly better than others. I prefer not to generalize, but in this case you pretty much get what you pay for. The more expensive Plössls typically maintain closer optical and mechanical tolerances and include multicoated optics as well as blackened lens edges and threads. Little things like these can make the difference between seeing a marginally visible object and not.

Tele Vue Plössls, generally considered to be the finest, are available in eight focal lengths, from 8 mm to 55 mm. All have 1.25-inch barrels except for the 55-mm, which requires a 2-inch focuser. Each has a 50° apparent field of view, except the 40-mm model, which has a 43° field. Each also includes a rubber eyeguard to help prevent unwanted light infiltration.

For those working on tighter budgets, less expensive Sirius Plössls by Orion Telescopes, Celestron Plössls, Meade Series 3000 Plössls, and Pro-Optic Plössls by Adorama are also worth strong consideration. Celestrons range from 6.3 mm to 40 mm in focal length; Sirius Plössls, from 7.5 mm to 40 mm; and Series 3000, from 5 mm to 40 mm; Pro-optic Plössls range from 6.3 mm to 40 mm. All feature 1.25-inch barrels and are a big step up from some of the shoddy eyepieces that are included as standard equipment with many telescopes. Price-wise, the Pro-Optics and Celestrons are the least expensive of the lot, being about half the price of Tele Vues. Sirius eyepieces are more expensive but also include built-in, foldable rubber eyeguards (aftermarket eyeguards may be added to the others). Owners of less expensive department-store telescopes should also note that all but the largest of the Sirius Plössls are also available with 0.965-inch barrels.

Meade's Series 4000 Super Plössls are also very good performers, although technically they are not conventional Plössls; instead, they are more of a four-element, hybrid design. Super Plössls come in focal lengths from 6.4 mm to 56 mm. All offer 52° fields except for the 40-mm, which has a 44° field, and all are designed to 1.25-inch focusers, except the 56-mm, which requires a 2-inch focusing mount. Super Plössls have excellent performance characteristics, including very good image sharpness across the field. Their foldable rubber eyeguards also make these attractive for amateurs in light-polluted environs.

Super-deluxe-extra-omni-ultra-maxi-mega-colossal ... Whatever happened to the words *standard* and *regular*? Are they still in the dictionary? Apparently not, judging by today's advertising. Every product, from refrigerators to pet food, is extra-special in some way. Top-of-the-line oculars are no different. Each is proclaimed by its manufacturer as something extraordinary. And do you know what? They really are quite good!

These super-duper eyepieces first came on the scene to meet the demanding needs of amateurs using the increasingly popular Schmidt-Cassegrains and mammoth Newtonians, but they also work equally well in refractors and reflectors with longer focal lengths. Most use multicoated lenses made from expensive glasses to minimize aberrations. With these eyepieces, the universe has never looked so good.

Not satisfied with the success of its Plössls, Tele Vue led the way in introducing oculars with extremely wide apparent fields of view. Tele Vue Nagler eyepieces (Figure 5.4h) come in four focal lengths; the 4.8-mm and 7-mm are designed to fit 1.25-inch focusers, while the 9-mm and 13-mm are skirted to fit both 1.25- and 2-inch focusers. All use a complex seven-element design to produce an incredible 82° apparent field while correcting for astigmatism, chromatic aberration, spherical aberration, coma, and just about every other optical fault.

This is not to say that Naglers are without their shortcomings; in fact, there are two. One is a minor loss of contrast when viewing bright objects such as the planets, a common problem with many multiple-element eyepieces. Their second drawback is known as the "kidney-bean effect" and is most evident in the 13-mm Nagler. As you shift your eye around to take in the eyepiece's full field of view, a dark, kidney-shaped area becomes noticeable to some.

From this latter problem was born the eight-element Tele Vue Nagler Type 2 design. Three eyepieces make up the Nagler 2 line, all with the same dramatic 82° apparent field of view. The 12-mm and 16-mm are designed to fit both 1.25- and 2-inch focusing mounts, while the 20-mm fits 2-inch focusers only. These longer focal lengths are ideal for deep-sky observing, especially for low-surface-brightness objects like diffuse nebulae and face-on spiral galaxies. Keep in mind that these are all *big* eyepieces, with weights listed in pounds rather than ounces. For many smaller telescopes, especially those on Dobsonian mounts, this may be too much to bear; owners would do better with other eyepieces. Price-wise, they are also measured in pounds (of dollars). But the view through a Nagler or a Nagler Type 2 is wondrous; you may actually have to move your eye or your head around to take in the whole view. Company literature, independent reviews, and consumers all liken the feeling to peering through a porthole of a spacecraft.

Tele Vue Panoptics are also heavyweights in the eyepiece arena, their six elements combining to yield unparalleled low-power views. Panoptics come in five focal lengths, each sporting a 68° apparent field of view. The 15-mm and 19-mm models feature 1.25-inch barrels, the 22-mm may be used in either 1.25-inch or 2-inch focusers, and the 27-mm and 35-mm fit 2-inch focusers only. Although their fields of view are not as large as the Naglers', eye relief is generally better, an important consideration for those who must wear eyeglasses. Consumers should note, however, that the eye relief of the 15-mm Panoptic is on the short side and might prove troublesome for some observers.

Many other companies offer wide-field eyepieces. Meade Super Wide Angle oculars employ six elements that provide an impressive apparent field of 67°, roughly comparable to a Panoptic's. The 13.8-mm, 18-mm, and 24.5-mm Super Wides are designed to fit 1.25-inch focusers, while the 32-mm and 40-mm require 2-inch focusers. The two shorter focal lengths yield tack-sharp star images across the field. The 18-mm is especially nice because of its comfortable eye relief and small size. The 24.5-mm also performs well, but images near the field's edge are not quite as crisp as those of the Tele Vue Panoptic 22-mm or 27-mm or of the Pentax XL 21-mm when used with telescopes with fast focal ratios. The two longer focal lengths show further signs of edge distortions, with out-of-focus "seagull" stars evident. Consumers should bear in mind, however, that Meade Super Wides are seventy to eighty dollars less expensive than Panoptics and Pentax XLs.

Meade Ultra Wide Angle oculars come in 4.7-mm, 6.7-mm, 8.8-mm, and 14-mm focal lengths. The two shorter ones are housed in standard 1.25-inch barrels, while the two longer ones feature 1.25-inch/2-inch step-down barrels. All use a combination of eight elements in five groups to produce a stated apparent field of 84°, approximately equal to a Nagler's. Owners expound the eyepieces' excellent performance. Some comments from the mailbag: "The 6.7-mm Ultra Wide has very short eye relief but is a very sharp, high-power eyepiece; works well even on the Astroscan; lighter than the 7-mm Nagler," and "The 14-mm Ultra Wide Angle is the eyepiece that I use most often and is probably the most common in our astronomy club." The Meade Ultra Wide Angle eyepieces are indeed right up there with the best available.

A newcomer on the eyepiece scene is the Japanese camera company Pentax, which makes several lines of oculars, although many have limited availability. In addition to 0.965-inch diameter Kellner and orthoscopic eyepieces, it also offers the premium-grade Pentax XL eyepieces, available in focal lengths of 5.2, 7, 10.5, 14, 21, 28, and 40 mm. All have a 65° apparent field of view, except the 28-mm, which has a 55° apparent field. The 40-mm has a 2-inch barrel, and the others fit only 1.25-inch focusers. (Don't be fooled by what appears to be a step-down barrel in photos of these eyepieces; the larger diameter is less than two inches.) All come with eyecups that are unlike any other available; although most eyecups rise about the eyepiece to shield the observer from stray light, Pentax eyecups look like, as one person put it, an upside-down funnel. To use them, the observer sticks his or her eye (which is slightly convex) *into* the eyecup. It takes a little getting used to, but I have heard of no complaints from those who use them.

For high-power applications, owners agree that the 5.2-, 7-, 10.5-, and 14-mm Pentax XL eyepieces are about as good as they get. Each features excellent eye relief and sharp, high-contrast views. Few seemed bothered by the fact that their apparent fields of view are not as wide as those of Tele Vue Naglers or Meade Ultra Wide Angle eyepieces. As one person put it, "The short focal length [Pentax XL] eyepieces are perhaps the most incredible eyepieces on the market, period."

Many commented, however, that the three longer-focal-length Pentax XLs fall a little short of Panoptic and Nagler performance. The 19-mm Panoptic, for instance, is just as sharp as the 21-mm Pentax XL but is much lighter and smaller.

Celestron Ultima eyepieces, manufactured in Japan by Vixen Optical Company, are available in eight focal lengths ranging from 5 mm to 42 mm. Depending on focal length, each incorporates between four and seven optical elements to create fields from a slim 36° in the 42-mm model to a more acceptable 51° in shorter-focal-length versions. Images are quite sharp, with excellent correction against aberrations. Eye relief is also quite good in all focal lengths except the two shortest, the 5-mm and the 7.5-mm, which measure only 4 mm and 5 mm, respectively.

If long eye relief is important to you, then Vixen Lanthanum LV eyepieces may be your salvation. With most designs, as focal length shrinks, so does eye relief. As a result, short-focal-length eyepieces require the observer to get uncomfortably close to the eye lens. Not so with these oculars. All Lanthanum LV eyepieces, ranging in focal length from 30 mm to 2.5 mm (the shortest focal length available, incidentally) feature a long 20-mm eye relief. As a result, the observer can stand back from the eyepiece and still take the whole scene in, making life at the eyepiece much more comfortable. Based on the Plössl, the Lanthanum LV adds a fifth lens element between the eye-lens and field-lens groups as well as an extra element(s) before the field lens. Apparent fields of view are in the 45° to 50° range. Image quality and contrast rank among the highest of any eyepiece, making Lanthanum's offerings popular among deep-sky and planetary observers alike.

Takahashi LE eyepieces are another proprietary design. Available in focal lengths from 5 mm to 30 mm (1.25-inch barrel) as well as 50 mm (2-inch barrel), Takahashi LE eyepieces are also highly regarded by amateur astronomers. One respondent to the first *Star Ware* survey described her 5-mm LE eyepiece as "the only lens that I use for planetary and lunar observing; very sharp, with no ghosting on bright objects like some of my other eyepieces; excellent eye relief." Several others commented on the similar performance characteristics of Takahashi LE eyepieces and Vixen's Lanthanum series. Which is better boils down to a largely subjective judgment; all will serve well.

Brandon eyepieces, offered exclusively by VERNONscope, yield sharp, crisp views with good image contrast across the entire field. Ghosting, prevalent in many lesser eyepieces, is effectively eliminated in four-element Brandons thanks to precise optical design and magnesium-fluoride coatings. Overall performance is comparable to Plössl and orthoscopic eyepieces for observing deep-sky objects but superior for observing the planets. Brandons come in six focal lengths from 8 mm to 48 mm; all have standard 1.25-inch barrels (except the 48-mm, which has a 2-inch barrel) and feature foldable rubber eyeguards. But take heed, those of you who plan on using screw-in filters: Brandon eyepieces do not have standard threads to accept other manufacturers' filters. They can use only VERNONscope filters.

Another company to jump on the bandwagon is the Telescope and Binocular Center (the company formerly known as Orion Telescope Center). In addition to selling many of the previously mentioned brands of eyepieces, they also feature Orion Ultrascopic oculars of their own design. Ultrascopics come in four "power ranges" based on their focal lengths, including 20-mm, 25-mm, and 30-mm low-power Ultrascopics; 15-mm and 20-mm moderate-power versions; 7.5-mm and 10-mm high-power variants; and 3.8-mm and 5-mm very-high-power models. Low- through high-power Ultrascopics incorporate five lens elements, while very-high-power versions add a two-element negative achromat (in effect, a built-in Barlow lens) as in many other hybrid eyepiece designs. Views are ghost-free, with pinpoint stars seen across the field. The general consensus among their owners is summed up by one who uses the eyepieces with a large-aperture reflector: "Ultra-scopics are excellent; very good images to the edge of the field, even in an f/5.2 Newtonian." All Ultrascopics come with rubber eyeguards and produce a reasonable 52° apparent field of view (the 35-mm has a 49° apparent field of view).

As previously mentioned, all of these super-deluxe-extra-omni-ultra-maxi-mega-colossal eyepieces offer excellent image quality as well as color correction, and are free of edge distortions. Their designers are to be congratulated for creating superb eyepieces for the backyard astronomer. On the negative side, however, I must point out that not only are their performances stellar, but their prices are, too! Some of these may actually cost more than an entire telescope. Are they worth the money? In all honesty, I have to say yes, especially if you own a fast, that is, low-focal-ratio telescope. These top-of-the-line eyepieces are great if you can afford them, but they are not absolutely necessary. Many hours of great enjoyment can be yours with less costly eyepieces.

Barlow lens. The Barlow lens was invented in 1834 by Peter Barlow, a mathematics professor at Britain's Royal Military Academy. He reasoned that by placing a negative lens between a telescope's objective or mirror and the eyepiece, just before the prime focus, the instrument's focal length could be increased (Figure 5.4i). The Barlow lens, therefore, is not an eyepiece at all but rather a focal-length amplifier.

Depending on the Barlow's location relative to the prime focus, the amplification factor can range up to 3×. For example, remember the 8-inch f/10 telescope and 40-mm ocular that I used to illustrate the concept of real field of view? If we insert the eyepiece into a 2× Barlow lens and then place both into the telescope, the combination's magnification will climb from 50× to 100×.

Why use a Barlow lens? The first reason should be obvious. By purchasing just one more item, the observer effectively doubles the number of eyepieces at his or her disposal. But the benefits of the Barlow lens go even deeper than this. Because a Barlow stretches a telescope's focal length, an eyepiece/Barlow configuration will yield consistently sharper images than an equivalent single eyepiece (provided that the Barlow is of high quality, of course). This becomes especially noticeable near the edge of the field. Another important advantage is increased eye relief, something that short-focal-length eyepieces always have in limited supply. Several of the super-deluxe-etc. eyepieces get their long eye relief and short focal lengths by incorporating a Barlow lens right in the eyepiece's barrel.

When shopping for a Barlow lens, make sure that it has fully coated optics; fully multicoated optics are even better. Avoid so-called variable-power Barlows, as they tend to perform less satisfactorily than fixed-power versions. Several high-quality Barlows are available in the 1.25-inch range, including those by Tele Vue, Meade, Celestron, and Orion. If you own 2-inch diameter eyepieces, the giant Barlows by Astro-Physics and Tele Vue are your only choices. But that's okay, because both are excellent; the Tele Vue is a little less expensive.

Focal reducers. Although the f/10 focal ratios of most Schmidt-Cassegrain telescopes have great appeal for observers who enjoy medium- to high-power sky views, their long focal lengths make it difficult to fit wide star fields into a single scene. To view large objects, such as the Andromeda Galaxy or the Orion Nebula, many amateurs use focal reducers to shrink a telescope's overall focal length by 37% (from f/10 to f/6.3), thereby increasing its field of view.

Modern focal reducers do more than just compress the focal length; they also give pinpoint star images across the entire field of view. Though beneficial to all observers, this is especially attractive to astrophotographers, as the reduction in focal length also cuts down exposures by a factor of 2.5. Not surprisingly, both Celestron and Meade offer focal reducers for their telescopes, each designed for one or the other specific brand. Consumers should note, however, that both will also fit each other's SCTs and work equally well, so the choice is yours.

Coma correctors. One look through chapter 4 and it should be clear that Newtonians with low focal ratios have become very common in telescope circles—and with them, unfortunately, so has coma. You'll recall that coma causes stars near the edge of an eyepiece's field to appear like tiny comet-shaped blobs instead of sharp points. Coma is an inborn trait of all Newtonians, although it becomes objectionable only when a telescope's focal ratio is f/5 or less.

To help counter coma's deleterious effect, many amateurs use coma correctors. At first glance, coma correctors look just like Barlow lenses; in fact, that is just how they are used. Simply slip an eyepiece into the corrector's barrel and then place the pair into the telescope's eyepiece holder. By refining the light exiting the telescope before it reaches the eyepiece, coma is effectively eliminated. As a side bonus, they also correct for off-axis astigmatism, another common problem with fast Newtonians. But bear in mind that coma correctors are not magic. They do nothing for bad optics. If your telescope has *on-axis* astigmatism, it is most likely caused by a poor-quality primary or secondary mirror.

Two companies, Tele Vue and Lumicon, manufacture coma correctors. Although both are basically the same, Lumicon uses two single lenses in its design, whereas Tele Vue's Paracorr is based around a pair of achromats for better correction. However, the Lumicon corrector does not affect the focal length of the system, while the Paracorr increases focal length by 15%. Both will fit only 2-inch focusing mounts, although they do come with adapters for 1.25-inch eyepieces.

Reticle eyepieces. For some applications, such as through-the-telescope guided photography, it can be useful (even necessary) to have an internal grid, or *reticle*, superimposed over an eyepiece's field of view. Reticle patterns, typically etched on thin, optically flat windows, come in a wide variety of designs depending on the intended purpose. The simplest have two perpendicular lines that cross in the center of view, while the most sophisticated display complex grids. One of the most versatile is Celestron's Micro Guide, shown in Figure 5.5. Built around a 12.5-mm orthoscopic eyepiece, the Micro Guide features a laser-etched reticle that includes a bull's-eye–style target for guiding during astrophotography and micrometer scales for measuring object size and the separations and position angles of double stars (but note that the scale is calibrated only for telescopes with 80-inch focal lengths).

Most reticle eyepieces come with strange-looking appendages sticking out one side of their barrels. These are illuminating devices used to make the reticle patterns visible to the observer. Ideally, an illuminator's light level should be adjustable so that the reticle is just bright enough to be seen but not so bright as to overpower the object in view. Illuminating devices use either small incandescent flashlight bulbs or light-emitting diodes (LEDs), both colored red. Celestron's Pulstar illuminator, rather than glowing steadily as other illuminators do, alternately flashes the reticle on and off. The advantage here is that by not keeping the reticle on continuously, fainter stars may be spotted and followed with greater ease. Rate and brightness of pulse are both adjustable by the user.

Figure 5.5 *Celestron's Micro Guide illuminated-reticle eyepiece. Photo courtesy of Celestron International.*

Putting the Puzzle Together

Now comes the moment of truth: Which eyepieces are best for you? It is always preferable to have a set of oculars that offers a variety of magnifications, became no one value is good for everything in the universe. Low power is best for large deep-sky objects such as the Pleiades star cluster or the Orion Nebula. Medium power works well for lunar sightseeing as well as for viewing smaller deep-sky targets such as most galaxies. Finally, high power is needed to spot subtle planetary detail or to split close-set double stars.

Who makes what? In appendix B I try to sort out the eyepiece marketplace by giving an account of what eyepieces are sold by which companies. This appendix also looks at all of the important criteria that should be weighed when deciding what to purchase. In an ideal world, eyepiece quality, eye relief, field of view, and optical coatings would be our chief considerations, but in the real world, most of us must also factor in cost. That is why eyepieces have been collected according to their costs (early 1998 prices).

As an aid to guide your selection, Table 5.3 also offers three different possibilities for four of today's most popular telescope sizes. Each lists eyepieces

Table 5.3 **Four Telescope/Eyepiece Alternatives**

Dream Outfit	Middle of the Road	Good and Cheap
4" f/10 refractor		
LP: 35-mm Panoptic	LP: 32-mm Plössl	LP: 25-mm RKE/Kellner
MP: 12-mm Nagler	MP: 12-mm Plössl	MP: 9-mm RKE/Kellner
HP: 7-mm Nagler	HP: 8-mm Plössl	HP:—
Barlow lens	Barlow lens	Barlow lens
8" f/6 Newtonian		
LP: 21-mm Pentax XL	LP: 25-mm Plössl	LP: 25-mm RKE/Kellner
MP: 14-mm Ultra Wide	MP: 10-mm Plössl	MP: 9-mm RKE/Kellner
HP: 6.7-mm Ultra Wide	HP:—	HP:—
Barlow lens	Barlow lens	Barlow lens
18" f/4.5 Newtonian		
LP: 27-mm Panoptic	LP: 25-mm Ultrascopic	LP: 25-mm Orthoscopic
MP: 12-mm Nagler	MP: 10-mm Plössl	MP: 9-mm Orthoscopic
HP: 7-mm Pentax XL	HP: 6-mm Orthoscopic	HP:—
2-inch Barlow lens	Barlow lens	Barlow lens
Coma corrector		
8" f/10 Schmidt-Cassegrain		
LP: 35-mm Panoptic	LP: 32-mm Plössl	LP: 25-mm RKE/Kellner
MP: 12-mm Nagler	MP: 12-mm Plössl	MP:—
HP: 7-mm Pentax XL	HP: 9-mm Orthoscopic	HP: 9-mm RKE/Kellner
Barlow lens	Barlow lens	Barlow lens
Reducer/compressor		

according to the magnification they would produce (LP = low power, MP = medium power, and HP = high power). The first is a dream outfit, where money is no object: the second offers a middle-of-the-road compromise between quality and cost: and the third represents the minimum expenditure required for a good range of eyepieces. Prices are not listed because they can change quickly and dramatically from dealer to dealer; be sure to shop around.

Keep in mind that these represent only three possibilities for each telescope. You are encouraged to flip back and forth among the eyepiece descriptions, appendix B, and Table 5.3 to substitute your own preferences. And remember, the only way to learn which oculars are really best for your particular situation is to try them out first. Once again, I strongly recommend that you seek out and join a local astronomy club and go to an observing session. Bring your telescope along and borrow as many different types of eyepieces as possible. Take each of them for a test drive. Then, and only then, will you know exactly what is right for you.

6

The Right Stuff

Congratulations on making it through what might be thought of as telescope and eyepiece obstacle courses. Although the range of choices was extensive, you should now have a fairly good idea about which telescopes and eyepieces best meet your personal needs.

A telescope alone, however, cannot simply be set up and used. First, it must be outfitted or combined with other things, such as a finderscope, maybe some filters, a few reference books, a star atlas, a flashlight, some bug spray . . . well, you get the idea. So take out your wish list and credit card once again. It is time to go shopping in the wide world of telescope-related paraphernalia!

Let me just take a moment to calm your fears at the thought of spending more money on this hobby. Sure, it is easy to draw up a list of must-have items as you look through astronomical catalogues and magazines. Everything could easily tally up to more than the cost of your telescope in the first place, but is this truly necessary? Happily, the answer is no. Before you buy another item, we must first explore the accessories that are absolutely mandatory, those that can wait for another day, and the ones that can be done without altogether.

A quick disclaimer before going on: So many accessories are available to entice the consumer that it would be impossible for a book such as the one you hold before you to list and evaluate every item made by every company. As a result, this chapter must limit its coverage to more readily available items. If, as you are reading this chapter, you feel that I have unjustifiably omitted something that you feel is the greatest invention since the telescope, then by all means share your enthusiasm with me. Write your own review and send it to me in care of the address listed in the acknowledgments at the beginning of this book. I will try to include mention of that item in a future edition.

Finderscopes

After eyepieces (and arguably, even before), the most important accessory in an amateur astronomer's bag of tricks is a finderscope. Finderscopes (Figure 6.1) are small, low-power, wide-field spotting scopes that are mounted piggyback on telescope tubes. Their sole purpose is to help the observer aim the main telescope toward its target. Most telescopes come with finders, but all are not created equally! What sets a good finder apart from a poor finder depends on how the finder is going to be used. First, some words of advice for readers who will be using finders only to supplement the use of setting circles for zeroing in on sky targets. If this applies to you (and I truly hope it does not, especially after reading my editorial in chapter 9), then the finder will probably be employed only for locating brighter Solar System objects, probably Polaris to align the equatorial mount to the celestial pole, and perhaps a few terrestrial objects. In this case, just about any finder will do.

If, however, star-hopping is your preferred method for locating sky objects (again, consult chapter 9), then finderscope selection is critical to your success as an observer. The three most important criteria by which to judge a finder are magnification, aperture, and field of view. Finderscopes are specified in the same manner as binoculars. A 10 × 50 finder, for example, has a 50-mm aperture and yields 10×. Most experienced observers agree that the smallest useful size for a finderscope is 8 × 50, although finders, like binoculars, should be sized to match sky conditions. Under suburban skies, an 8 × 50 finder (with a 6-mm exit pupil) will penetrate to about 8th magnitude, which is roughly comparable to the magnitude limit of many popular star atlases. Rural astronomers with darker skies may prefer giant 10 × 70 and 11 × 80 finders. These reveal stars a magnitude or two fainter but suffer from smaller fields of view. Meade Instruments; the Telescope and Binocular Center (formerly Orion Telescope Center); Roger W. Tuthill, Inc.; and Lumicon offer good selections of finders from which to choose.

Figure 6.1 An 8×50 finderscope.

Some telescopes come with right-angle finders. In these, a mirror- or prism-based star diagonal is built into the finder to turn the eyepiece at a 90° angle. Sure, a right-angle prism can make looking through the finder more comfortable and convenient, but is it really a good idea? To my way of thinking, *no!* There are two big drawbacks to right-angle finders. First, although the view through a right-angle finder is upright, the prism also flips everything left to right. This mirror-image effect matches the view through a telescope using a star diagonal (Schmidt-Cassegrain, refractor, etc.) but makes it very difficult to compare the field of view with a star atlas. By comparison, straight-through finders flip the view upside down but do not swap left and right. Personally, I find it easier to turn a star chart upside down to match an inverted view than to turn it inside out to matched a mirrored image! (Special prisms, called Amici prisms, can be used in place of an ordinary star diagonal to cancel out the mirroring effect but are supplied with relatively few finders, probably because they also cause images to dim.)

A straight-through finder also permits the use of both eyes when initially aiming the telescope. By overlapping the naked-eye view with the one seen through the finder, an observer may point the telescope quickly and accurately toward the desired part of the sky. If you keep both eyes open when using a right-angle finder, one eye will see the sky, and the other will see the ground!

Deep down inside, most observers agree that right-angle finders are poor substitutes for their straight-through counterparts. At least I assume they must, given the incredible popularity of 1× aiming devices in recent years. By far, the most common of these sighting contraptions is the Telrad (Figure 6.2), invented by the late amateur telescope maker Steve Kufeld. The Telrad is described as a "reflex sight." Using a pair of AA batteries and a red LED, a bull's-eye target of three rings (calibrated at 0.5°, 2°, and 4°) is projected onto a clear piece of glass set at a 45° angle. The observer then sights through the

Figure 6.2 *A one-power Telrad aiming device by Steve Kufeld, one of the most popular accessories among today's amateur astronomers.*

glass, which acts as a beamsplitter, to see the reflected target rings as well as stars shining through. The brightness of the rings is controlled by a side-mounted rheostat that also acts as an on/off switch.

The Telrad is not meant to be used in lieu of a finder but only to supplement its use. The biggest advantage of the Telrad is that it allows telescopes to be aimed easily without any need to flip or twist star charts. Its only disadvantage is that the window tends to dew over quickly in damp environs, but because it is not an optical device, the window may be wiped clear with a finger, sleeve, or paper towel. A more permanent solution to the dewing problem is to simply make a roof over the glass window by bending a large file card that has been painted black over the Telrad from side to side and secure it in place with masking tape.

Rigel Systems offers the QuickFinder, which looks and works like a vertical Telrad. The QuickFinder projects an adjustable-brightness 0.5°-diameter red circle onto an angled window. Unlike any of the other 1× aiming devices, the QuickFinder's red circle can be set to pulse off and on. This allows the observer to keep the circle's brightness relatively high while aiming at dim stars, a helpful feature. The QuickFinder weighs just 3.4 ounces (as compared to Telrad's 11 ounces) and stands about four and a half inches tall. Its small footprint, just 1.38 inches, is the most compact of any 1× aiming device currently made, although it also makes me wonder if it might be easily snapped off. To prevent this from happening, the QuickFinder can and should be removed from its base and stored separately whenever the telescope is moved.

Two other 1× aiming devices, Celestron Star-Pointer and Tele Vue Qwik-Point, are direct adaptations of an aiming device that Daisy Manufacturing Company markets for its BB guns and pistols (as is the homemade aiming device detailed in chapter 7). Both are made from black plastic and project single red dots onto partially reflective curved windows that measure about three-eighths of an inch in diameter. The observer-side surfaces of the windows are coated with a clear reflective material, while the outer surface is uncoated. Both surfaces reflect images of the red dot, which merge into one when the observer's eye is on axis. Power to the units is provided by thin button batteries mounted on the bottoms of the units. Both models can be adjusted for brightness, although doing so requires a small screwdriver. The Tele Vue unit's dimmer control is on an exposed circuit board that protrudes in front, while the Celestron's dimmer is built in. Each comes with a mounting base (Tele Vue's is plastic, Celestron's is aluminum) for mounting on its as well as others' telescopes.

In principle, Orion's EZ-Finder is the same as the Tele Vue and Celestron models. Its long tube, however, makes it a little more difficult to sight through when out under the stars. Like the others, its brightness is fully adjustable, in this case by turning a small side-mounted potentiometer.

Leanest and sexiest of all is Tele Vue's Star Beam. Like the Celestron Star-Pointer, Orion EZ-Finder, and Tele Vue Qwik-Point, the Star Beam projects a red LED dot onto a partially reflective window through which the observer looks when aiming the telescope. However, the others are made of plastic, while the slim body of the Star Beam is crafted from machined aluminum. Not surprisingly, the Star Beam costs about four times as much as the others. Beauty may be only skin deep, but cost cuts right to the bone, so the Star Beam has never achieved the popularity of some of its ugly duckling cousins.

Finally, I'll say a word or two (OK, a paragraph) on finder mounts. Good finders are secured to telescopes by a pair of rings with six adjustment screws that allow the finder to be aligned precisely with the main instrument. Let the buyer beware, however, that many smaller finders (primarily 5 × 24 and 6 × 30 models) use single mounting rings with only three adjustment screws. These are notoriously difficult to adjust and even more difficult to keep in alignment. If there is a choice between a single ring or a pair of rings for mounting a finder, always select the pair.

Filters

For years, photographers have known the importance of using filters to change and enhance the quality and tone of photographs in ways that would be impossible under natural lighting conditions. Today, more and more amateur astronomers are also discovering that viewing the universe can also be greatly enhanced by using filters with their telescopes and binoculars. Some heighten subtle, normally invisible planetary detail; others suppress the ever-growing effect of light pollution; and still others permit the safe study of our star, the Sun. Which filter or filters, if any, are right for you depends largely on what you are looking at and from where you are doing the looking.

Light-Pollution Reduction (LPR) Filters

As all amateur astronomers are painfully aware, the problem of light pollution is rapidly swallowing up our skies. In regions across the country and around the world, the onslaught of civilization has reduced the flood of starlight to a mere trickle of what it once was. Are we powerless against this beast? Not entirely.

While chapter 9 discusses the dilemma of light pollution in more specific detail, this section examines some filters that may be used to help counteract part of the problem. First, let's briefly define *light pollution*: unwanted illumination of the night sky caused largely by poorly designed or poorly aimed artificial lighting fixtures. Rather than illuminating only their intended targets, many fixtures scatter their light in all directions, including up.

There are two kinds of light pollution: *local* and *general*. Local light pollution shines directly into the observer's eyes and may be caused by anything from a nearby streetlight to an inconsiderate neighbor. No currently available filter will counteract this sort of interference. LPR filters are much more effective against general light pollution, or *sky glow.* The most destructive type of light pollution, sky glow is the collective glare from untold hundreds or even thousands of distant lighting fixtures. It can turn a clear, blue daytime sky into a yellowish, hazy night sky of limited usefulness.

Although modern technology caused the problem in the first place, it also offers partial redemption. As indicated in Figure 6.3, many sources of light pollution shine in the yellow region of the visible spectrum, between 550 nm and 630 nm [nm is short for *nanometer,* a very small unit of measure; one nanometer is equal to 10^{-9} meters, or 10 ångstrom (Å) units]. At the same time, many nonstellar sky objects (planetary nebulae, emission nebulae, and comets) emit most of their light in the blue-green region of the spectrum. Emission nebulae, for example, glow primarily in the hydrogen-beta (486 nm) and oxygen-III (496 nm and 501 nm) spectral regions. In theory, if the yellow wavelengths could somehow be suppressed while the blue-green wavelengths were allowed to pass, then the effect of light pollution would be greatly reduced.

What exactly do light-pollution reduction filters do? A popular misconception is that LPR filters (Figure 6.4) make faint objects look brighter. Not true! LPR filters consist of thin pieces of optically flat glass that are coated in multiple, microscopically thin layers of special optical material. They are designed to block specific wavelengths of light while letting others pass. The

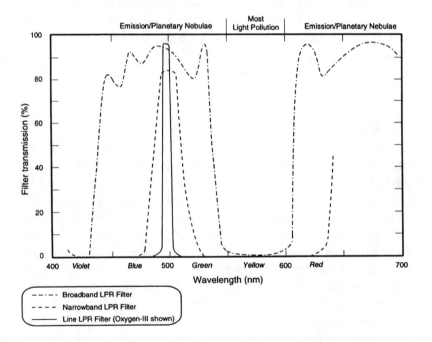

Figure 6.3 *Transmission characteristics of light-pollution reduction (LPR) filters. Chart based on data supplied by Lumicon, Inc.*

Figure 6.4 *An assortment of LPR filters. Photo courtesy of Lumicon, Inc.*

observer need only attach the filter to his or her telescope (usually either by screwing the filter into the field end of an eyepiece or, in the case of many Schmidt-Cassegrains, inserting it between the telescope and star diagonal). The net result is increased contrast between the object under observation and the filter-darkened background. (Many deep-sky objects also shine in the deep-red hydrogen-alpha [656nm] portion of the spectrum, a region that is all but invisible to the eye under dim light conditions. Still, most LPR filters transmit these wavelengths as well for photographic purposes.)

Light-pollution reduction filters (also called *nebula filters*) come in three varieties: *broadband, narrowband,* and *line filters.* The biggest difference is in their application. As shown in Figure 6.3, broadband filters pass a wide swath of the visible spectrum, from 430 nm to 550 nm or so. Narrowband filters transmit wavelengths of light between 480 nm and 520 nm. Line filters have an extremely narrow window, allowing only one or two specific wavelengths of light to pass. No one type of filter is best for everything. Some enhance the object under observation, some have little or no effect, while others may actually make the target fade or disappear completely! Which LPR filter is best under what circumstances remains one of the hottest topics of debate among amateur astronomers today. Table 6.1 may help sort things out. Each filter is rated on a four-star scale based on performance. One star indicates either negligible or negative results, two stars connote minimal positive effect, three stars indicate a noticeable improvement in appearance, and four stars denote a vast improvement in object visibility and/or detail.

Should you even buy a filter? If you are interested at all in deep-sky observing, then the answer is an unqualified yes. *Which* filter you should buy, however, is not so easy to figure out. The answer depends on where you do your observing. If you are a captive of a city, then you ought to consider a broadband filter. Though the improvement in deep-sky objects is marginal, broadband filters do help increase contrast with the surrounding sky, making

Table 6.1   **Light-Pollution Reduction (LPR) Filters: A Comparison**

	Open Clusters	Globular Clusters	Emission Nebulae	Reflection Nebulae	Planetary Nebulae	Galaxies	Comets
Broadband							
Celestron LPR	*	*	**	**	***	*	**
Lumicon Deep Sky	**	**	**	**	***	**	**
Meade 908B	*	**	**	**	***	*	**
Orion Skyglow	**	**	**	**	***	**	**
1000 Oaks LP-1	**	**	**	**	***	**	**
Narrowband							
DayStar 300	*	**	***	**	****	*	*
Lumicon UHC	*	*	****	**	****	*	*
Meade 908N	*	*	***	**	****	*	*
Orion Ultrablock	*	*	****	**	****	*	*
1000 Oaks LP-2	*	*	***	**	****	*	*
Line							
Lumicon H-Beta	*	*	***	*	*	*	*
Lumicon OIII	*	*	***	**	****	*	*
Lumicon Swan Band	*	**	**	**	***	*	****
1000 Oaks LP-3 (OIII)	*	*	****	**	****	*	*
1000 Oaks LP-4 (H-Beta)	*	*	***	*	*	*	*

the view more aesthetically pleasing. All are very close to one another in performance, and you will probably be happy with any of those detailed here. But if I had to rank their performance, the Orion SkyGlow filter seems to work the best, followed closely by the Lumicon Deep Sky, Thousand Oaks LP-1 broadband filter, Meade 908B, and Celestron LPR filters in that order (the Meade and Celestron are virtually identical). From suburban or rural sites, however, broadband filters are nearly worthless.

Observers interested in spotting fine detail in emission and planetary nebulae, regardless of where they are viewing, would do better by purchasing either a narrowband or line filter. Each is a vast improvement over an unfiltered telescope as well as any of the broadband filters. And as an added bonus, each narrowband filter also improves the views of Mars and Jupiter, making them more versatile than perhaps even their manufacturers realize! Which is best? Honestly, all work well, but if I had to rate them in order of preference, I would give a slight nod to the Orion Ultrablock, Lumicon UHC, and Thousand Oaks LP-2, followed closely by the DayStar 300 and Meade 908N. Personally, I would recommend that you let price and availability be your guide; you should be happy with any of them. One exception to this is the Meade 908N, which seems to outperform the others by a noticeable margin when used with refractors (although I don't know why).

Finally, what about the line filters? Owners of medium- and large-aperture instruments should also give serious consideration to an O-III filter. Although

I personally prefer the narrowband filters, oxygen-III (O-III) filters have a well-deserved and loyal following. Frequently, they will reveal subtle features in nebulae that remain invisible through the others. Indeed, for many bright emission nebulae, O-III filters outperformed all but the Orion UltraBlock and Lumicon UHC filters. My only complaint is that O-IIIs tend to darken the background sky too much. Both the Thousand Oaks LP-3 and the Lumicon O-III work well, but I'll cast my vote for the LP-3.

Lumicon's H-Beta filter and Thousand Oaks's LP-4 hydrogen-beta filter work well on a few objects, notably the Horsehead and California Nebulae (two notoriously difficult deep-sky objects), but on most objects they will actually have a negative impact! Finally, comets, because of their unique spectra, are best observed using Lumicon's Swan Band filter, designed with comets in mind.

The bottom line is this: LPR filters are wonderful assets for the amateur astronomical community. With them, observers can spot nonstellar objects that were considered impossible to find just a few years ago. But remember, no filter can deliver the beauty of the universe as well as a truly dark, light-free sky.

Lunar and Planetary Filters

Most seasoned observers agree that fine details on both the Moon as well as the five planets visible to the naked eye can be significantly enhanced by viewing through color filters. This improvement occurs for two reasons. First, they reduce irradiation, the distortion of the boundary between a lighter area and a darker region on the Moon's or a planet's surface. This effect is usually caused by either turbulence in the Earth's atmosphere or by the human eye being overwhelmed by the dazzling image.

Filters also help to increase subtle contrasts between two adjacent regions of a planet or the Moon by transmitting (brightening) one color while absorbing (darkening) some or all other colors. For instance, an observer's eye alone may not be able to distinguish between a white region and a bordering beige region on Jupiter. By using filters of different colors, the contrast between the zones may be increased until their individuality becomes apparent.

Which filters are best for which objects in the sky? Table 6.2 compares different heavenly sights with the results that may be expected when they are viewed through a variety of color filters, which are listed according to their Wratten number, as well as their color. The Wratten series of color filters was created by Eastman Kodak and contains over one hundred different shades and hues, each of which is assigned a number; today, the Wratten system is the industry's standard way to refer to a filter's precise color. (The entire series was revamped not too many years ago, which is why some older books may refer to, say, a deep yellow filter as a K3 instead of referring to it as a Number [#]15.)

For those just starting out, choose basic colors, such as deep yellow (#15), orange (#21), red (#23), green (#58), and blue (#80A). Although you are free to

Table 6.2 **Color Filters: A Comparison**

Object	Filter	Result
Moon	Moon filter (Neutral density)	Reduces brightness of Moon evenly across the spectrum, making observations easier without introducing false color
	#15 (Deep yellow)	Enhances contrast of lunar surface
	#58 (Green)	Like deep yellow, green will enhance contrast and detail in some lunar features
	#80A (Light blue)	Reduces glare
	Polarizer	Like the neutral density filter, reduces brightness without introducing false colors
Mercury	#21 (Orange)	Helps to see the planets' phases
	#23A (Red)	Increases contrast of the planets against blue sky, aiding in daytime or bright twilight observation
	#25 (Deep red)	Same as #23A, but deeper color
	#80A (Light blue)	Improves view of Mercury against bright orange twilight sky
	Polarizer	Darkens sky background to increase contrast of planet; helpful for determining phase of Mercury
Venus	#25 (Deep red)	Darkens background to reduce glare; some say they also help reveal subtle cloud markings
	#80A (Light blue)	Improves view of Venus against bright orange twilight sky
	Polarizer	Reduces glare without adding artificial color (especially important for viewing the planet through larger telescopes)
Mars	#21 (Orange)	Penetrates atmosphere to reveal reddish areas and highlight surface features such as plains and maria (best choice for small apertures)
	#23A (Light red)	Same as #21, but deeper color (best choice for medium and large apertures)
	#25 (Deep red)	Same effect as #23A, but deeper color (best choice of the three for very large apertures)
	#38A (Deep blue)	Brings out dust storms on the surface of Mars
	#58 (Green)	Accentuates "melt lines" around polar caps.
	#80A (Light blue)	Accentuates polar caps and high clouds, especially near the planet's limb
Jupiter	#11 (Yellow-green)	Reveals fine details in cloud bands
	#21 (Orange)	Accentuates cloud bands
	#56 (Light green)	Accentuates reddish features such as the Red Spot
	#58 (Green)	Same as #56, but has deeper color
	#80A (Light blue)	Highlights details in orange and purple belts as well as white ovals
	#82A (Very light blue)	Similar effect as #80A, though not as pronounced
Saturn	#15 (Deep yellow)	Helps to reveal cloud bands
	#21 (Orange)	Similar effect as #11, but deeper color
Comets	#80A (Light blue)	Increases contrast of some comets' tails
Other	#15 (Deep yellow)	Helps block ultraviolet light when doing black-and-white astrophotography
	#25 (Red)	Reduces impact of light pollution on long-exposure black-and-white photographs taken from light-polluted areas
	#58 (Green)	Same as #25; works well for emission nebulae
	#82A (Very light blue)	Suppresses chromatic aberration in refractors
	Minus Violet	Reduces impact of light pollution without dramatically distorting color; cheaper than broadband LPR filters

use photographic filters sold in camera stores, most amateurs prefer color filters designed to screw into the field end of eyepieces. Many telescope manufacturers and suppliers, including Orion, Meade, and Celestron, offer a wide assortment of standard-threaded color filters.

Solar Filters

Monitoring the ever-changing surface of the Sun is an aspect of the hobby that is enjoyed by many. Before an amateur dares look at the Sun, however, he or she must be aware of the extreme danger of gazing at our star. **Viewing the Sun without proper precautions, even for the briefest moment, may result in permanent vision impairment or even blindness.** This damage is caused primarily by the Sun's ultraviolet rays, the same rays that cause sunburn. It may take many minutes before the effect of sunburn is felt on the skin, but the Sun's intense radiation will burn the eye's retina in a fraction of a second.

There are two ways to view the Sun safely: either by projecting it through a telescope or binoculars onto a white screen or piece of paper or by using a special filter. Sun filters come in a couple of different varieties. Some fit in front of the telescope, and others attach to the eyepiece. ***NEVER*** use the latter ... that is, the eyepiece variety. They can easily crack under the intense heat of the Sun (focused by the telescope, as is the Sun's image), leading tragically to blindness. Happily, I know of no new telescope that is supplied with an eyepiece solar filter, but many were in the past.

Safe solar filters fit securely in front of a telescope or binoculars (Figure 6.5 shows one example). This way, the dangerously intense solar rays (and accompanying heat) are reduced to a safe level prior to entering the optical instrument. Be sure to use only specially designed solar filters; *do not* use photographic neutral-density filters, smoked glass, overexposed photographic film, or other makeshift materials that may pass invisible ultraviolet or infrared light.

Full-aperture solar filters are made from either Mylar or glass. Let's take a look at each option. The most popular glass solar filters are sold by Thousand Oaks Optical Company, the Telescope and Binocular Center (formerly Orion Telescope Center) and JMB, Inc. All three brands are about comparable in performance, displaying the Sun as a yellow-orange disk. Thousand Oaks and JMB solar filters come in three types. Thousand Oaks's Type 2 filters are triple-coated with nickel, chrome, and stainless steel and come mounted in an aluminum cell. The new Type 2+ filters are coated with Solar II Plus, a coating that the manufacturer claims is more durable. Both reduce the Sun's brightness to 0.001% of its original intensity and are designed for visual and photographic use. Type 3 filters reduce the Sun's brilliance to 0.01% and are intended for photographic use only. JMB also offers three different solar filters, its Class A, B, and C being roughly equivalent to the Thousand Oaks lineup. The Orion filters are most similar to the Type 2 variants.

Figure 6.5 *Full-aperture solar filter. Photo courtesy of Thousand Oaks Optical.*

Roger W. Tuthill, Inc., has been selling the popular Solar Skreen filters since 1973. Solar Skreen filters are made from two tissue-paper-thin pieces of aluminized Mylar that are sandwiched together. Like most of the glass models, Solar Skreen filters reduce the Sun's brightness to 0.001% of its unfiltered intensity. There are some pluses and minuses to Solar Skreen filters. One drawback is that they turn the Sun blue. This can be a little disconcerting at first, but most people get used to it after a while. If you must see a yellow Sun (when doing color solar photography, for instance), add a Wratten #15 (deep yellow) or #21 (orange) filter to the eyepiece or camera. Mylar filters also are not as durable as glass filters and are prone to pinholes, although with proper care they can last for years. On the plus side, inch for inch, Solar Skreen filters are less expensive than their glass rivals. And unlike glass filters, which may warp slightly under the heat of the Sun, causing image distortion and focus shift, Mylar filters retain their optical properties.

Both glass and Mylar filters may be purchased either already mounted in a cell designed to slip over the front end of a specific telescope or unmounted. Because these are full-aperture filters, the larger the telescope, the more expensive the filter. Keep in mind, however, that a telescope's full aperture is not required for observations. To save a few dollars, purchase a small, unmounted filter and then mount it yourself using an off-axis mask similar to that detailed in chapter 8 for testing telescope optics. Of course, although the price will be lower, definition may be a little lower, too, owing to the smaller aperture.

All of the solar filters just discussed might be thought of as broadband filters, because they filter the entire visible spectrum (and you can't get much broader than that). There are also special narrowband solar filters that allow observers to see our star in a completely different light! These filters block all of the light from the Sun with the exception of one distinctive wavelength: 656 nm. This is a frequency of glowing hydrogen called hydrogen-alpha (H-alpha). Viewing the Sun with an H-alpha filter allows the observer to monitor ruby-red solar prominences, bright white filaments, and intricate surface granulation.

Hydrogen-alpha filters typically consist of two separate pieces: an energy-rejection filter (ERF) that fits over the front end of a telescope to prevent overheating and the H-alpha filter itself, which fits between the telescope and eyepiece. The filters are available in several different, extremely narrow bandwidths that are usually expressed in angstroms. The narrower the bandwidth, the higher the contrast but fainter the image. Bandwidths of either 0.8 or 0.9 angstrom are the most popular because these offer the best compromise between brightness and contrast.

Unfortunately, H-alpha solar filters are expensive. DayStar Filter Corporation is one of the leading sources for these special accessories. The most expensive of the DayStar filters is the University model, which, as the name implies, is geared toward the rigid requirements of professional institutions. The ATM filter, on the other hand, is designed with the serious amateur solar astronomer in mind. Both the University and ATM filters need to be heated in order to function properly and therefore require an external source of AC power. The least expensive H-alpha filter of all DayStar models is the T-Scanner. Unlike the other two, the T-Scanner requires no external power supply and so is completely portable, making it even more attractive. Its only drawback is that the bandwidth passed is a little wider than is passed by the others, lowering contrast some. Still, the T-Scanner is the most popular hydrogen-alpha filter available today.

Lumicon also manufactures H-alpha solar filters for the amateur marketplace. The Lumicon Solar Prominence Filter works on the same no-heater principle as the T-Scanner. Its 1.5-angstrom bandpass, wider than that of any of the DayStar filters, renders good views of prominences but is not designed for viewing flares, filaments, and granulation. Complete with ERF prefilter and all required adapters and hardware, the Lumicon filter costs hundreds of dollars less than the T-Scanner and is an excellent value for anyone living within a budget who wishes to get into this phase of solar observing.

Other Eyepiece Accessories

Collimation tools. A telescope will deliver only poor-quality, lackluster images unless it is in proper optical alignment, or collimation. Several companies manufacture collimating tools for amateur telescopes, with some more useful for certain optical designs than others. Typically, refractors should never have to be collimated, while reflectors and Schmidt-Cassegrains should be checked regularly. Rich-field reflectors (that is, those with focal ratios less than f/6) should be checked each time they are set up.

The least expensive collimation tool sold today, at less than $10, is the Aline from Rigel Systems. The Aline is simply a black plastic cap with a centrally drilled peephole and a reflective surface on the cap's inside surface. Place the Aline into your telescope's 1.25-inch focuser and look through the hole. Adjust the telescope's diagonal and primary mirrors until both appear centered. The Aline is a handy tool for slower-focal-ratio reflectors, but faster instruments require some of the more sophisticated tools discussed in the following paragraphs.

The three most popular collimation tools remain the sight tube, the Cheshire eyepiece, and the Autocollimator. The sight tube is helpful for approximating the collimation of just about any telescope. At one end lies a small peephole used by the observer to check collimation, while at the other, a pair of thin crosshairs serves as a reference for centering optical components. (Collimation procedures are outlined in detail in chapter 8.) Tectron Instruments; AstroSystems, and the Telescope and Binocular Center (Orion) all sell 1.25-inch-diameter sight tubes that will fit most amateur telescopes. The Cheshire eyepiece (Figure 6.6), invented 75 years ago by Professor F. J. Cheshire at a British university, is intended primarily for in-the-field adjustments of fast Newtonians. Not an optical assembly, a Cheshire eyepiece is a variation on the sight tube with a part of one side cut out and a 45° mirrored surface inside. Tectron, AstroSystems, and the Telescope and Binocular Center all sell Cheshire eyepieces. Those sold by the former two include sight-tube adapters to create handy all-in-one sight tube/Cheshire eyepiece combinations. Last and certainly not least, the Autocollimator is the most sensitive of the three tools for collimating a fast Newtonian reflector. At first glance, it looks like a short sight tube, that is, a hollow tube capped with a piece of metal that has a tiny hole drilled in the center. The difference becomes evident only when you look inside the Autocollimator. There, encircling the peephole and perpendicular to the eyepiece mount's optical axis, a flat mirror serves to reflect all light from the primary mirror back down the telescope. Only Tectron sells the Autocollimator.

Are all three necessary? The sight tube is a *must* for anyone who owns a telescope and is concerned with optimizing its performance. I also consider the Cheshire eyepiece a requirement for owners of faster Newtonian reflectors.

Figure 6.6   *A Cheshire eyepiece, a must for collimating fast Newtonian reflectors.*

Don't leave home without it! The Autocollimator, though a handy tool for fine-tuning the alignment of fast Newtonians, is not absolutely necessary.

Recently, laser collimators have become all the rage in certain circles. Lasers emit precise, narrow beams of light that do not expand or elongate as they travel away from their source. Within the past few years, small, low-power units have been created and incorporated into many diverse applications, including telescopes. To aid telescope enthusiasts, AstroSystems has created the AstroBeam laser collimator.

Do laser collimators work as advertised? The short answer is yes, but with some qualification. These accessories are not designed to be used with refractors or catadioptric instruments; they are intended only for short-focus, rich-field Newtonian reflectors. If your Newtonian has a longer focal ratio than this, then the three previously mentioned collimation tools are just fine.

The principle behind a laser collimator is simple enough. Place the laser into the eyepiece holder and turn on the beam. If the telescope's optics are properly collimated, the beam should bounce from the diagonal mirror to the primary mirror and then exactly back on itself. You then look through the front of the telescope tube at the bottom of the collimator protruding out of the focusing mount. If you see an off-center image of the beam on the collimator, the optics are not properly aligned.

Laser collimators are not without their drawbacks. First, all of the preceding discussion assumes that the laser collimator itself is collimated. To its credit, AstroSystems offers a check-and-clean service for the lifetime of its collimators if you suspect that something is amiss. Secondly, although the AstroBeam is designed to fit both 1.25- and 2-inch focusers, the snout of the collimator does not protrude enough to make it easily visible in some focusing mounts. Finally, there is the cost: At more than one hundred and sixty dollars, a laser collimator is far more expensive than other collimation tools.

Rubber eyecups. To help shield an observer's eye from extraneous light, many eyepieces sold today come with collapsible rubber eyecups, and they certainly can make a big difference!

In the previous chapter, I tried to point out whenever an eyepiece came with a built-in eyecup. But just because an eyepiece is cupless doesn't mean it is not worth consideration. Most can be retrofitted with aftermarket eyecups that work just as well. The Telescope and Binocular Center, Edmund, Tele Vue, and several other companies sell rubber eyecups designed to fit both their own particular lines of oculars as well as others'. Always match the proper size eyecup to your eyepiece before ordering, because a too-tight or too-loose eyecup isn't of much value.

Binocular viewers. Research has shown that when it comes to viewing the night sky, binocular (two-eyed) vision is definitely better than monocular

Figure 6.7 *Binocular viewing attachment for telescopes. Photo courtesy of Tele Vue Optics, Inc.*

(one-eyed) vision. Our power of resolution and ability to detect faint objects are dramatically improved by using both eyes. Some people enjoy up to a 40% increase in perception of faint, diffuse objects by viewing them with both eyes instead of just one. In addition, color perception and contrast enhancement also benefit.

Few experts will argue against the benefits of using binoculars rather than a monocular telescope of comparable size, but the advantages of binocular viewing attachments for conventional telescopes are not as clear. Binocular viewers (Figure 6.7) customarily either screw onto a telescope (as with Schmidt-Cassegrain telescopes) or slide into a telescope's focusing mount. Inside, a beamsplitter cuts the telescope's light into two paths of equal intensity, sending the light toward two eyepiece holders.

Are these rather costly accessories worth the price? Well, yes and no. First, they must be used with two absolutely identical eyepieces, raising the total investment even higher. Even then, however, they do not work as well as true binoculars. This should come as no surprise when you stop and think about it. In essence, binoculars are two independent telescopes strapped together. Binocular viewers, on the other hand, must rely on the light-gathering power of a single telescope. As such, the images perceived will be dimmer than those obtained through the same telescope outfitted with a single eyepiece.

Binocular viewers come into their own when viewing the brighter planets and, especially, the Moon. Our satellite seems to look three-dimensional (actually, an optical illusion), an effect that cannot be duplicated with a traditional monocular telescope. In addition, an increase in planetary detail can also be expected with a binocular viewer thanks to the aforementioned improvement in subtle contrast and color perception.

Several companies offer binocular viewers, but most use prisms that are too small for the intended purpose, causing field vignetting, especially at lower magnifications. In fact, of those sold, the only one that works well is that by

Tele Vue. As one owner summed up his experiences: "When I have compared Saturn [through an Astro-Physics 180EDT apochromatic refractor] with and without the Tele Vue binocular viewer at similar magnifications, I find the difference in illumination almost unnoticeable. Markings on the disk of Saturn, however, are decidedly more apparent with the binocular viewer than without. Admittedly, however, some dimming of faint objects is detectable." To my eye, the dimming amounts to half a magnitude, give or take.

Consumers should note that not all telescopes (especially Newtonian reflectors) have enough range in their eyepiece focusing mounts to accommodate the long path that light must follow inside a binocular viewer. Another owner found that the only way that he could get his 18-inch Newtonian to focus with a binocular viewer was to cut 5 inches from the instrument's truss poles! He notes that "while that part was easy, it created yet another difficulty ... with the diagonal mirror that much closer to the primary, the light cone at that point was larger than the diagonal mirror, causing an effective loss in aperture." He ultimately had to use a larger diagonal mirror built into a new upper tube/eyepiece assembly exclusively with the binocular viewer.

Star Atlases, Periodicals, Books, and Computer Software

Not long ago, I read that no pastime has more new books published about it each year than amateur astronomy. Although that's bad news for us authors (too much competition!), it is good for the hobby. In fact, with so many excellent books, periodicals, and software packages available, it is difficult to draw up a short list. Here is a *brief* listing of some of the better resources on the market today.

Periodicals

Amateur Astronomy, 3544 Oak Grove Drive, Sarasota, FL 34243. *Amateur Astronomy* is a quarterly magazine that offers a nice mix of stories on observing projects, homemade telescopes, and recaps of astronomy conventions. Although it lacks the gloss and color of the larger magazines, it offers an alternative to amateurs who long for more "backyard astronomy" articles.

Astronomy, P.O. Box 1612, Waukesha, WI 53187. This is North America's best-selling astronomical periodical. Each issue of *Astronomy* includes several feature articles covering a wide range of topics as well as a large monthly star map, regular columns on sky events and observing projects, book and product reviews, and occasionally details on how to construct some useful do-it-yourself projects. A showcase of amateur astrophotographs is also highlighted, with a cash prize awarded to "the best" each month.

Astronomy Now, **193 Uxbridge Road, London W12 9RA, Great Britain.** From England comes this popular-level monthly periodical. Each issue of *Astronomy Now* contains several interesting feature articles, test reports, book reviews, and columns highlighting sky events for the month to come.

ATM Journal, **17606 28th Avenue SE, Bothell, WA 90012.** Like the sorely missed magazine *Telescope Making,* *ATM Journal* offers amateur telescope makers a forum to display and describe homemade telescopes, observatories, and other accessories. Published quarterly but has suffered some inconsistency in the past.

Griffith Observer, **Griffith Observatory, 2800 East Observatory Road, Los Angeles, CA 90027.** Published bimonthly, the *Griffith Observer* features one or two articles on anything from astronomical history to the latest discoveries as well as a simple star map and sky data.

Sky & Telescope, **P.O. Box 9111, Belmont, MA 02178.** Like *Astronomy, Sky & Telescope* includes feature articles geared for beginning and intermediate amateurs as well as book and software review columns and brief highlights of late-breaking astronomical news. After the centerfold star map of the monthly sky, regular departments highlight sky events occurring during the month.

Sky Calendar, **Abrams Planetarium, Michigan State University, East Lansing, MI 48824.** Here is the cheapest way to learn what's up in the sky. Published monthly, this single-sheet publication features a calendar-style format that highlights one or more interesting naked-eye sights for most nights of the month. On the reverse side is a monthly star map for constellation identification. *Sky Calendar* is highly recommended for beginners.

Sky News, **National Museum of Science and Technology, P.O. Box 9724, Station T, Ottawa, Ontario K1G 5A3 Canada.** This new bimonthly magazine comes as one of the benefits of joining Canada's National Museum of Science and Technology as an "astronomy associate." Inside each issue, readers will find several feature stories, equipment reviews, and a sky chart and calendar covering the issue's two-month period. *Sky News* is edited by Terence Dickinson, one of Canada's most prolific astronomy authors.

Stardate, **University of Texas, McDonald Observatory, Austin, TX 78712.** Each bimonthly issue of *Stardate* features a lead article or articles as well as a star map and descriptions of celestial events for the months to come.

Starry Messenger, **P.O. Box 6552, Ithaca, NY 14851.** The original used-equipment periodical, listing hundreds of classified advertisements for all sorts of telescopes, binoculars, and related equipment.

Annual Publications

Astronomical Calendar, **Ottewell, G.; Furman University.** The abundantly illustrated *Astronomical Calendar* describes celestial events visible around the world for the particular year of issue, making it one of the most useful publications in an amateur's library.

Exploring the Universe; Kalmbach Publishing. This magazine-style annual publication from *Astronomy* highlights the year's meteor showers, eclipses, planets, and comets as well as the four seasonal skies. An introduction to astrophotography and an overview of astronomy software is also included. *Exploring the Universe* is aimed primarily at beginning stargazers.

Observer's Handbook; Percy, R., et al.; Royal Astronomical Society of Canada. Like the *Astronomical Calendar*, the *Observer's Handbook* details all of the predicted celestial events that will occur during its year of publication, with events for each month listed chronologically. Special departments describe in detail upcoming eclipses, meteor showers, comets, and much more.

Sky Watch; Sky Publishing Corporation. *Sky Watch* itemizes the year's events and gives some rudimentary advice on using star maps and viewing the sky. Like *Astronomy* magazine's *Exploring the Universe, Sky Watch* is intended for novice sky watchers who might be intimidated by the *RASC Observer's Handbook* or *Astronomical Calendar*.

Star Atlases

Cambridge Star Atlas 2000.0; Tirion, W.; Cambridge University Press, 1996. An excellent first star atlas! The *Cambridge Star Atlas 2000.0* shows all stars down to magnitude 6.5 as well as 866 deep-sky objects. A list accompanies each colorful chart tabulating the finest objects found in that particular slice of sky.

Edmund Mag 6 Star Atlas; Dickinson, T., V. Costanzo, and G. Chaple; Edmund Scientific Company, Barrington, NJ, 1982. Another collection of star charts identifying stars down to magnitude 6.2 and hundreds of telescopic targets. Facing each chart is a list of interesting sights located within that area of sky. Also included is an excellent introduction to astronomy and observing.

Herald-Bobroff Astroatlas; Herald, D., and P. Bobroff; HB2000 Publications, 1994. One of the most sophisticated star atlases ever published, the *H-B Astroatlas* is designed to grow as its user becomes a more seasoned observer. How? By including not one but three sets of black-stars-on-white-background charts in one 12-by-16-inch volume. The first set of charts details the complete sky to magnitude 6.5, an ideal level for those just starting out with binoculars and smaller telescopes, and nearly three thousand deep-sky objects. The second set of charts, designed for those with 6- to 10-inch instruments, shows the sky to magnitude 9 and more than thirteen thousand deep-sky objects to magnitude 14. For overly crowded regions of the sky, these are augmented with a third set of charts that show stars as faint as magnitude 11.5 and deep-sky objects to magnitude 15. These are further augmented with even more detailed charts showing stars to magnitude 14 for selected regions, such as the Coma-Virgo Realm of Galaxies and the Large Magellanic Cloud. Spiral-bound to lie flat, the *H-B Astroatlas* is the closest thing to a perfect star atlas yet.

Millennium Star Atlas; Sinnott, R., and M. Perryman; Sky Publishing, 1997. The Millennium Star Atlas is the largest, most expensive star atlas for the amateur astronomer ever published. Scattered among its 516 9.25-by-13-inch charts are over one million stars to magnitude 11 and more than ten thousand deep-sky objects. The *Millennium* atlas is published in three hardbound volumes (though they are sold only as a complete set), with each covering an eight-hour slice of sky from pole to pole. Though thorough, its girth makes it difficult to manage at the telescope.

Norton's 2000.0; Ridpath, I., et al.; Longman Scientific and Technical, 1989. Now in its 18th edition, *Norton's 2000.0* (formerly known as *Norton's Star Atlas*) is the granddaddy of all modern star atlases. Data tables adjacent to each atlas page list details for many of the objects found in that particular part of the sky. Extensive discussions on telescopes and observing techniques make this more a reference book of amateur astronomy than just a star atlas.

1000+; Lorenzin, T., with T. Sechler; Lorenzin, 1987. Like *Norton's, 1000+* is much more than just a star atlas. The book is centered around descriptions of more than one thousand deep-sky objects, each described as seen through the author's 8-inch Schmidt-Cassegrain and plotted on the accompanying atlas, which shows stars to the 6th magnitude.

Sky Atlas 2000.0; Tirion, W.; Sky Publishing Corporation, 1981. The *Sky Atlas 2000.0* plots all stars in the sky down to magnitude 8 and 2,500 deep-sky objects on 26 charts. The *Sky Atlas 2000.0* is available in three editions: an unbound "field edition" (showing white stars on a black background), "desk edition" (black stars on white), and a spiral-bound "deluxe color edition" (black stars, red galaxies, etc.).

Uranometria 2000.0; Tirion, W., B. Rappaport, and G. Lovi; Willmann-Bell, 1987. The *Uranometria 2000.0* plots more than three hundred and thirty thousand stars to magnitude 9.5 and 10,000 deep-sky objects on 472 9-by-12-inch charts; it is sold in two volumes that may be purchased separately. The first volume covers the northern sky from declination +90° to −6°, while the second includes +6° to −90° (the overlap is intentional). This atlas is recommended for advanced amateurs only.

Introductory Books

Nightwatch; Dickinson, T.; Camden House, 1989. A favorite book among novice stargazers, *Nightwatch* covers a wide breadth of topics in a clear, easy-to-understand manner. Chapters cover the constellations, telescopic equipment, and observing techniques. Beginners will find the simplified star atlas in the back a terrific tool for learning their way around the telescopic sky.

Peterson's Field Guide to the Stars and Planets; Pasachoff, J., and D. Menzel; Houghton Mifflin, 1992. The *Field Guide to the Stars and Planets* is one of the finest introductory books to the science and hobby of astronomy. Though small in physical size, it contains more useful information than most books four times as large. The current edition includes an all-sky star atlas to magnitude 7.

Sky Observer's Guide; Mayall, N., et al.; Western Publishing, 1985. First published in 1959, the *Sky Observer's Guide* includes sections on observing the

Sun, Moon, planets, comets, meteors, asteroids, satellites, deep-sky objects, and just about anything visible in small telescopes.

***365 Starry Nights*; Raymo, C.; Fireside Press, 1990.** An excellent introductory guide for the whole family. Features a different sky activity for each night of the year, building the reader's knowledge with the passage of time.

Observing Guides

(Note that although a few of these guides include finder charts, most require the use of a separate star atlas.)

***Binocular Stargazer*; Peltier, L.; Kalmbach, 1995.** Several books on observing with binoculars are currently in print. My favorite is written for the intermediate-to-advanced amateur (see below), but this book is the best for those just starting out with a pair of small binoculars and a curiosity. Includes some simple star charts and a wonderful text written by the consummate amateur astronomer.

***Burnham's Celestial Handbook* (Volumes 1, 2, and 3); Burnham, R., Jr.; Dover, 1978.** Throughout its 2,138 pages, *Burnham's Celestial Handbook* breaks down the sky by constellation, covering just about everything in the known universe. Although much of its technical information is now outdated, *Burnham's* still offers a great guide for those wondering what deep-sky objects to look at on the next clear night.

***Eclipse!*; Harrington, P.; John Wiley & Sons, 1997.** It is hard to give an unbiased assessment of a book that I wrote, so instead, I'll quote from some comments made by eclipse expert Joe Rao, author of *Guide to the July 1991 Solar Eclipse* and *Guide to the January 1992 Solar Eclipse:* "This is a richly informative book on the phenomena of solar and lunar eclipses. Eclipsophiles will certainly find it to be a useful tool well into the 21st century, as it probably answers most every question you might have had about upcoming eclipses as well as some you probably hadn't thought of."

***Exploring the Moon Through Binoculars and Small Telescopes*; Cherrington, E.; Dover, 1983.** An essential guide for the devout lunatic (Moon watcher). Divided into 28 day-by-day chapters, the author describes the appearance and location of just about every major feature visible on our nearest neighbor.

***Hartung's Astronomical Objects for Southern Telescopes*; Malin, D.; Cambridge University Press, 1995.** An updated version of one of the best guides to the southern sky ever compiled. This book lists and describes hundreds of objects visible both above and below the celestial equator (focusing mainly on the latter) through amateur-size telescopes and also includes some terrific photographs.

***Messier Marathon Observer's Guide*; Maccholz, D.; MakeWood Products, P.O. Box 1716, Colfax, CA 95713, 1994.** Each March, many amateurs enjoy an annual ritual of trying to see all of the Messier objects in a single sunset-to-sunset "Messier Marathon." This guide is designed to help the reader plan and successfully run the marathon. Included within are specially prepared finder charts, a sequential checklist, and a table of the best dates for the marathon between now and 2019.

Observing Handbook and Catalogue of Deep-Sky Objects; Luginbuhl, C., and B. Skiff; Cambridge University Press, 1990. An excellent reference for the intermediate-to-advanced amateur. After brief introductory comments, the book describes over two thousand objects visible through medium-size telescopes. Sketches and photographs of some of the objects are included, as are computer-generated charts for some of the more crowded fields.

Touring the Universe Through Binoculars; Harrington, P.; John Wiley & Sons, 1990. Now, this is a book! (You expected me to say anything less?) *Touring the Universe Through Binoculars* is unique among binocular books, as it was written for the intermediate or advanced amateur who has made a conscious decision to use binoculars for observing the night sky. The book contains information on observing the Moon, Sun, planets, minor members of our Solar System, and over one thousand deep-sky objects.

Turn Left at Orion; Cosmolmagno, G., and D. Davis; Cambridge University Press, 1995. One of the best books for the beginning observer. This recently revised guide helps steer the reader to 100 of the most interesting night-sky objects by using clearly drawn charts, concise descriptions, and realistic eyepiece sketches. Also included are tables of lunar and planetary information through 2006.

Universe From Your Backyard; Eicher, D.; Cambridge University Press, 1988. From *Astronomy* magazine comes this collection of reprinted articles describing 690 of the finest deep-sky objects for backyard astronomers. Simple finder charts, photographs, and drawings combine to make this an excellent introduction to deep-sky observing.

Astrophotography

Astrophotography Basics (Publication P150); Eastman Kodak; Kodak, 1988. As the name implies, *Astrophotography Basics* provides entry-level instruction to help the beginner capture heavenly bodies successfully on film. And best of all, it can be had for free (in single quantities only). Call the Kodak Customer Service hotline at (800) 242-2424 and ask for a copy of Publication P150.

Astrophotography for the Amateur; Covington, M.; Cambridge University Press, 1991. This remains one of the finest books written for the amateur looking to break into the challenging field of astrophotography. Features good discussions on capturing on film just about everything in the sky, from the simplest star trails to advanced through-the-telescope techniques.

Choosing and Using a CCD Camera; Berry, R.; Willmann-Bell, 1992. A well-written book on the fundamentals of charge-coupled devices (CCDs) for the amateur astronomer. Within its 96 pages are basics on CCD chips and imagers, adapting telescopes to CCD imaging, and how to take and process images.

Telescope Making and Optics

Amateur Telescope Making (volumes 1, 2, and 3); Ingalls, A., et al.; Willmann-Bell, 1996. *The* classic telescope-making reference returns to print! Volume 1 describes techniques for grinding and testing mirrors of Newtonian reflectors and discusses observatories; volume 2 discusses refractors, binoculars, Schmidt cameras, and telescope adjustments; and volume 3 includes details about eyepieces, optical coatings, and other optical instruments. Each volume has been expanded and brought up to date.

***Backyard Astronomer's Guide,* Dickinson, T., and A. Dyer; Camden House, 1994.** A terrific book for the amateur astronomer. Divided into three sections, the *Backyard Astronomer's Guide* addresses telescopes and equipment, observing the night sky, and simple astrophotography. Several guest experts contribute short sections to help round out the presentation.

***Build Your Own Telescope;* Berry, R.; Willmann-Bell, 1995.** An excellent first book on constructing telescopes. Includes complete plans for five refractors and Newtonian reflectors, ranging in aperture from 4 to 10 inches, all with full instructions, plans, and numerous photographs showing their construction.

***Making and Enjoying Telescopes;* Miller, R., and K. Wilson; Sterling, 1995.** Like *Build Your Own Telescope,* this book offers clear descriptions and plans for constructing five different Newtonian reflectors plus a variation of the Scotch mount/barn-door camera tracker. Advice on purchasing components and caring for and using telescopes is also offered.

***Perspectives on Collimation;* Menard, V., and T. D'Auria; self-published, 1993.** If you want to get the most out of your telescope, this booklet is for you. Messrs. Menard and D'Auria take the reader through the complete process of optical collimation, an area addressed, albeit briefly, in chapter 8 of the book you are now holding in your hands. Because the book is self-published, you'll have to order it directly from the authors (Vic Menard, 2311 23rd Ave. West, Bradenton, FL 34205, or Tippy D'Auria, 1051 NW 145th St., Miami, FL 33168).

***Star Testing Astronomical Telescopes;* Suiter, H.; Willmann-Bell, 1994.** One of the most eye-opening optical tests that anyone can perform on his or her telescope is called the *star test*. Although chapter 8 of the book you're reading offers guidance on performing the star test, Suiter's book provides an in-depth discussion of the many fine nuances that one might encounter.

References and Catalogues

***Deep Sky Field Guide;* Cragin, M., J. Lucyk, and B. Rappaport; Willmann-Bell, Inc., 1993.** A companion for the advanced deep-sky observer, this itemized catalogue details the coordinates, magnitudes, apparent dimensions, and other data for all of the deep-sky objects plotted in the *Uranometria 2000.0* star atlas. All data is arranged by chart number in the *Uranometria* for handy reference.

***Sky Atlas 2000.0 Companion,* Strong, R.; R.A. Strong, 1994.** This privately published work contains data on each of the 2,500 objects plotted in the *Sky Atlas 2000.0.* Objects are itemized twice, first alphabetically and then by chart number. Data given include the object type, right ascension and declination, apparent size, and magnitude—everything needed to make this an ideal resource for amateurs with 6- to 10-inch telescopes.

Miscellaneous Reading

Alvan Clark and Sons: Artists in Optics; **Warner, D., and R. Ariail; Willmann-Bell, 1995.** Here is the story of America's consummate telescope maker, Alvan Clark and Sons. Even in light of today's technology, Clark refractors remain some of the finest astronomical instruments ever made. If you enjoy the historical aspects of telescopes, this book offers a enchanting read.

Computer Software

Astronomical software can be divided into three basic categories based on their intent: observing programs (which include planetarium simulations and, often, telescope motion control), image-processing programs (typically used in conjunction with CCD cameras, but they may also be used to enhance scanned-in conventional photographs), and special-purpose programs (covering a wide range of subjects, from optical design to predicting eclipses). Because most amateur astronomers are primarily interested in observing the sky, I have restricted the listing here to some of the best observing programs.

Deep Space (CD-ROM, DOS); David Chandler Company. A fine program for intermediate and advanced amateurs, Deep Space includes a huge database of more than eighteen million stars, ten thousand deep-sky objects, and ten thousand comets and asteroids. Like most other programs of this type, Deep Space will print finder charts, but what makes these charts unique is the way a user can move titles and labels around to minimize overlapping. Deep Space will also steer many digitally controlled telescopes. The only drawback is that because Deep Space is a DOS-based rather than Windows-based program, some users may find it a little less friendly to use. Requires an IBM-compatible 386 or better, 640K RAM plus 1 MB extended memory, 13 MB hard-drive space, EGA or better monitor, MS-DOS 5.0 or higher, and a CD-ROM drive. A math coprocessor is recommended.

Guide (CD-ROM, Windows); Project Pluto. An excellent program for the deep-sky observer at a terrific price! Guide plots over eighteen million stars, over twenty thousand asteroids, hundreds of comets, and seventy-five thousand deep-sky objects from the Messier, NGC, and IC lists as well as many comparatively unknown catalogs. Like many of the other programs listed here, Guide can also be used to pilot a Meade LX200 instrument around the sky. Requires an IBM-compatible 286 or better, 640K RAM, Windows 3.1/95, EGA or better monitor, and CD-ROM drive.

Megastar (CD-ROM, Windows); ELB Software. The single best program ever created for the diehard deep-sky observer. Megastar plots over eighteen million stars to magnitude 16, an incredible 110,000 deep-sky objects, and more than thirteen thousand archived asteroids and comets, which amounts to more viewable objects than just about any other program currently available. Megastar will also generate lists of deep-sky objects for a night's observing, print high-quality charts, and navigate a Meade LX200 mounting around the sky by simply pointing and clicking the computer's mouse. Requires an IBM-compatible 486DX or better, 4 MB (Windows 3.1) or 8 MB (Windows 95) RAM, 5 MB hard-disk space, Windows 3.1/95, VGA display, CD-ROM drive, and mouse.

Realsky (CD-ROM, Windows or Macintosh), Astronomical Society of the Pacific. The Palomar Observatory Sky Survey, an invaluable photographic survey long held in high regard by professional astronomers, is now available in a nine-disk CD-ROM set for either Windows or Macintosh systems. *Realsky North* covers the northern sky to declination −15°, while *RealSky South* extends to the South Celestial Pole. Both display stars as faint as 19th magnitude as well as innumerable deep-sky objects. The software lets the user zoom the field of view from 1° to 5 arc-minutes and then print out a finder chart (more correctly, a finder photograph) for use at the telescope. Viewing software is supplied with both CD-ROMs, but other software packages, such as *The Sky* or *Guide*, can be used to manipulate *RealSky* more effectively. Requires IBM-compatible 486-66 or better, 16 MB RAM, and Windows 3.1/95, or for Macintosh, LCII+ (68040—25 mHz) or PowerMac, 16 MB RAM, System 7.0 or later; SVGA display, 2× CD-ROM drive, and mouse.

Starry Night Deluxe (CD-ROM, Macintosh), Sienna Software. The most widely acclaimed astronomy program offered exclusively for the Macintosh. Starry Night Deluxe accurately plots more than nineteen million stars and deep-sky objects from the Hubble Guide Star Catalog, displays the sky from any planet or moon in the Solar System at any time, prints customized star charts, and aims and controls a Meade LX200 telescope. And the graphics are terrific! Requires System 7.0 or later, 8 MB RAM, and CD-ROM drive.

The Sky (CD-ROM and/or 3.5-inch disk, Windows); Software Bisque. *The Sky* is one of the finest sky-simulation and telescope-control programs available. With this program, you can show the Sun, Moon, and planets against a myriad of background stars for any time, any date, any place on Earth. With an optional interface cable, the program can also slew a Meade LX200 or Celestron Ultima 2000 to any object in its extensive database. *The Sky* comes in either Level II (CD-ROM plus 3.5-inch disks), Level III (CD-ROM only), or Level IV (CD-ROM only), depending on the number of objects stored in its database. The flagship Level IV versions plots 19 million stars and nonstellar objects from the Hubble Guide Star Catalog plus more than one hundred thousand objects from various deep-sky catalogs. Requires IBM-compatible 386DX or better, 8 MB RAM, 10 MB hard-disk space, Windows 3.1/95, VGA display, and mouse.

Voyager II (CD-ROM and/or 3.5-inch disk, Macintosh); Carina Software. Another outstanding sky-simulation program for Macintosh users. *Voyager II* displays 259,000 stars; 50,000 deep-sky objects; and many comets, asteroids, and Earth-orbiting satellites. With it, the sky from any place, any time, may be shown easily by a click of the mouse button. Animation compresses time to show eclipses or planets orbiting the Sun. Requires MacPlus to Quadras and PowerMacs, System 7.0 or later, 2.5 MB RAM, 10 MB hard-disk space, and 3.5-inch disk or CD-ROM drive.

In addition to these commercial programs, there are many shareware and freeware packages available that have many of the same advanced features as their more expensive competitors. Some of the best for Windows-based computers include *Deepsky 97* (Steven S. Tuma), *Earth-Centered Universe* (Nova Astronomics), *Home Planet* (John Walker, excellent freeware from Switzerland), *SkyChart 2000* (Southern Stars Software), and *SkyMap* (SkyMap Software, considered the best astronomical shareware by many). All are downloadable off the Internet; see Appendix C for their snail-mail addresses and Uniform Resource Locators (URLs).

Electronic Gadgetry

As with just about every other aspect of our lives, amateur astronomy is becoming more sophisticated thanks to tremendous advances in electronics and computerization. Here is a short review of some of the more popular electronic equipment on the market today.

Digital Setting Circles

One of the biggest complaints that amateur astronomers have had for years is that the setting circles that come with many equatorial mounts are useless. Because of their small size and gross calibration, they are little more than decoration. All this changed with the invention of electronic digital setting circles, which make it possible to aim a telescope accurately to within a small fraction of a degree. And forget polar aligning (although it is still usually required to use the clock drive); just aim the telescope at two of the many stars stored in memory, tell the unit which ones they are, and the built-in algorithms do the rest. This ease of operation means that the setting circles can be attached to just about any kind of telescope mount—either equatorial or alt-azimuth. Don't know what to look at? That's OK, because many of these units come with a whole catalog of thousands of sky objects from which to choose. Simply move the telescope until the LED prompt announces that you have hit the preselected target, look in the eyepiece, and there it is! Want one? Here is a quick rundown of some of the more popular digital setting circles on the market today.

The Lumicon Sky Vector is a neat little unit that mounts unobtrusively to just about any telescope. Two encoders attach to the instrument's axes and are connected to the "brain" of the outfit by means of a pair of thin wires. Once properly secured and calibrated, the digital setting circles will automatically keep track of the passage of time as well as where the telescope is aimed. The LED is accurate to within 10 arc-minutes in declination and 1 minute of right ascension. The Sky Vector I includes 220 sky objects, the Sky Vector II has stored information on 10,000 objects, and the Sky Vector III library includes 12,000 objects and can be connected to a personal computer via a serial port, which allows the observer to link to various software packages like *Megastar* or *The Sky*.

The Magellan I and Magellan II systems from Meade Instruments are designed specifically for its own telescopes. The Magellan I, for the Meade 8-inch LX-10 Schmidt-Cassegrain and Starfinder reflectors, includes a library of more than twelve thousand objects and may be updated with more advanced data that may become available in the future. Each Magellan I kit comes with all necessary encoders, cables, and hardware. The Magellan II, made expressly for Meade's LX-50 catadioptric instruments and Starfinder equatorial reflectors, includes all features of the Magellan I and has a built-in drive corrector for photographic guiding. The Magellan II Starfinder kit also includes the required declination motor assembly and a new control panel that replaces the one originally supplied with the instrument. One caveat for owners of older AC Starfinder equatorial telescopes: Sorry to say, but neither Magellan system will work with your instrument's drive.

Celestron's Advanced Astro Master works in pretty much the same way as the Sky Vector. The Celestron unit is designed specifically for Celestron-brand telescopes supported on either fork or Super/Great Polaris equatorial mounts, although it may be adapted to other brands and mounting styles with a drill (careful!) and a little creativity. It, too, features a ten-thousand-plus object library from which the user may browse, and it may be connected to a personal computer via an RS-232 serial interface cable.

Simplest of the three digital-setting-circle units sold by Jim's Mobile, Inc. (JMI), is the NGC-microMAX, which provides accurate telescope positions and includes data on 245 sky objects. Next up, the NGC-miniMAX (Figure 6.8) holds data on more than three thousand nine hundred deep-sky objects. Finally, the

Figure 6.8 *The Meade ETX Maksutov telescope outfitted with NGC-Max digital setting circles and Motofocus electric focusing, both manufactured by JMI. Photo courtesy of JMI, Inc.*

NGC-MAX includes a library of more than twelve thousand objects and can link directly to a computer's serial port. The JMI MAX family is designed to fit almost all popular telescopes around and is also adaptable to homemade instruments.

Similar to the JMI units is Orion Telescope's Sky Wizard, sold exclusively through its Telescope and Binocular Center retail and catalog outlets. The Sky Wizard comes in three varieties. Sky Wizard 1 contains data on all of the Messier objects, 28 stars, and all of the planets; Sky Wizard 2 adds 2,000 deep-sky objects and has room for 27 of your favorite objects; Sky Wizard 3 contains a 10,000-object database and may be connected to a PC through a serial interface. As with the others, Sky Wizard works equally well with both equatorial and alt-azimuth mounts.

All of these computerized telescope-aiming systems require that a pair of encoders be mounted to the telescope's axes. Many offer customized packages for attaching the encoders to fork-mounted Celestron or Meade Schmidt-Cassegrains simply by unscrewing and replacing one screw on both axes. Other telescopes, however, may not have it so easy, requiring some drilling and tapping in order to fit the encoders to their mountings. I strongly urge you to contact the manufacturer to find out what is required to attach the encoders to your particular telescope before you purchase any of them.

Another important question for readers considering the purchase of one of these units for their 8-inch Schmidt-Cassegrain telescope: Can the telescope swing into its storage position without hitting the right-ascension encoder? The encoders engage the right-ascension axis at its pivot point, smack dab in the middle of the fork. If the encoder sticks up too high, the front of the telescope tube will not clear. This forces the user to detach the encoder whenever the instrument is disassembled for storage in its footlocker. Be sure to ask the manufacturer before you buy!

Digital Setting Circles: A Commentary

Back in the 1980s, one manufacturer advertised that an observer could see "100 galaxies per hour" with its computer-aided digital circles. Sure, they are a great way to increase your productivity as an observer, if productivity is only measured in terms of sheer numbers. But since when are we out to break the land-speed record? The idea is not to whiz across the universe at warp speed but rather to get to know it intimately. Without becoming too philosophical, I cannot recommend digital setting circles, especially if you are just starting out. It would be like giving a calculator to someone who does not know how to multiply. Sure, the calculator will give the right answer, but without it, the person is lost. Unless you are involved in an advanced observing program, such as variable-star observing or hunting for supernovae in other galaxies; run frequent public observing sessions; or are forced to observe from an area so overwhelmed by light pollution that you can't see enough to star-hop by, resist the impulse until you finish the section of chapter 9 entitled "Finding Your Way."

Clock-Drive Correctors

Most of today's top-of-the-line clock-driven telescopes are powered by DC motors and feature hand controllers that not only allow fine adjustment of the tracking rate but also permit relatively fast slewing across the sky. Older instruments use AC synchronous motors in their clock drives and therefore require external drive correctors. Because the speed of synchronous motors will not vary with changes in voltage (above their threshold voltage, they run; below, they stall), drive correctors alter the frequency of the supply voltage.

Drive correctors come in two basic styles: single-axis or dual-axis. The simpler, single-axis correctors adjust only the speed of the right-ascension clock drive motor, while dual-axis correctors also control the declination slow-motion motor.

The only manufacturer currently offering external clock-drive correctors is Jim's Mobile, Inc. Its Mototrak quartz-controlled drive correctors enjoy some interesting and useful features, including dual-axis operation, as well as optional declination motor and electronic Motofocus focusing control (discussed in just a bit).

A drive corrector is a must for long-exposure astrophotography, but if all you want to do is run the clock drive from a car battery, then you really don't need one. Instead, save some money by purchasing an AC/DC inverter, which will plug into a standard 12-volt automobile cigarette lighter to produce 110 volts AC. Though the current is too low to power an appliance, it is adequate to run a clock drive. Inverters are available from many auto parts stores and camping outlets.

Motorized Focusing Devices

Apart from poor sky conditions, the greatest hurdle to overcome when making high-powered observations is achieving a sharp focus. As magnification rises, precise focusing becomes extremely critical. At the highest powers, even the slightest turn of the focusing knob will vibrate most telescopes, blurring the image. Although not required for observations made at low or medium power, electric focus motors are handy for the crucial focusing required for high-power visual and photographic observations.

Some premier telescopes come with motorized focusing as standard equipment, but many such devices are sold as options. Most of these are dedicated units designed to fit only one or two specific telescope models. As an aftermarket solution, Jim's Mobile, Inc., offers the Motofocus electric focusing motor, with models to fit many of the most popular telescopes. The Motofocus attaches to the telescope without drilling and allows either coarse or fine focus adjustment. The more adventuresome might even wish to make their own motorized focusers. The next chapter illustrates what one amateur telescope maker accomplished for less money than a comparable commercial unit would have cost.

Smile ... Say Pleiades

Chapter 10 concludes with an introduction to the art of astrophotography, one of amateur astronomy's most popular pastimes. As most soon discover, however, there is a lot more to taking a good picture of the night sky than one might suspect at first. Patience is the most important requirement of the astrophotographer, followed closely by equipment. It's tough to bottle the former for sale, but lots of companies are looking to sell you the latter!

Cameras

Although quality astrophotographs can be taken with many different types of cameras, most amateurs prefer the 35-mm single-lens reflex (SLR) camera. Single-lens reflexes allow the photographer to look directly through the lens of the camera itself, a critical feature for aligning the image, especially when photographing through a telescope (with most other cameras, the photographer is viewing through a separate viewfinder). SLRs also offer the maximum flexibility in terms of film and lens availability, both of which will be discussed later.

Not all 35-mm SLRs are suitable. For astrophotography, a camera must have a removable lens with manually adjustable focus; provisions for attaching a cable release to the camera and the camera to a tripod; a manually set, mechanical shutter with a "B" (bulb) setting; mirror lockup; and interchangeable focusing screens. Unfortunately, few of today's 35-mm SLRs fit this bill. In an attempt to attract more weekend photographers, most camera manufacturers offer cameras with automatic everything, from focus to exposure to flash control. All of these are nice for taking pictures of the family picnic but are of no use to astrophotographers. Quite to the contrary, the long exposures required for astrophotos (usually measured in minutes, even hours) will quickly drain the power from expensive camera batteries. When that happens, the camera shuts down and becomes useless until a fresh set of batteries is inserted.

Which cameras are best for astrophotography? Table 6.3 lists several excellent alternatives, both past and present.

Expensive does not necessarily mean "better for astrophotography." All of these cameras will work well for wide-field constellation shots as well as through-the-telescope photos of the Moon and Sun (the latter requiring safety precautions outlined in chapter 10), but their differences will become more apparent when taking long-exposure telescopic shots. Here, the benefit of interchangeable focusing screens and mirror lockup will become apparent. Most subjects photographed through telescopes are very faint, making it difficult to line up and focus the shot when viewing through the majority of standard focusing screens. A simple ground-glass screen will provide the brightest possible images, a great aid in focusing and composing (see further discussion under "Focusing Aids" later in this chapter). Mirror lockup is essential for reducing *mirror slap*, which occurs every time the shutter is tripped and the camera mir-

Table 6.3 **Suggested Cameras for Astrophotography**

Model	Nonbattery "Bulb" Setting	Manual Lens	Interchangeable Focus Screen	Mirror Lockup
Today's Best				
Contax S2	Y	Y	Y	N
Leica R6.2	Y	Y	Y	Y
Nikon F3HP	Y	Y	Y	Y
Nikon FM2	Y	Y	Y	N
Olympus OM-2000	Y	Y	N	N
Olympus OM-4T	Y	Y	Y	N
Pentax K1000	Y	Y	N	N
Pentax LX	Y	Y	Y	Y
Ricoh KR-5 Super II	Y	Y	N	N
Yashica FX-3 Super 2000	Y	Y	N	N
A Few of Yesterday's Best				
Canon F-1	Y	Y	Y	N
Canon FTb	Y	Y	N	Y
Minolta SRT-101	Y	Y	N	Y
Miranda G	Y	Y	Y	Y
Nikon F, F2	Y	Y	Y	Y
Olympus OM-1, OM-2	Y	Y	Y	Y

ror pivots out of the way. Swinging the mirror out of the way before the shutter is opened eliminates most vibration, reducing the chances for blurred images.

Lenses

Just as all cameras are not suitable for photographing the sky, neither are all lenses. But before an educated choice can be made, the photographer must first determine what he or she wants to photograph. For wide-field photography, either with the camera attached to a fixed tripod or guided with the stars, the standard lens may be all that is needed. Most 35-mm SLR cameras are supplied with lenses of 50 mm to 55 mm focal length; these cover an area of sky 28° × 40°. If a wider field is desired, then either a 28-mm or 35-mm lens would be a good choice; they cover 50° × 74° and 40° × 59°, respectively. On the other hand, if a magnified view is what you want, try an 85-mm, 135-mm, or 200-mm telephoto lens, with 16° × 24°, 10° × 15°, and 6.9° × 10.3° fields of view, respectively.

In general, an astrophotographer wants to use as fast a lens as possible, because the faster the lens, the shorter the required exposure. Photographers refer to the speed of a lens just as astronomers talk about the focal ratio (the "f/" number) of a telescope. These terms mean the same thing.

Quite simply, the faster the lens, the lower the focal ratio. Holding the focal length constant, the only way to lower the focal ratio is to increase the lens's aperture. For instance, most 50-mm and 55-mm camera lenses have f-ratios between f/2 and f/1.2, resulting in apertures that range between about one and two inches. The larger the aperture, the greater the light-gathering ability of the lens, and, therefore, the shorter the required exposure.

Years ago, lens quality varied dramatically from one company to the next, but today, design and manufacturing procedures have been so perfected that most lenses will produce fine results (of course, every photographer thinks his or her favorite brand is best). Flare and distortion, two of the biggest problems in older lenses, have been all but eliminated thanks to optical multicoatings. In general, most lenses made by reputable companies (all of the camera manufacturers listed above, as well as those by Vivitar, Sigma, and Tamron, to name a few) are fine for astrophotography. A *big* exception to all this is the zoom lens. Although extremely popular for everyday photography, zoom lenses almost always produce results that are inferior to those of their fixed-focal-length counterparts.

Film

The film industry has seen advances in the past two decades the likes of which have never been witnessed before. Without a doubt, the wide variety of films that are readily available today is a great boon to astrophotography. The selection is so vast, in fact, that it can leave the photographer confused. "Which film shall I use?" is a question often posed even by veterans.

To help alleviate some of this bewilderment, Table 6.4 lists some currently popular films. The table is broken into three broad categories by film type (black-and-white, color negative, and color slide) and then sorted by film speed, or ISO value. ISO (International Standards Organization) is the modern equivalent of the older ASA (American Standards Association) designation. Basically, the greater the numerical ISO value, the faster the film records light. For instance, an ISO 400 film will record the same amount of light in one-quarter of the time required by ISO 100 film. High-ISO-value films allow shorter exposures and, therefore, less chance of photographer error (accidentally kicking or hitting the telescope, tracking error due to polar misalignment, etc.).

Why, then, would anyone consider using slower films? If a frame of film is studied close up, it will be found to be made up of a pebbly surface called *grain*. It is one of those irrefutable laws of nature that fast film has larger grain structure than slower film. Larger grain means lower resolution and, therefore, poorer image quality.

Another reason to use slower film is something called *reciprocity failure*. This is one of those terms bounced around by most photographers but probably understood by few. As an illustration, think back to the example of ISO 100 film versus ISO 400 film. The films differ in ISO value by a ratio of 1 to 4, while their speed differs by a ratio of 4 to 1. Because these two ratios are reciprocals of each other, this concept is called reciprocity. But reciprocity does not remain constant as exposure times increase; instead, the film's ability to record light falters and finally ceases. After that, no further light buildup will occur regardless of length of exposure, hence the term reciprocity failure.

Table 6.4 **Film Comparisons**

Black-and-White Film

Film (Manufacturer's Code)	ISO	Grain
Kodak Tech Pan 2415 (TP-2415)	25[1]	Extra fine
Kodak T-Max 100 (TMX 5052)	100	Very Fine
Kodak Tech Pan 2415 (hypered) (TP-2415)	200	Very fine
Kodak T-Max 400 (TMY 5053)	400	Moderate
Kodak T-Max 3200 (TMZ 5054)	3200	Very coarse

Color-Negative Films

Film (Manufacturer's Code)	ISO	Grain
Fujicolor Super G Plus 400[1] (CH)	400	Moderate
Kodak Pro 400 (PPF)[1]	400	Fine
Kodak Ektapress Multispeed 640 (PJM-2)	640	Fine
Fujicolor Super G Plus 800 (CZ-1)[1]	800	Moderate
Fujicolor Super HG 1600 (CU)	1600	Coarse
Konica SR-G 3200[1] (SR-G 3200)	3200	Very Coarse

Color-Slide Films

Film (Manufacturer's Code)	ISO	Grain	Sky color (background)
Ektachrome 100 Professional (EPN)	100	Very fine	Neutral
Fujichrome Provia 100 (RDP II)[1]	100	Fine	Green
Ektachrome Elite II 200 (ED)	200	Moderate	Neutral
Fujichrome Sensia 400 (RH)	400	Moderate	Green
Ektachrome Elite II 400 (EL)	400	Moderate	Neutral
Ektachrome P1600 Professional (EPH 5040)[1]	1600[2]	Moderate	Neutral
Fujichrome Provia 1600	1600[2]	Moderate	Neutral

Notes:
1. Also available hypersensitized
2. May be processed at various speeds

There are ways to diminish a film's reciprocity failure while also increasing its ISO speed. The most popular technique is called *gas hypersensitizing*. The film is baked for days in a special oven containing a combination of nitrogen and hydrogen called *forming gas*. Although some amateurs prefer to cook their own, gas-hypersensitized film may be purchased from many mail-order sources, such as Lumicon. Users should note that not all films respond well to "hypering" and that hypered film has an effective shelf life of about a month. All hypered film should be sealed in an airtight package (such as the plastic film can it came in) and stored in a freezer before and after use.

Tripods

If you will be affixing your camera to a tripod (as opposed to shooting through a telescope), then pay close attention to the tripod you will be using. Many less expensive tripods sold in department stores and other mass-market outlets are just not sturdy enough to support a camera steadily for any length of time. If the tripod is shaky, then the photographs will be hopelessly blurred. It makes no sense mounting a camera outfit costing hundreds, even thousands of dollars on a cheap tripod!

Here are a few things to look for when purchasing a camera tripod. First, the legs should be extendable so that the camera may be raised to a comfortable height. Make certain, however, that the tripod remains steady when fully extended. Sturdier models feature braces that bridge the gap between the tripod's legs and the center elevator post. Next, take a look at the footpads. Better tripods have convertible pads that feature both a rubber pad for use on a solid surface as well as a spike for softer surfaces like grass or dirt. (A tip: When using a tripod on sand, place a plastic coffee can lid under each foot for added rigidity.)

Of all the tripods made, most photographers agree that the sturdiest are manufactured in Italy by Manfrotto and marketed in the United States under the Bogen name. For instance, the Bogen model 3036 is sturdy enough to hold a 4-inch refractor even with its legs fully extended; lesser tripods would collapse under such a load. Other brands worth considering are Tiltall, Star-D, Velbon, Vivitar, and Slik.

Camera-to-Telescope Adapters

For prime-focus photography (if the term is foreign to you, a preview of chapter 10's introduction to astrophotography would be in order here), the most common way to affix a camera to a telescope is a two-piece T-ring/adapter combination. The T-ring attaches to the camera in place of its lens, while an adapter attaches to the telescope. The ends of the adapter and T-ring are then screwed together to form a single unit.

Different cameras require different T-rings. For instance, a T-ring for a Minolta will not fit a Canon. Likewise, different adapters are required for different telescopes. In the case of most catadioptric telescopes, a device called a *T-adapter* screws onto the back of the instrument in place of the visual back that holds the star diagonal and eyepiece. Most refractors and reflectors, on the other hand, use a different item called a *universal camera adapter*, which is inserted into the eyepiece holder.

Positive-projection astrophotography, commonly used when shooting the planets or lunar close-ups, requires that an eyepiece be inserted between the lensless camera and the telescope. Most camera adapters, such as the one shown in Figure 6.9, come with an extension tube for this purpose. The eyepiece is inserted into the tube, and the tube is then screwed in between the adapter and T-ring.

Celestron, Meade, Questar, and some other telescope manufacturers offer camera adapters that custom-fit onto their telescopes. Aftermarket brands, often less expensive, are also available. For instance, the Telescope and Binocular Center (Orion) sells several different adapters to fit most popular tele-

Figure 6.9 *Camera-to-telescope adapter system. Photo courtesy of Orion Telescope Center.*

scopes. None are supplied with camera T-rings, which must be purchased separately. Better photographic supply stores carry T-rings for most common single-lens reflex cameras, as do many astronomical mail-order companies.

Off-Axis Guiders

For long-exposure, through-the-telescope astrophotography, photographers have no choice but to visually monitor their telescopes' tracking. To do this, they must either peer through a side-mounted guide scope or an "off-axis" guider or use an autotracker, outlined later under "CCD Cameras." Mounting a guide scope onto the side of the main instrument can be both clumsy and expensive, leading many astrophotographers to choose an off-axis guider.

An off-axis guider looks like the letter **T**, with two hollow tubes attached to each other at a 90° angle. The main body of the guider fits between the telescope and camera, while the perpendicular leg contains a tiny prism that is used to divert a small amount of starlight toward an illuminated-reticle eyepiece that fits into the guider. To use the guider, one plugs or screws it into the telescope's eyepiece holder between the telescope and the camera and aligns the crosshairs of the eyepiece with a star. During the exposure, monitor the guide star through the reticle eyepiece to make sure it never leaves the crosshairs.

Off-axis guiders are available from a number of different sources. The simplest and, therefore, cheapest feature rigid prisms, while more expensive models come with prisms that may be rotated around the field. In practice, the latter are easier to use because the freedom of prism movement permits a much wider choice of guide stars. The Orion UltraGuider, Celestron Radial Guider, Lumicon Easy-Guider, and Spectra Astro System SpectraGuide are among the best off-axis guiders for the amateur astrophotographer.

Focusing Aids

Focusing a telescope for the eye alone is easy, but achieving a sharp focus when looking through a camera viewfinder can be quite another matter. Although standard focusing screens work well under bright conditions, they produce dim, ill-defined images when used at night. Many astrophotographers swap their camera's replaceable focusing screen for a clear ground-glass matte screen. Special viewfinder screens called Intenscreens are available from Beattie Systems, Inc., and deliver images up to four times as bright as those obtained with standard screens. (Some camera light meters that take readings off the focusing screen might need their ASA/ISO speed dials adjusted to compensate for the Intenscreen's extra brightness. Although you aren't likely to use a light meter for astrophotography, it is something to keep in mind if you use the same camera for terrestrial photography.)

But Intenscreens are not available for every camera, and even with a clear matte screen in place, getting a crisp image is still tough to do. Three focusing devices that have won favor among astrophotographers are the SureSharp and PointSource, both by Spectra Astro Systems, and Celestron's Multi Function Focal Tester. The SureSharp, introduced in 1986, consists of two parts: a modified T-ring that attaches to either a telescope's camera adapter or off-axis guider and a conical housing containing a Ronchi grating (a clear glass window with evenly spaced, opaque, etched lines) at the exact distance as the film plane of a 35-mm SLR camera. To use the SureSharp, aim the telescope towards a 4th-magnitude or brighter star and look at its image through the grating. If the star is out of focus, it will appear as a disk crossed by dark gray bands. By turning the focus, the number of bands will decrease until the lines disappear at the point of focus.

The PointSource works on the same principle as the SureSharp. Screw the PointSource into any camera adapter that uses a T-ring for mounting onto a camera. Attach the camera adapter to the telescope, look at the Ronchi grating, and focus the telescope as described in the previous paragraph. The biggest advantage of the PointSource is that it can be used with just about any T-ring and may be shared with different camera brands. Celestron's Multi Function Focal Tester (MFFT-55) also screws onto a standard T-adapter in place of a camera. Just aim the telescope at a bright star and focus its image on the ground-glass screen of the MFFT-55. That's all there is to it. Remove the MFFT-55 (carefully, so as not to disturb the focusing), attach the camera, and the telescope is set for in-focus photography. Instructions also explain how the MFFT-55 can be used for a variety of other tests, such as telescope collimation and checking the squareness of the camera adapter to the optical axis.

If you are interested in photographing bright, extended objects like the Sun (warning: Read chapter 10 before you try solar photography!), Moon, planets, or maybe the brightest deep-sky objects, then one of these focusing devices may not be needed. To get a sharp focus first time, every time, make a mask for your telescope by cutting a circular piece of cardboard of the same diameter as its aperture. Now cut two smaller circles in the mask directly opposite each other, making certain that they are not cut off by the telescope's secondary mirror (if it's a reflector or catadioptric). Two-inch diameters work well for 8-inch and larger telescopes. With the mask secured in front of the telescope, attach the camera, aim at the target, and look through the viewfinder. If the telescope is out of focus, you will see two images. Turn the focusing in or out until the two slowly blend into one. When only a single image is seen, the telescope is properly focused. This works with all telescopes but only for extended objects (objects with perceptible width) like the Moon and planets, not stars. A commercial version of this device, called Kwik Focus, is marketed by Jim Kendrick Studios.

CCD Cameras

The pursuit of astrophotography has changed dramatically in the past decade. The availability of super-sensitive charge-coupled devices, or CCDs for short, now makes it possible to take photographs of the Moon, planets, and deep-sky objects using exposures many times shorter than those that would be required with conventional film. A CCD camera uses a silicon chip made up of thousands of light-sensitive areas called *pixels* to convert light into electrons during an exposure. At the end of the exposure, each pixel converts its stored electrons into a numeric value and then downloads that result to a computer. The computer then converts the values received from each pixel into a shade of gray and combines them into an image displayed on the computer's monitor.

Exposures as short as 30 seconds with a CCD camera (Figure 6.10) will record the same detail as a half-hour exposure using conventional film. (Estimates indicate that these cameras have an equivalent ISO speed value in

Figure 6.10 *The Lynxx-PC CCD camera in action. Note the image of Jupiter on computer screen.*

excess of 20,000 after computerized image processing!) In addition, the effects of light pollution can be eliminated by computer processing, making it possible to capture great deep-sky images from within cities. Fantastic!

But there is much more to this than just buying a CCD camera and hooking it up to your telescope. Besides having to buy a computer (laptop and notebook models seem to be the most popular choice) as well as a sturdy equatorial mounting, you will also need to have some idea of what kind of imaging you are interested in doing. The ideal setup for planetary work may not necessarily be appropriate for deep-sky imaging and vice versa. Planetary images are typically bright and small, requiring short exposures. The amount of sky coverage is usually not critical; indeed, eyepiece projection is almost always used to enlarge the planet's image. Deep-sky objects, on the other hand, may cover wide expanses and are comparatively very dim, requiring larger, more expensive chips and longer exposures. Figure 6.11 graphically compares the area captured by some of the more common CCD chips in use today to that of a 35-mm negative. The comparison may surprise you!

As exposure grows, so does the background electronic noise, or *dark current*, generated by the CCD camera. Dark current can be reduced by cooling the CCD chip, but some chips require more cooling than others to bring this noise level down to workable levels.

CCD chips are listed as being 8-bit, 12-bit, or 16-bit. These numbers refer to the analog-to-digital (abbreviated as A/D) converter used for storing and processing images. Simply put, the larger the number, the greater the dynamic range in image brightness and, therefore, the better the image quality. But keep in mind that a 12-bit converter will require about 50% more memory space to store and is

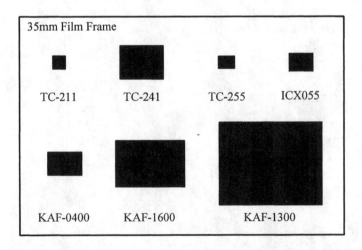

Figure 6.11 *A comparison of the sky coverage offered by today's most popular CCD chips compared to that of 35-mm film. This comparison assumes that each is used at the same focal length.*

slower to process than an 8-bit converter, while a 16-bit requires twice the allotted memory of an 8-bit. Equally critical is the number of picture elements, or pixels, for this will dictate the amount of sky covered at any one time.

All of the CCD cameras listed in this section image in black-and-white only, yet color CCD photography is possible thanks to a variation on the tricolor photographic technique introduced by James Maxwell in 1861. Three separate exposures are made with red, green, and blue filters over the camera and then combined electronically to produce a true-color image. Some CCD cameras have optional built-in color-filter wheels, while others require a separate attachment.

Several companies market CCD cameras for today's amateur astronomer. Some of the most highly regarded are manufactured by Santa Barbara Instrument Group, or SBIG for short. The least expensive model in its collection is the ST-4, an 8-bit imager that uses the Texas Instruments TC211 chip. The ST-4 is both an imaging camera as well as an autotracker. In the latter mode, it will automatically follow a selected star across the sky when coupled to a properly equipped telescope, such as a Meade LX200 or Celestron Ultima 2000. The observer can then mount a camera piggyback on the telescope to take perfectly tracked photos of the sky. The ST-4 can also be placed at the prime focus of a telescope for CCD imaging, although the area included is very limited (4 arc-minutes in an 8-inch f/10 telescope) because of the chip's small size. The imaging mode requires that the ST-4 be attached to a computer, while the autotracker mode does not. The two modes cannot be combined, however, which is not the case in more sophisticated CCD units.

The SBIG ST-6 has received rave reviews from all who have tried it. Using Texas Instruments' TC241 CCD chip, the ST-6 covers a 16' × 12' area with an 8-inch f/10 telescope. (That's 16 arc-minutes by 12 arc-minutes— ' can be an abbreviation for arc-minutes or feet, just as " can be an abbreviation for arc-seconds or inches. In this book, if the context is celestial, I'm expressing measurements in arc-seconds or -minutes; if terrestrial, in feet or inches.) Image quality is outstanding, approaching that of photographic film quality, thanks to 16-bit technology. The ST-6 also includes a feature that SBIG calls "Track and Accumulate." This allows the camera to spend most of its time imaging, but once in a while it switches briefly to autotrack mode to make sure the telescope is tracking properly, thereby freeing the observer from the tedious task of manually checking that the instrument is still on target. The only apparent drawback to the ST-6 is its overall size. Although it weighs only 2.25 pounds, the camera's head is large enough to interfere with the mounting of fork-mounted Schmidt-Cassegrain telescopes.

As impressive as the ST-6 is, SBIG's ST-7 and ST-8 cameras are in a completely different league. Like the ST-6, the ST-7 and ST-8 have 16-bit A/D converters but they are built around much larger CCD chips from Kodak; the ST-7 uses the KAF-0400 chip, while the ST-8 uses the KAF-1600. Through an 8-inch f/10 instrument, the ST-7's effective field of view spans 8' × 12', while the ST-8's is double that still. But perhaps their most unique feature is that both cameras also feature dual-CCD autoguiding; that is, each has a separate TC211 chip just for autoguiding, rather than burdening the one chip for both functions.

SBIG recently teamed up with Celestron to create the Celestron Fastar 8, an 8-inch f/10 Schmidt-Cassegrain instrument designed to have its conventional secondary mirror replaced with a CCD camera manufactured for Celestron by SBIG. By placing the camera at the prime focus of the primary mirror, the Fastar operates at an incredible f/1.95 for great wide-field potential. (See chapter 4 for more on the Fastar.) The PixCel 255 camera used with the Fastar is also available separately for more conventional applications. Based around the Texas Instruments TC255 chip (240 × 320 pixel array), the PixCel covers a 5.4 × 4.1 arc-minute area of sky when used with an 8-inch f/10 telescope (27.7' × 21.0' with the Fastar at f/1.95).

Meade sells a number of CCD products, the least expensive of which is the Pictor 201XT autoguider. This is not an imaging device; rather, the 201XT is connected between a telescope and its dual-axis clock drive to follow a selected star automatically, eliminating the need to adjust the instrument's tracking rate manually during a conventional photographic exposure. No personal computer is needed. Several owners report trouble trying to lock the 201XT onto a star, but once this is accomplished, the autotracker works properly. Personally, however, I would recommend that prospective consumers should invest a little more money and purchase a CCD imager/autotracker, such as the Meade 208XT.

The 208XT is one of four CCD imagers offered by Meade. Both it and the 216XT use the Texas Instruments TC255 chip. The difference between them? Sophistication. The 208XT uses an 8-bit A/D converter, while the 216XT enjoys a wider dynamic image range thanks to its selectable 12- or 16-bit imaging, and will also attach to the #616 color filter system for tricolor imaging. Both may also be used as separate autoguiders or in a sophisticated "shift-and-combine" mode where the imager pauses momentarily during an exposure to autoguide.

The two most advanced CCD cameras in the Meade line are the 416XT and 1616XT. Both are based on CCD chips from Kodak, the KAF-0400 and KAF-1600, respectively. As already noted in Figure 6.11, these chips provide more resolution than the more common Texas Instruments chips by a wide margin while also covering a larger area of sky. The KAF-1600, for instance, will span nearly half of the Moon's disk when used in an 8-inch f/10 instrument. Meade recommends coupling these cameras to computers with a SCSI (Small Computer System Interface) card for fastest image downloading (not all computers permit this). With a SCSI interface, image download time is 1 and 4 seconds for the 416XT and 1616XT, respectively; with the more common serial interface, download times climb to 65 and 260 seconds, respectively!

Apogee Instruments sells a number of top-of-the-line CCD imagers for industrial and research institutions. Recently, it introduced a comparatively low-cost series of three 16-bit CCD imagers aimed at the amateur astronomical market. All use CCD chips from Kodak for wide sky coverage: The AM4 uses the KAF-0400 chip, the AM16 contains the KAF-1600, and the AM13 utilizes the KAF-1300. All yield excellent image quality thanks to low noise and precise chip-cooling control. Although they are more expensive than some competing models, none are capable of autoguiding.

SpectraSource Instruments continues to make its line of Lynxx CCD cameras. The Lynxx PC uses the same 0.1-inch-square Texas Instruments TC211 chip as the ST-4, resulting in very small fields. The more advanced Lynxx 2000 CCD camera features an expanded format of 324 × 242 pixels by using the popular TC-255 chip. The associated software is available for both IBM-compatible and Macintosh computers.

A relatively new arrival on the CCD scene is the Starlight Xpress MX5 CCD camera by Functional Design & Engineering of England. The MX5, available in either 12- or 16-bit versions, is one of the smallest CCD cameras made—only 2 inches in diameter by 4 inches in length. This lets it fit directly into a 2-inch focusing mount, reaching focal planes in some telescopes that might otherwise be inaccessible (a 1.25-inch adapter is also included). The MX5 is also threaded to accept Pentax screw-mount camera lenses for wide-field astrophotography. The heart of the MX5 is the Sony ICX055 CCD chip, which measures 500 × 290 pixels, or about twice the area of the Texas Instruments TC255. The camera is designed to connect to a computer's parallel port via the included 15-foot cable. System requirements include a 486-66 computer and an SVGA monitor.

If you are looking for a challenge, consider making your own CCD camera. The *CCD Camera Cookbook* by Richard Berry, Veikko Kanto, and John Munger outlines how anyone who is knowledgeable in electronics can make not one but two different CCD cameras. The simpler of the two is based on the Texas Instruments TC211 chip, while the other uses the larger TC245 chip. Both work quite well when assembled correctly.

Although still more expensive than 35-mm SLR cameras, CCD cameras continue to drop in price just as CCD-based home-video camcorder prices have fallen as the technology becomes cheaper to produce. Of course, all imagers require a personal computer and high-resolution monitor, adding further to the start-up cost, yet many amateurs think this is a small price to pay for the outstanding quality that can be achieved with CCDs.

Unfortunately, the area of CCD photography is far too complex to address adequately in the small space provided here. If you are new to this burgeoning field, I recommend that you consult one of the books listed earlier in this chapter before purchasing *anything*.

Video Cameras

Speaking of video astronomy, Adirondack Video Astronomy sells a trio of terrific little video cameras that are ideal for mating with a telescope. Each of their AVA

Astrovid CCD videocameras is small enough to fit into a jacket pocket yet powerful enough to get some terrific views of the Moon and brighter planets through just about any telescope on any mounting. The Astrovid 400 weighs only 9.25 ounces (260 grams) and is rated at 0.03 lux (*lux* being the unit of measurement of the amount of illumination on a particular surface or object), while the more advanced Astrovid 2000 weighs 11 ounces (300 grams). The Astrovid 2000 offers higher resolution than the 400, with 600 lines of horizontal resolution, and enjoys an increase in sensitivity to 0.01 lux. Although these cameras are designed primarily for lunar and planetary photography, the manufacturer states that both are sensitive enough to record 9th-magnitude stars through an 8-inch telescope. Finally, if you are looking for color, consider the Astrovid 7000. Though not as sensitive as the black-and-white cameras, the 1.5-lux Astrovid 7000 is more than sensitive enough to record the true colors of Jupiter, Saturn, and Mars as well as brighter double and multiple stars. The Astrovid 7000 weighs 16 ounces (450 grams) and measures 5.75 inches (145 mm) long, making it the largest of the three. All come complete with power supply, cables, a T-C adapter (needed to couple the camera to a telescope), and instructions. All you need to add is a telescope and a video recorder (for field use, it might be best to attach the video camera to a battery-powered camcorder rather than a conventional VCR).

The Great Outdoors

An area that few manuals of amateur astronomy address is the environment around the observer. Sure, most books complain about excessive light pollution and the need for good sky conditions, but there is so much more to enjoying the night sky than just the sky.

Baby, It's Cold Outside

The old saying that clothing makes the man (excuse me, person) is certainly true in astronomy. Nothing can take the enjoyment out of observing more quickly than weather-related discomfort. Although this is usually not a problem during summer, it certainly can be at other times of the year. Even the sturdiest telescope mount will wobble if the observer using it is shivering!

It goes without saying that the clearest nights occur after a high-pressure weather front sweeps the atmosphere of clouds, haze, and smog. Unfortunately, the clear atmosphere also causes the Earth to lose a great deal of the heat that it has built up during the day. Many amateurs decide to stay indoors on these cold nights, but by doing so they are missing some of the clearest skies of the season. Others try to brave the cold by wearing their usual overcoat and a thin pair of gloves, but they soon return inside, teeth chattering and fingers numb. Is this any way to enjoy the wonder of the universe?

Most hardy souls agree that layering clothes works best. For temperatures above about 20°F wear (from the inside out) a T-shirt, flannel shirt, sweater, and parka above the waist, while underwear, long underwear, and heavy pants should be worn below. In colder temperatures, or when the wind is howling, replace the sweater with a one-piece worksuit, snowmobile suit, or insulated coveralls. These provide a good, windproof barrier between you and the cold, cruel world.

These items should keep you warm enough in moderately cold conditions, but they also can make you stiff as a board! Multilayered clothing can make it difficult to pick up that pencil that was dropped on the ground or even to bend for a peek through the eyepiece. There must be a better way! Happily, thanks to modern synthetic fabrics, it is now possible to stay outside even in subzero temperatures in relative comfort and with full freedom of movement. Increasing numbers of observers are joining other outdoor enthusiasts who wear clothing made of advanced materials such as Dupont's Thermax and Thinsulate. Both have amazing heat-retention properties yet are thin and light enough to permit the wearer to bend with ease.

The best selections of cold-weather apparel are found at either local sporting goods retailers or in national mail-order catalogs such as L.L. Bean, Campmor, and Damart. Unlikely as it may sound, I have gotten much of my cold-weather clothing either from local bike shops or from two mail-order outlets: Bike Nashbar and Performance Bicycle Shop. (I'm an avid bicyclist by day.)

Keeping the extremities warm is the most critical part of your cold-weather regimen. In less extreme temperatures, a pair of thick socks and work boots for the feet, a hat for the head, and a pair of gloves for the hands should do the trick. Under colder conditions, the head is best protected with a silk or wool balaclava. Looking like a full-face ski mask, a balaclava is thin enough to hear through yet warm enough that it may make a heavy hood or hat unnecessary.

For the hands, try a pair of ski mittens stuffed with Thinsulate or a similar material. Unfortunately, though they are warmer than gloves, mittens can make it difficult even to focus the telescope. Some mittens come with a thin insulating glove that may be worn separately, which is handy when changing eyepieces but still a problem when taking notes or making drawings. One other possibility is to wear a separate thin glove liner inside each mitten. Once again, many outdoor outlets sell mittens and glove liners that work quite well.

Nothing is more painful than frozen feet. I have seen some people walk out in 10°F weather wearing a heavy parka, down vest, long underwear, thick hood, heavy gloves, and a pair of sneakers! They didn't last long.

Work boots and so-called sorels and moon boots have excellent heat-retention qualities but only when used with a good pair of socks. The two-pair strategy usually works best. Wear a thin pair on the inside and a thicker, thermal pair on the outside. For truly frigid weather, even the best insulated gloves and boots may not do the trick. Although some outdoorsfolk use heated hand warmers that run on cigarette lighter fluid, I get nervous keeping an incendiary device near clothing. (Do you smell something burning?)

No doubt about it, battery-powered socks will keep your feet warm, but they fail on several different fronts. First, all of those that I have tried tend to warm my feet as a microwave oven warms a bowl of leftover spaghetti—unevenly! Invariably, part of my foot was too hot, while another part was too cold. Then, too, you must decide what to do with the battery pack. In most cases, it must be strapped onto your leg, adding to the discomfort. Some people love electric socks, which is fine, but I prefer another approach.

A safer means for dispelling the cold are nontoxic chemical hand- and feet-warming pads, such as those manufactured by Mycoal Warmers Company in Japan and sold in North America by Grabber of Grand Rapids, Michigan. Once removed from their packaging and exposed to the air, these nontoxic heaters maintain a temperature of about 140°F for several hours. (I have found the advertised claim that they last "seven hours or more" to be a little optimistic. From actual use, four or five hours is probably a better estimate.) Afterwards, simply discard the pad. It is important to note, however, that the warmer must be wrapped in a cloth or other protective material before use. Burns can result if these heaters are left in direct contact with the skin for long. Other companies make similar heat pads, but I have found those by Mycoal to be the best.

Don't Bug Me!

All that talk about cold weather makes me long for spring and warmer nights. But, of course, with warmer weather come things that go buzz in the night. Mosquitos, gnats, and blackflies can prove more annoying than the cold. Can anything be done to ward off these nighttime pesties? Different solutions, ranging from voodoo-like rituals to toxic chemical brews, have been advanced over the years with varying degrees of success.

The best way to avoid insect bites is a combination of long-sleeved clothing, an observing site that is high and dry, and a good insect repellent. Studies show that the most effective brands use N, N-diethyl-3-methylbenzamide, better known as DEET. Most commercial repellents specify a DEET concentration of no more than 25% to 30% because higher amounts are potentially harmful. Common repellents that use DEET include Cutter, Off, Deep-Woods Off, and 6-12 Plus.

Exercise caution whenever applying an insect repellent, DEET may be applied directly to the skin and clothing, but be sure to use it far from your telescope or other equipment. Although it works well at warding off insects, DEET also acts as a wonderful solvent when sprayed onto vinyl, plastic, painted surfaces, and optical coatings! DEET should also not be used on infants or young children.

A repellent growing rapidly in popularity is actually not a repellent at all. For years, Avon has been selling Skin-So-Soft lotion and bath oil as a way to promote youthful skin; according to modern-day folklore, it works as a mosquito repellent, too. Tests in *Consumer Reports* magazine and by others have found that it works marginally, if at all. Maybe that's why so many astronomers look so young!

Still More Paraphernalia

Flashlights

Every astronomer has an opinion of what makes a flashlight astronomically worthy. Some prefer pocketable penlights, others like focusable halogen models, and a few favor dual-bulb models (which provide a built-in replacement just in case one bulb burns out). Most agree that the best are small enough to fit into a pocket but large enough not to get lost at night. White or brightly colored housings are also preferred to black or dark models, because they are easier to find if dropped.

Regardless of the style or design, a flashlight must be covered with a red filter to lessen its blinding impact on an observer's night vision. There are many different ways of turning a white light red. Some of the more common methods include painting the bulb with red fingernail polish or using red tissue paper or transparent red cellophane from stationery and party supply stores. These treatments tend to chip, tear, or fade with time, forcing repeated filter renewal or replacement. A more permanent solution is to use red gelatin filter material sold in art supply or camera stores. One or two layers of Wratten gelatin filter #25 (yes, the same classification system as eyepiece filters) works especially well. But perhaps the most versatile red filter material is sold by auto parts stores as repair tape for car taillights. Sold in rolls of several feet, its adhesive backing makes it ideal for sticking onto a flashlight.

Far better are flashlights that use a light-emitting diode (LED) instead of a conventional bulb. These are growing in popularity thanks to their deep, pure red color and low power drain. Many astronomical suppliers sell LED "astronomy flashlights," with one of my favorites being the Starlite by Rigel Systems. The Starlite features an adjustable-brightness LED and runs on a single 9-volt battery.

Dew Caps and Dew Guns

One of the most frustrating things that we, as amateur astronomers, are forced to deal with is dew. The formation of dew on a telescope objective, finder, or eyepiece can end an observing session as abruptly as the onset of a cloud bank.

Dew forms on any surface whenever that surface becomes colder than the *dew point* temperature, which varies greatly with both air temperature and humidity. To illustrate this, think of a can of cold soda. In the refrigerator, the can's exterior is dry because its surface temperature is above the dew point of the surrounding refrigerated air. Now take the can out of the refrigerator. Almost immediately, its surface becomes laden with moisture because it is now colder than the warmer air's dew point. Under a clear sky, objects radiate heat into space and soon become colder than the surrounding air.

Nothing, neither telescopes, binoculars, eyepieces, star atlases, nor cameras, is impervious to the assault of dew, but there are ways to slow the whole process down (no, one way is *not* to give up and go inside—although I have been tempted at times!). One option is to install a dew cap on the telescope. A *dew cap* is a tube extension that protrudes in front of a telescope tube to shield the optics from wide exposure to the cold air, thus slowing the cooling process. Binoculars, refractors, and catadioptric telescopes stand to gain the most from dew caps because their objectives and corrector plates lie so near the front of the telescope tube; reflectors usually do not need a dew cap because their primary mirrors lie at the bottom of the tube (which itself acts as a dew cap). The only exceptions to this are if the secondary mirror dews over (only in exceptionally damp conditions) or if the reflector has an open-truss "tube." The former situation can be slowed by installing battery-powered heater elements around the secondary, a technique that will be described later. For the latter, many amateurs wrap the truss with cloth, effectively shielding the mirror from radiational cooling.

Dew caps merely slow the cooling of an objective lens or corrector plate and, therefore, only delay dew from forming. Depending on the humidity, this may be enough, but to be effective, a dew cap must extend in front of the objective or corrector at least 1.5 times (preferably 2 to 3 times) the telescope's diameter. Although most refractors built during this century have been supplied with dew caps, most binoculars and catadioptrics are not (the lone exception being the Questar Maksutovs), so their owners must either purchase or construct dew caps.

Surprisingly few companies offer dew caps. Of those available, most are made of molded plastic and designed to slip on and off the telescope as needed. For instance, the Telescope and Binocular Center supplies Orion FlexiShield dew caps. Made of a thin, durable material called Kydex, FlexiShields are sold as flat rectangles. They are formed into tubes by wrapping the ends around and sealing the full-width, permanently attached Velcro closure. After the observing session, simply pull the Velcro seal apart, and the FlexiShield lies flat for easy storage. FlexiShields come in sizes to accommodate telescopes from 3.5 to 14 inches in diameter.

Roger W. Tuthill, Inc., also offers a complete line of No-Du dew caps (Figure 6.12). The No-Du caps are rigid, plastic cylinders, making them more difficult to store than the Orion caps. Each is lined with black felt, which is claimed to increase the cap's effectiveness. Tuthill's larger dew caps also feature built-in cylindrical heating elements. By warming the corrector plate and the air inside the cap *ever so slightly* by convection, dewing can be effectively prevented even in high humidity. The cap's heating element emits between 10 and 20 watts of heat, depending on telescope size, and requires a 12-volt DC power source, such as a car battery, DC power supply (useful only if a 110-volt AC outlet is nearby), or a rechargeable battery with at least 5 (preferably 10 or more) ampere-hour capacity. No-Du caps are available for telescopes ranging from 3.5 to 14 inches in diameter, but heated versions are sold only for 8-inch and larger scopes.

Figure 6.12   *Tuthill No-Du Cap.*

Jim Kendrick Studios sells dew caps that seem to be a combination of Orion FlexiShields and Tuthill unheated No-Du caps. Kendrick's models all have sewn-in Velcro strips and felt linings, and they unfold flat for easy storage. They are available for most popular-size telescopes.

For 8- through 11-inch Schmidt-Cassegrains, the Telescope and Binocular Center also offers a different style of contact heating element called the Orion Dew Zapper, which wraps around the front end of the telescope tube to heat the corrector plate by conduction to just above the air's dew point. Although it may be used independently of a dew cap, the Dew Zapper is more effective with a cap in place. Both 110-volt AC and 12-volt DC versions are available; the former delivers up to 25 watts of heat, while the latter supplies about 12 watts at 1 amp.

Perhaps the most flexible dew-prevention system is sold by Canadian amateur astronomer Jim Kendrick. The Kendrick Dew Remover works equally well on all types of telescopes as well as eyepieces, finderscopes, camera lenses, and just about any other optical surface likely to be turned skyward by today's stargazers. The Kendrick system consists of a small control box that is powered by a 12-volt DC source, such as a car battery. Kendrick recommends using, at a minimum, a 12 amp-hour battery. The system may also be run off house current by plugging the controller into an AC/DC converter of adequate amperage.

The Kendrick heater elements are wrapped in black nylon strips of various lengths and widths, each with elastic straps and Velcro pads for fastening. They come in various shapes and sizes, some for wrapping around tubes and others for laying under primary and secondary mirrors. Depending on just how damp the conditions are, power to the heating elements can be set at either low, medium, or high and will cycle off automatically to prevent overheating.

Some find the Kendrick system, Orion Dew Zapper, and Tuthill No-Du cap inconvenient to use because they all require external power sources, which means each comes with a long electric cord. Cords are easy to snag and trip over, especially at night, but because of the required power draw, there seems to be no way around this.

Right about now, if you listen carefully, you can hear the traditionalists in the crowd jumping up and down, screaming that all this is blasphemy. Many amateur astronomers believe that imparting any heat to a telescope will cause optical distortion. True enough, heat will upset the delicate figures of optics, which is why some telescopes must be left outside for an hour or more before observing to let the optics adjust to the outdoor temperature. In practice, however, the disturbance from contact heaters will be minimal as long as the heating is done in moderation. By definition, the dew point can only be less than or equal to the air temperature, never greater (under most clear-sky conditions, the dew point is going to be much less than the air temperature). The purpose of a contact heater is to raise the telescope's temperature not above that of the air but only above the dew point. Therefore, the telescope should never feel warm to the touch, and overheating should never become a problem.

A third way to wage war against dew is to use a low-power hair dryer or heat gun. By blowing a steady stream of warm air across an optical surface, dew may be done away with, albeit temporarily. If an AC outlet is within reach, a portable hair dryer at its lowest setting makes a good dew remover, but if you are out in the bush away from such amenities, use a heat gun designed to run off a car battery (DC heat guns draw too much power to be used with rechargeable batteries). A wide variety of sources sell basically the same 12-volt heat gun at a wide variety of prices for a wide variety of uses. Although the Telescope and Binocular Center sells it as a "dew remover gun," auto parts stores call it a "windshield defroster," and camping equipment outlets offer it as a "mobile hair dryer." Call them what you will, all are pretty much the same. The gun puts out about one hundred fifty watts of heat—not exactly enough to dry hair or defrost a windshield but plenty to remove dew from an eyepiece.

Although I never go observing without a heat gun, they do have some drawbacks. First, although they are fine for undewing finderscopes and eyepieces, their small size limits their effectiveness for objectives and corrector plates. (If the lens or mirror is much larger than about three inches, it is likely that the entire surface will not be cleared before a portion becomes fogged again.) A second shortcoming is that, sadly, the dew will return as soon as the surface cools below the dew point, making it necessary to halt whatever you are doing and undew the optic all over again. If dew is a big problem from where you observe, the Orion Dew Zapper and, especially, the Kendrick system are hard to beat; in drier environs, you can probably make do with a dew cap and heat gun. Never wait for dew to form; always turn the heater on before observing begins. Put the dew cap on (remember the finder, too) and have the heat gun at the ready to clear any fog that may form on the eyepieces. By following this three-step program, optical fogging should be minimized, if not eliminated.

Observing Chairs

It is a well-proven fact that faint objects and subtle detail will be missed if an observer is fatigued or uncomfortable, yet many amateurs spend hours outside at their telescopes without ever sitting down. Observing is supposed to be fun, not a marathon of agony!

Observing chairs help relieve the stress and strain associated with hours of concentrated effort at the telescope. The best have padded seats and may be raised or lowered with the eyepiece (not always possible with long-focus refractors or large-aperture Newtonians).

Musician's stools or drafting-table chairs work splendidly for this purpose. Check the offerings at local music shops and drafting/art supplies stores.

For the astronomical market, the Telescope and Binocular Center offers three observing seats. Two of these differ little from piano stools; the standard model adjusts between 19 and 25 inches, while the deluxe chair will get you anywhere between 21 and 27 inches above the ground. The third one is an innovative design that is also sold by Pocono Mountain Optics as the Starbound Viewing Chair. The padded seat can be set at any height between 9 to 32 inches by sliding it along the chair's framework of black metal tubes, making it usable with most amateur telescopes. Although much more flexible than a stool, the Orion/Starbound chair is also considerably more expensive.

If that proves a little rich for your blood, consider the all-oak observer's chair from K.C. Woodsmith. Similar in design to the Starbound chair, the K.C. Woodsmith chair also stands about thirty-six inches tall but will need a seat cushion for those long viewing sessions. Both it and the Starbound chair fold flat for storage.

Finally, StarMaster Telescopes offers a unique chair/stepladder combination (Figure 6.13). Two aluminum side rails, each standing 38 inches tall and placed at right angles to each other, support a triangular platform of oak, which

Figure 6.13 *The StarMaster observing chair/stepstool. Photo courtesy of StarMaster Telescopes.*

in turn holds the side rails together. The platform can be locked in four height positions; the standard chair sits the observer as high as 29 inches, while a deluxe version raises the platform another 6 inches. Unlike the others described here, the StarMaster chair may also be used as a short stepladder, although the manufacturer warns that this should be done only with the platform set in either of its two lowest positions. The entire chair disassembles for flat storage.

Telescope Mounts

When the Beach Boys sang "Good Vibrations," they certainly were not singing about telescope mounts! One of the biggest complaints that my informal survey of telescope owners revealed is dissatisfaction with their mountings. In an effort to remedy this situation, many retrofit their instruments with substantially larger, sturdier support systems. Some of the finest are manufactured by Astro-Physics and Takahashi, discussed in chapter 4 under those companies' respective headings. Here are a few other companies that manufacture some excellent telescope mounts.

For small refractors (that is, up to 3.1 inches or 80 mm in aperture), Tele Vue sells the Up-Swing mount head. Based on Tele Vue's Panoramic alt-azimuth mounting, the Up-Swing acts as a go-between to mount the telescope onto a camera tripod. To use the Up-Swing, the telescope must have a 0.25-20 threaded tripod socket, and the tripod must be sturdy enough to hold both it and the Up-Swing assembly.

Some of the world's finest telescope mounts are manufactured by Hollywood General Machining Company and carry the Losmandy name on their sides. Losmandy product line includes a number of add-on features to make an existing mount a little easier to use as well as complete mounts for a wide variety of telescopes. Especially impressive is the G-11 German equatorial mount, previously noted in chapter 4's discussion of the Celestron CG-11 and CG-14 SCTs. The G-11 may be purchased separately and is ideal for telescopes up to about sixty pounds, while a smaller version, the Losmandy GM-8, is designed for instruments up to about thirty pounds. Both come with well-designed dual-axis drive systems that feature built-in quartz correctors and periodic error correction as well as optional polar-alignment scopes.

Another giant in the telescope-mounting industry is Edward R. Byers Company. Like Losmandy, Byers is famous for its mountings' rigidity and precise tracking ability. Chiefly a build-to-suit custom house, Byers features mountings designed primarily for medium- to large-aperture amateur telescopes and also manufactures some of the finest clock drives available on the market. Especially popular among serious astrophotographers is its Celestron 14 Star-Master retrofit drive system for older, fork-mounted models.

Parallax Instruments offers three equatorial mountings suitable for its instruments as well as other manufacturers' models. The Centaurus is a nicely crafted German-style equatorial mounting on a solid ash tripod. Both axes move smoothly, the right-ascension axis supported by a tapered roller bearing and the declination axis on a radial ball bearing. The mounting comes outfitted with the same, sophisticated, dual-axis drive package found on the Losmandy G11, which includes DC stepper motors, periodic-error-correction circuitry, and four drive rates. The only thing missing is a built-in polar-alignment scope, which is a nice feature to have but not an absolute necessity. The equatorial head, minus counterweights, weighs 29 pounds, while the tripod weighs 30 pounds. The manufacturer states that the Centaurus is capable of supporting up to 7-inch refractors, 12.5-inch Newtonian, and 14-inch Cassegrains and catadioptrics, although making this claim for the latter two seems a little optimistic.

For larger instruments, the Parallax HD150 and HD200 German equatorial mounts are good choices. Both feature stainless-steel axes and counterweights held in cast housings and mounted onto steel piers. The HD150 will support telescopes up to 100 pounds, while the HD200 will handle up to 160 pounds. Both are designed more for permanent setup in an observatory than for portability.

One of the biggest drawbacks to Dobsonian telescopes is that they do not track the stars automatically; instead, observers have to nudge them to keep up with the sky. One way around this is to put the telescope, mounting and all, onto an equatorial table. Equatorial tables are designed to pivot along with the sky just like an equatorial mounting, bringing anything mounted on top of it along for the ride. Two companies that offer equatorial tables for Dobsonians are TL Systems and Equatorial Platforms. The TL Systems unit is sold only as a kit. All parts except the plywood for the platform itself are supplied, although you also need a few basic power tools such as a drill and saw as well as a rudimentary knowledge of carpentry. If you would prefer instant gratification (but at a much higher price), the Equatorial Platforms mounting comes fully constructed from finely crafted wood and well-machined components. Various models are available for different apertures. Both the TL Systems and Equatorial Platforms models are powered by 12-volt motors that work very well, although polar alignment can prove more challenging with them than with more conventional equatorial mounts.

Finally, if you want your Dobsonian telescope to track the sky but can't afford an equatorial table, you might consider the Dob Driver II from Tech2000. The Dob Driver II is a dual-axis drive system run by a sophisticated microcomputer that lets any alt-azimuth mount (although it was designed primarily for Dobsonians) track the sky automatically. Two 12-volt DC motors drive the mounting's axes by means of stainless steel wheels and/or timing pulleys and belts, depending on the design of your telescope. Once set up and engaged, the Dob Driver will automatically step the telescope up or down, left or right, to stay centered on the selected field. The user can then choose from four operation modes—track, pan, guide, and seek—from the small hand controller that is supplied as part of the package. In the track mode, you teach the computer by centering the object of interest and then letting it drift. By using the hand controller to correct the drift, the computer figures out just how the object is arcing across the sky and will track it from then on. In pan mode, you move the telescope around the sky by pressing the directional buttons on the hand controller. The guide mode amounts to a dual-axis drive corrector, which allows some limited guided astrophotography. Because this is not a true equatorial mount, however, exposures are limited because the field will slowly rotate as the telescope tracks the sky. Finally, the seek mode helps you find an elusive object, once the telescope has been pointed in the object's general direction, by slowly moving the telescope in an expanding square spiral pattern. This way, you can concentrate on looking for the object rather than worry about moving the instrument too far or too fast. One caveat: Installing the Dob Driver II will require some surgery on your telescope mount, including cutting and drilling. Instructions are provided, of course, but some people might still find it a little too challenging. Depending on how accurately the telescope mount is assembled, further modifications may be needed.

Binocular Mounts

After purchasing a pair of giant binoculars, most people suddenly come to the realization that the binoculars are too heavy to hold by hand and must be attached to some sort of external support. The favorite choice is the trusty camera tripod, but as was discussed earlier, not all tripods are sturdy enough to do the job. All of the brands mentioned in the earlier section are strong enough to support binoculars, but most are too short to permit the user to view anywhere near the zenith comfortably (the bigger Bogen/Manfrotto models being an exception to this). What are the alternatives? Basically only one: Use a special mount designed specifically for the purpose.

Some of the best binocular mounts around are sold by Virgo Astronomics (Figure 6.14). Working like a swing-arm desk lamp, the binoculars are attached to an arm that may be raised and lowered without affecting the aim of the binoculars. This lets observers both short and tall view comfortably without having to re-aim the glasses each time. In early 1998, Virgo Astronomics introduced their new Sky/mount. Made of machined aluminum that is finished with a black powder-coat, Virgo states that the Sky/mount is designed to hold binoculars as heavy as 5 pounds, an average weight for 80-mm giant glasses. This might be a little taxing for the mounting, which is made from aluminum square tubing (hollow with 0.0625-inch-thick walls). Open tubes tend to flex and twist more than solid beams when under a torquing load, such as might be produced when moving heavy binoculars balanced at one end. But for smaller glasses (say, to 63- or perhaps 70-mm aperture), the mounting will perform well.

One reason for the hollow tubes is to accommodate the Sky/mount's telescoping counterweight shaft, which can be set at 18 or 24 inches. This flexibility lets the Sky/mount counterbalance a wide range of binocular weights with just a single 5-pound counterweight. Motions are smooth in both altitude and

Figure 6.14   *The Sky/mount binocular mount, available through Virgo Astronomics. Photo courtesy of Virgo Astronomics.*

azimuth thanks to nylon pads that are placed at all points between the beams as well as between the two parts of the mount's base. A large knob is used to lock the binoculars in altitude after they are aimed at a target.

Vibration Dampers

These make great stocking stuffers for any amateur who owns a tripod-mounted telescope. One of the biggest problems facing amateurs using such telescopes is how to combat vibration. Vibration is caused by a wide range of sources; anything from nearby automobile traffic or another person moving about to the wind can be the culprit.

Nothing can guarantee shake-free viewing, but Celestron has introduced a simple little gadget that can make a big difference. Its vibration dampers are made of a hard outer resin and soft inner rubber polymer, isolated from each other by an aluminum ring. By simply placing one under each tripod leg, you ensure that any vibration will be absorbed by the inner polymer before it can be transmitted to the telescope. Such a simple yet effective idea—why didn't someone think of this before?

The list of available accessories for today's amateur astronomer could run on for pages and pages, but I must cut it off somewhere. To help put all of this in perspective, Table 6.5 lists what the well-groomed amateur astronomer is

Table 6.5 **Must-Have Accessories for the Well-Groomed Astronomer**

Commercial
- A good red flashlight
- Star atlas
- Narrow-band LPR filter
- Amici prism for right-angle finder
- Adjustable observer's chair (see also chapter 7 for plans on how to construct one)
- Warm clothes
- Hand and foot warmers
- Bug repellent (Cutter, Deep-Woods Off)
- Guidescope or off-axis guider (long-exposure photography only)
- Drive corrector (if telescope comes equipped with an AC clock drive)
- Camera-to-telescope adapter

Home-made
- Eyepiece and accessory tray attached to telescope
- Two old briefcases, one for charts, maps, and flashlights, and the other lined with high-density foam for eyepieces, filters, etc.
- Observation record forms
- Cloaking device (see chapter 9)

wearing these days. Some of the items may be readily purchased from any of a number of different suppliers; others can be made at home. A few may even be lying in your basement, attic, or garage right now.

The most important accessory that an amateur astronomer can have is someone else with whom to share the universe. Although many observers prefer to go it alone, there is something special about observing with friends; even though you may be looking at completely different things, sharing the experience with someone else is always nice. If you have not already done so, seek out and join a local astronomy club. For the names and addresses of clubs near you, contact a local museum or planetarium, the Astronomical League, or, if you have access to the Internet, check out the World Wide Web; you will find addresses and URLs aplenty in appendix E. If you have an astronomy buddy, then you have the most important accessory of all, and one that money cannot buy!

7

The Homemade Astronomer

Amateur astronomers are an innovative lot. Although manufacturers offer a tremendous variety of telescopes and accessories for sale, many hobbyists prefer to build much of their equipment themselves. Indeed, some of the finest and most useful equipment is not even available commercially, making it necessary for the amateur to go it alone.

Many books and magazines have published plans for building complete telescopes. Rather than reinvent that wheel here, I thought it might be fun to include plans for a variety of useful astronomical accessories. For this second edition, here are ten new projects, ranging from the very simple and inexpensive to the advanced and costly. Their common thread is that they were created by amateurs to enhance their enjoyment of the universe. These projects are just a few samples of the genius of the amateur astronomer.

Portable Tablecase

One look through the previous chapters and you can see that being an amateur astronomer carries with it a lot of baggage. Eyepieces, charts, notebooks, flashlights, bug spray ... the list goes on and on. Where do you put it all, and what do you do with it once you get to your observing site? For Terry Alford of Gray, Tennessee, the answer came in the form of a tablecase (Figure 7.1). The same size as a large briefcase, Alford's tablecase doubles as both a carrying case for all of his observing essentials as well as a portable table for spreading out his gear. The legs for the table are held to the underside of the case with a bracket when not in use.

Figure 7.1 *Terry Alford and his tablecase, both ready for a night of observing from the Winter Star Party. Photo by Terry Alford.*

The tablecase is based on a 18″ × 24″ × 5″ box constructed from plywood. As Figure 7.2 shows, its bottom is made from 0.5-inch plywood, while the sides and top are constructed from 0.25-inch plywood. Alford notes that 0.75-inch plywood might be easier to work with during construction, but it would cause the case to increase substantially in weight. You, the reader, should feel free to vary the dimensions as you see fit; just make sure they are consistent.

The parts list for the tablecase is given in Table 7.1.

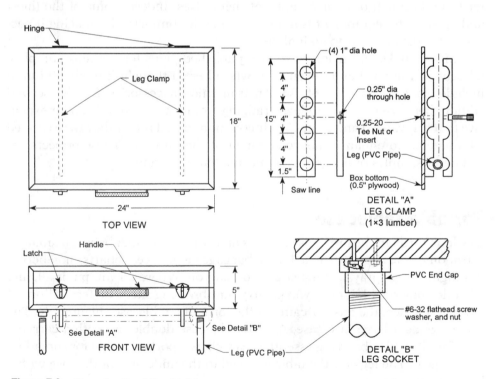

Figure 7.2 *Plans for the Alford tablecase.*

Table 7.1   **Tablecase Parts List**

Description	Quantity
Top, 18″ × 24″ (0.5″ plywood)	1
Bottom, 18″ × 24″ (0.25″ plywood)	1
Side, 5″ × 24″ (0.25″ plywood)	2
Side, 5″ × 17.5″ (0.25″ plywood)	2
Leg clamp, 15″ long (1″ × 3″ lumber)	2
Hinges	2
1″-diameter, Schedule 40 PVC pipe (thick-walled type)	8′
1″ PVC end caps	8
Latch	2
Chain, approximately 6″ long each	2
Handle	1
6d finishing nails	—
Carpenter's glue	—
Paint or stain	—

The case can be constructed in two ways: either by constructing two separate halves to be joined later or by making a single box that will be cut in half lengthwise afterwards for the case's top and bottom. Alford suggests the latter approach to ensure that the top and bottom match each other identically.

Begin your tablecase by cutting the box's panels and leg clamps to the dimensions in the parts list and shown on the drawing. Once all are cut, construct the box using some 6d finishing nails (about one inch long) and woodworking glue. Be especially careful to make sure that all of the corners are squared and then set the box aside to let the glue dry thoroughly.

When the glue has set, use a carpenter's square to divide the box in half lengthwise (in effect, creating two 18″ × 24″ × 2.5″ half boxes) and then cut it in half along that line. Although this is ideally done on a table saw, a circular saw or sabre saw may also be used; just be sure to use a saw fence or guide to keep the edges as straight as possible.

To make each leg clamp, drill a single, slightly oversized 0.25-inch hole in the center of each as shown in Detail "A" of Figure 7.2. Next, bore four 1-inch holes spaced approximately as in the figure. Split the clamps in half as shown, and glue and nail one half of each clamp to the bottom of the case. When the glue has set, drill through the centered hole in each clamp into the bottom of the case's box. Working from inside of the tablecase box, set a 0.25-20 tee nut or threaded insert into the hole as shown.

When the box and leg clamps are complete, add some edge molding to the edges both to protect it and to keep round objects like pencils from rolling off. Stain or paint the box to protect the wood against warping from inevitable nighttime moisture and then set it aside to dry.

Cut the table's four 22-inch legs from 1-inch diameter Schedule 40 thick-walled PVC pipe, commonly available from hardware, plumbing, and home supply stores. Attach a PVC end cap onto one end of each leg. Fasten the other four PVC end caps onto the four corners of the case's bottom with a pair of small screws and nuts as shown in Detail "B" of Figure 7.2. Screw them in flush to the surface so that they will not interfere with the legs when they are inserted.

Once the paint or stain has dried, install the hinges, latches, handle, and restraining chains to the case, and you're done.

To set up the table, insert the four legs into the PVC end caps on the bottom of the case. Dismantle the table by removing the four legs and storing them using the two leg clamps. To customize your case, you might want to add a small container to hold pens and pencils or a felt-lined compartment to hold eyepieces. Alford also suggests that more electronically minded astronomers might even install one or more red LEDs to help illuminate star charts.

Alford has found his tablecase to be an indispensable accessory. He adds that "once you have used a tablecase, you will wonder how you ever got along without it."

Aiming Device

As already mentioned in chapter 6, the single most often mentioned accessory for a telescope in my survey was a 1× aiming device like the Telrad. Nearly everyone, even those with computer-driven telescopes, raved about how useful these are for aiming instruments toward specific regions of the sky. Although the Telrad broke new ground when the late Steve Kufeld first introduced it in the 1980s, several similar devices have subsequently appeared on the market. All have one thing in common: a price tag of $40 or more. This is not a lot of money in today's astronomical marketplace, but the industrious amateur can easily make a similar gadget for about half that much money in a single afternoon.

Like several of the commercial versions, this homemade unit is based around the Electronic Point Sight, which is manufactured in China and sold by the Daisy Manufacturing Company of Rogers, Arkansas. You will find it for sale in sporting goods stores, discount department stores like Wal-Mart and K-Mart, and directly from Daisy Manufacturing (see appendix C for the address and telephone number). The Electronic Point Sight projects a red point of light from an LED onto a small, round, glass window. Viewing along the Electronic Point Sight, the eye sees the intended target through the window as well as the red point's reflection.

The only problem with the Daisy Sight is that because it is designed primarily for daytime use, its LED is very bright. For astronomical hunting, the LED has to be dimmed. Some people have simply glued gray filter material (such as a small piece of photographic negative) in front of the LED, but there is a better, electronic way of accomplishing this. Simply lowering the voltage of the battery will not work, because LEDs do not dim when the applied voltage drops; they will

Table 7.2 **Aiming Device Parts List**

Description	Part Number	Quantity
Daisy Electronic Point Sight	Daisy #7809	1
Daisy Point Sight adapter	Daisy #7853	1
Archer Horizontal PCB-Mount Micro Potentiometer	Radio Shack #271–282	1
Solder		—
Double-backed tape		—
Epoxy glue		—

not light below a certain threshold voltage (unlike traditional light bulbs, which may be dimmed). Instead, to change the brightness of an LED, the circuit's amperage must be reduced by inserting a resistor in line with the battery.

Many amateurs have undoubtedly concocted a whole host of ways to incorporate a resistor into the Electronic Point Sight, but I think the method offered by David Goldberg of Springfield, Virginia, is the cleanest and simplest. And although he originally specified a 100-ohm resistor, I personally prefer a variable resistor (also known as a potentiometer) so that the LED's brightness may be adjusted in the field. Table 7.2 lists the items you will need.

Begin surgery on the sight by unscrewing the two adjustment screws, one on the top and one on the side of its body. These screws, used to adjust the aim of the sight one it is mounted in place, are held by two crimped nuts. To unscrew them, grip the crimped nuts with a pair of pliers or vice grips. Don't be afraid to really lean on the screwdriver, as the screws will come out only with some effort. Once they are removed, the body of the sight can be separated as shown in Figure 7.3.

Carefully disassemble the sight, taking care not to pull or break the thin wires that make up the sight's circuitry. Note how the black (ground) wire goes from the battery clip to the LED, while the red (positive) lead goes from battery to the on/off switch and then on to the LED. Snip the black wire in half, offsetting the cut slightly toward the battery. Strip about a quarter inch of the

Figure 7.3 *Adding a tiny dimmer control onto the Daisy Sight.*

wire insulation away from either end. The specified potentiometer has three contacts on it. Solder the "top" contact onto one end of the wire and either of the "bottom" contacts to the other end. Once the solder has cooled, turn the switch on to make sure the LED still lights. Using a small screwdriver, twist the potentiometer's adjustable dial to the left and right. Does the LED change in brightness? If so, good, but if not, check all connections and soldered points. Also make sure that the battery hasn't slipped out of its holder.

Carefully slip the sight back together (without the screws) and mark where the potentiometer lines up with the side of the sight's housing (either side is okay, but I chose the side opposite the on/off switch). Drill a 0.25-inch hole in the sight's body at that point. Also mark where the potentiometer's leads hit the bottom and side of the sight's T-shaped "beam." Using the hacksaw carefully cut away that part of the **T** as I did in the figure. You'll note that my cut is far from neat which is not a problem because this part of the sight will not be seen once reassembled.

With all of the cuts made and the hole drilled, glue the potentiometer in place so that its adjustable dial protrudes through the hole in the sight's body. Just be careful not to get any epoxy on the dial itself! Then set the whole thing aside and let it dry fully before proceeding.

Before reassembling the sight, turn the switch on and be sure that the LED lights. With all working, put the sight back together, again using the pliers or vice grips to hold the crimped nut in place.

Attach the sight to your telescope tube using the Daisy adapter mentioned in Table 7.2. Although the adapter can be drilled and screwed onto the tube, the sight is light enough that double-backed adhesive tape should do the trick. Try to mount the adapter as straight as possible on the tube, but remember that the sight can be adjusted both in elevation and azimuth using the two adjustment screws.

The Daisy Sight also incorporates a neutral-gray filter (possibly a polarizer) into the round window through which you stare when aiming. That may be fine for daytime use, but the shading will dim things noticeably at night. To correct this, pop the window out by using the eraser end of a pencil to carefully press against it from the back. (Don't push too hard, as the window may be glued in place. In those cases, use a modeling knife to cut along the edge of the retaining ring in front of the window.)

The windows of some Daisy Electronic Point Sights are coated with a thin layer of reflective material on one side, while others include a second window that is tinted all the way through. If yours is the latter (and there is no way to tell before disassembly), I'm afraid you're stuck with a dim view, although you might try reassembling the window with only the single, clear lens. That will help increase image brightness but may also distort the view.

For those whose Daisy Sights include a single, coated lens, try the following technique suggested by Charlie Dilks of Newark, Delaware. First, find out which lens surface is coated by touching each side with a pencil point. The pencil's reflection will appear to touch the coated surface but will be offset on the other. Next, per Dilks, massage the coated surface with a Lead-Wipe Gun-Cleaning Cloth by Rig Products (P.O. Box 1990, Sparks, NV 89432-1990). After about ten to fifteen minutes, the lens should be clear. Dilks warns, however, that because of the chemicals impregnated in the cloth, rubber gloves should always be worn. Afterwards, reassemble the window by pressing it and the retaining ring back into place.

Telescopes on the Move

As an amateur astronomer grows into the hobby, it is not unexpected that his or her telescopes may grow in size as well. The question then becomes "Where will I store my telescope?" Some people build observatories to house their instruments, but for those of us who use our homes or other buildings for storage, one of the biggest drawbacks can be moving the telescope around and setting it up. Many large-aperture, Dobsonian-mounted Newtonians incorporate casters or wheelbarrow handles into their bases to make them more manageable, but let's face it, even an 8-inch Schmidt-Cassegrain telescope can be a burden to set up and move after a long day at work or school. To help solve this dilemma, many amateurs have devised ways of rolling their telescopes from their storage locations to their observing sites. Here are just a couple of alternatives.

To help move his 11-inch Schmidt-Cassegrain, Don Fox of Jacksonville, Florida, built an inexpensive wooden platform. Table 7.3 gives its short parts list.

Measure the footprint of the telescope's tripod when fully set up as well as the width of the smallest doorway through which you will have to wheel the telescope. The smaller of the two dimensions will dictate the width of your platform. Transfer that dimension onto a piece of plywood. (If the doorway is the limiting factor, reduce the dimension of the wood by 2 inches to allow for clearance.) Once laid out on the wood, cut the triangular platform from the plywood. Fox found that a 4' × 4' piece of 0.75-inch plywood was all that was needed for his scope. After cutting out the wooden triangle, attach three 3-inch-diameter rubber lockable casters to each corner of the base, securing each with four bolts through its predrilled base, and then square off the triangle's three points to eliminate the sharp corners. Not only do these make it easy to roll the fully

Table 7.3   *Fox Telescope Platform Parts List*

Description	Quantity
Plywood, exterior grade (0.75″ thick minimum)	1 (sized to suit the telescope's footprint)
Casters, locking, 3″ diameter minimum	3
Hardware (nuts, bolts, etc.)	12 sets
	(depends on the casters, probably 4 sets per wheel)

assembled telescope out for an impromptu observing session, but the platform also makes a handy place to store some accessories, such as an eyepiece case, dew cap, and portable battery. Fox finds the platform to be surprisingly steady, with vibrations dampening in one to three seconds.

Ed Stewart of Austin, Texas, created a sled for his 8-inch Schmidt-Cassegrain. The telescope is left permanently set up on the sled, even when in storage. To move the instrument, Stewart slips a hand truck under the sled, tilts the telescope back, and walks it around to his chosen site (Figure 7.4). Figure 7.5 shows his design: Table 7.4 lists the sled's parts.

Begin by cutting the horizontal and vertical beams of the sled. As with Fox's design, the breadth of your sled will depend on the span of the tripod's legs as well as the width of the smallest doorway through which the scope will pass. If a doorway's width is the critical dimension, allow yourself 2 inches for clearance. In other words, if the width of the door through which the telescope must pass through is 30 inches, cut a 28-inch-long horizontal beam. On the other hand, if the limiting dimension is the spread of the ends of your tripod legs and that measures 41 inches, cut the beam 43 inches long. This will allow for a little give-and-take when drilling the holes for the tripod's feet.

Next, cut the vertical beam. The tip of the vertical beam must measure an equal distance away from both ends of the horizontal beam, as shown in Figure 7.5. Set up the tripod with two of its legs on the horizontal beam and slide a length of 2″ × 4″ under the third leg to create the **T** shown in the drawing. Note where the third leg rests on the vertical beam and cut the beam 1 inch longer. (All of this is assuming that the two beams will be joined by a simple butt joint, i.e., with one beam butting up against the other. If you are a skilled woodworker and wish to use a more complex joint, be sure to allow for additional length.)

Figure 7.4 *Ed Stewart using his telescope sled to move his Schmidt-Cassegrain telescope. Photo courtesy of Ed Stewart.*

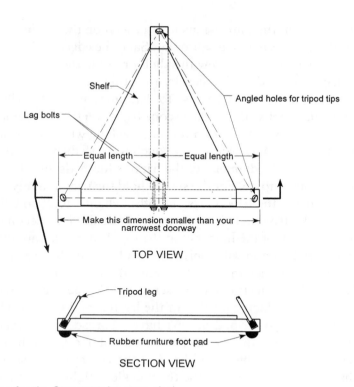

Shelf

Lag bolts

Angled holes for tripod tips

← Equal length → ← Equal length →

← Make this dimension smaller than your →
narrowest doorway

TOP VIEW

Tripod leg

Rubber furniture foot pad

SECTION VIEW

Figure 7.5 *Plans for the Stewart telescope sled.*

Lay the vertical and horizontal beams flat on the ground to form a **T,** as shown in the drawing. Centrally drill two pilot holes through the horizontal beam into the vertical member for two 6-inch-long lag bolts. Coat the two surfaces to be joined with carpenter's glue and screw in the lag bolts to make the sled's main **T** shape. After everything is together, Stewart recommends rounding off the corners of the 2″ × 4″s with a sander. He also used a router with a 0.25-inch round-over bit along the T-frame's edges to give a more finished look.

Now decide how large you want the shelf, remembering to allow enough room for things like a portable battery, eyepiece case, and other observing

Table 7.4 **Stewart T-Sled Parts List**

Description	Quantity
2″ × 4″ beam (length to suit)	2
0.5″-thick plywood shelf (cut to suit)	1
Lag screws, 6 inches long	2
Furniture foot	3
Hand truck	1
Foam insulation	—
Bungee cord	1 or 2 depending on their length

necessities. Lay out the dimensions for the shelf on the 0.5-inch-thick plywood and cut out the pattern with a sabre saw. Sand the edges and smooth the shelf's top surface; then glue and screw the shelf onto the top of the 2" × 4" T-frame.

The tripod is held to the sled by inserting the tips of its legs into three matching holes in the sled's beams. To determine where to drill the holes, set the sled on a flat floor with your telescope tripod on top, but remember to open the tripod's legs only to the width that they will be when the tripod is installed on the finished sled. Mark where each touches the sled's beams. Drill three holes slightly larger than the diameter of the tripod's footpads into the wood; a 1.375-inch hole saw ought to do. For the sled to hold the tripod tightly, the holes must be angled to match the tripod legs and offset slightly (0.25 to 0.5 inch) toward the center of the tripod. This is critical so that the legs become fully spread when at the bottom of the hole, not at the top. If the holes are drilled properly, the tripod spreader knob will apply outward force on the legs to lock their tips into the holes. (Another method to secure the legs is to make three roughly 2" × 0.75" simple metal straps with a screw hole in each end that can be curved around the outward-facing portion of the leg tips and screwed into the wood.)

If your telescope's tripod doesn't have a spreader assembly (such as the Meade LX10 or Celestron's Great Polaris mounts), attach a heavy gauge wire from the underside of the tripod head to the center of the sled, with a turnbuckle halfway between. When the turnbuckle is tightened, it will pull the tripod's legs into the holes, locking them in place.

Nail or screw a rubber furniture footpad under each end of the beams. Judge the height of the footpads according to your hand truck; they must raise the sled off the ground enough to let the front lip of the hand truck get under the sled. Finally, paint, varnish, or otherwise seal the sled to protect the wood against moisture.

The choice of hand truck is not critical, although one with soft inflatable tires will not transfer shocks to the telescope as readily as those with hard rubber tires. When transporting your telescope, it is best to secure the top of the tripod to the sled using an elastic bungee cord or two, just in case you run into an errant bump when wheeling the telescope around. It is also recommended that you add some foam insulation to the frame of the hand truck to protect the telescope's finish from getting scratched.

Stewart leaves his telescope permanently set up on its tripod sled, thus keeping setup time down to less than one minute. He writes, "I just wheel it out onto the deck, turn the power on, get my eyepieces and observing guides, and observe."

Homemade Collimation Tools

Nothing can degrade the performance of a telescope more readily than improperly collimated optics. In the course of writing this book, I have heard from many amateurs who, at first, thought their telescopes were producing poor results because of inferior optics. In many of these cases, however, what

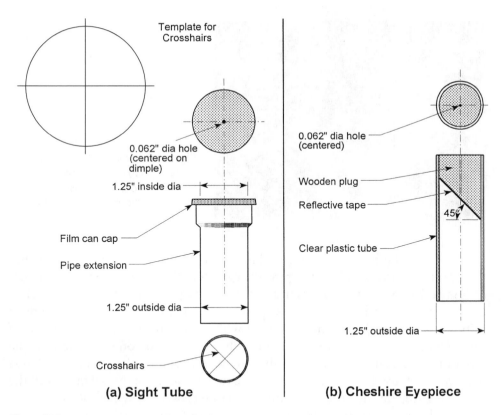

Figure 7.6 *Plans for a sight tube (left) and Cheshire eyepiece (right).*

proved to be inferior was not the optics but rather the optical collimation. Chapter 6 discussed what collimation tools are available and where they may be purchased, while chapter 8 outlines how to use them properly. Here are plans for making two of them at home.

First, let's make the sight tube (Figure 7.6a). Table 7.5 lists the few parts you will need to buy.

Your local plumbing supply house will have the 1.25-inch brass drainpipe. The drainpipe necks down from a 1.25-inch inside diameter to a 1.25-inch outside diameter. Check several samples, selecting the roundest and straightest of the lot. Once back home, check to see that the tube fits snugly into the eyepiece holder; if it doesn't, return it for another piece.

Table 7.5 **Sight Tube Parts List**

Description	Quantity
Brass bathroom drainpipe, 1.25″ diameter	1
Cap from a 35-mm film canister	1
White nylon thread	As needed
Glue	

Figure 7.7 *Rob O'Toole's home-made Cheshire eyepiece in the focusing mount of his Newtonian reflector. Photo courtesy of Rob O'Toole.*

A sight tube has a pair of crosshairs marking its exact center. To ensure their precise location, place the pipe over the template in the upper left corner of Figure 7.6. Mark the four points of the cross pattern onto the brass pipe. Now, using a thin hacksaw blade across the tube, cut four 0.063-inch-deep slots at the marks. Lay a short piece of white nylon thread across one pair of slots. Place a spot of epoxy or other household cement on one slot and tape the thread to hold it in place. Stretch the thread across the tube's center and glue the other end into its slot; once again, use tape to hold the thread taut until the glue dries. Repeat the procedure for the second crosshair. When the glue is dry, remove the tape and trim away excess thread and glue.

The sight tube's cap is easily made using the plastic lid from a Kodak 35-mm film can. Drill a tiny hole (no larger than 0.063 inch across) into the lid's central dimple. Put the lid over the large end of the tube and glue it in place to complete the sight tube.

Next, from Rob O'Toole of Chester, New Hampshire, comes this simple-to-make plastic-and-wood Cheshire eyepiece (Figure 7.7). Table 7.6 offers an itemized parts list.

Purchase the acrylic tube first. You will find acrylic tubing at a local plastics retailer (look under "Plastics" in your telephone directory's Yellow Pages) or possibly even from a hardware or home improvement store. Be sure to bring along an eyepiece from your telescope to check the diameter of the barrel against the outside diameter of the tube; the match should be as close as

Table 7.6 **Cheshire Eyepiece Parts List**

Description	Quantity
Clear acrylic tube, 1.25″ outside diameter	About 6 inches
Hardwood dowel, diameter to match inside diameter of acrylic tube	About 1.5 inches
Reflective aluminum tape	As needed

possible. Next, purchase the hardwood dowel. This time, bring along a slice of the acrylic tube, because the dowel needs to fit snugly inside. Finally, purchase a roll of reflective aluminized tape, available at better hardware stores and home improvement centers.

Once you have all of the parts, begin by cutting a 6-inch length of acrylic tube using a miter-box saw or table saw (Figure 7.6b). The exact length is not critical, although the cut should be as square as possible. If you are using a hand saw, select one with the finest teeth available, because coarse teeth tend to rip rather than cut. Next, cut a 1.5-inch length of the hardwood dowel—again, an exact length is not critical.

The next step however, is critical. For the Cheshire eyepiece to perform properly, the center peephole must be *exactly* in the center of the dowel. Use the techniques found in chapter 8 for locating the center of a mirror to find the exact midpoint of the dowel. Mark one end and then do the same for the other. Now drill a 0.0625-inch hole through the dowel lengthwise. A drill press would be best for this, but a hand drill and a good eye will work as well. Check your accuracy by comparing where the drill popped out to the center mark. If you were right on the mark, great; otherwise, try again until you succeed.

From one end of the rod, widen the center hole, using a 0.25-inch drill bit, to within 0.25 inch of the other end. Cut off a section of the dowel at a 45° angle, using either the miter-box saw or a table saw. Again, use a fine-toothed blade to keep the wood from ripping.

Cover the diagonal face of the dowel with a piece of the reflective tape. The tape should be as smooth as possible, although a few ripples will not affect the tool's function. Don't forget to punch through the tape at the center hole. Complete your Cheshire eyepiece by inserting the dowel into one end of the acrylic tube and pushing it in until it is flush with the end of the tube.

Let's Go to the Videotape

Nowadays, many people use video camera recorders, or camcorders, to capture memories of a child's first day at school, a fifth-grade play, Cousin Lester's wedding, or that weeklong trip to Walt Disney World. Camcorders are as popular today as Super-8 color movies were 25 years ago and actually cost much less to use. And although they are popular for recording terrestrial sights and sounds, few amateur astronomers seem to realize that camcorders are also ideal for capturing brighter celestial objects through telescopes.

Most camcorders come with permanently attached zoom lenses. Because of this configuration, amateur astrovideographers are restricted to photographing sky objects using the afocal method (see chapter 10). This means that both the camera's lens and telescope's eyepiece must be in place to focus the image onto the camcorder's screen. The size of the recorded image can be varied either by using eyepieces of different focal lengths or by zooming the

Figure 7.8 *Dan Ward's camcorder bracket attached to an 8-inch Schmidt-Cassegrain telescope.*

camcorder's lens in or out. Exposures can be adjusted using the camcorder's high-speed shutter settings, if it is so equipped.

One of the biggest challenges facing astrovideographers is aiming and holding up the camcorder to the eyepiece. Let's face it—keeping a telescope trained even on something as large as the Moon while also trying to point a handheld camcorder into the telescope's eyepiece is difficult! Some choose to mount the camcorder on a separate tripod, but this, too, can be daunting because the telescope and camera tripod must both keep up with the moving sky.

Here's a better solution, if you own a small 8-mm or VHS-C camcorder and a Schmidt-Cassegrain telescope. Dan Ward of Ridgefield, Connecticut, has created the simple bracket shown in Figure 7.8 for mounting his camcorder behind the eyepiece of his Celestron 8. This same design should be easily adaptable to any telescope whose eyepiece is located at the back of the tube, provided that the instrument is supported on a sturdy mounting. Table 7.7 offers the bracket's short parts list.

Cut one of the brace beams so that it will span from the front of the tube to the back plus another 12 inches behind. For example, this brace measures 27 inches long on Ward's Celestron 8. Once cut, drill two 0.19-inch-diameter holes through the beam as shown in Figure 7.9. Later, when the bracket is assembled, a pair of screws will be used to attach the braces into two accessory holes on the telescope tube.

Cut a second brace beam to extend 2.5 inches behind the telescope. In the case of Ward's Celestron 8, the second brace beam measures 12 inches altogether. A hole located as shown on the drawing lets the second brace attach to

Table 7.7 **Camcorder Bracket Parts List**

Description	Quantity
Brace beam: Aluminum angle, 0.75" wide × 0.125" thick	2
Platform plates: Aluminum flat bar, 1" wide × 0.25" thick	5
Ball-head tripod head	1
Screws, nuts, bolts	As needed

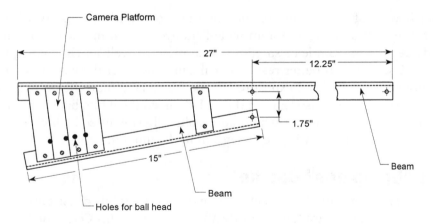

Figure 7.9 *Plans for the Ward camcorder bracket.*

another accessory hole at the back of the scope. Note that the two brace beams taper outward behind the telescope; Ward found that by using a triangular bracket, the support was sturdier than if the beams were parallel to one another.

For the camcorder's platform, cut four crosspieces from 1-inch-wide by 0.25-inch-thick flat aluminum bar stock (alternatively, cut the platform from a single piece of 4-inch by 0.25-inch-wide aluminum plate). The platform will bridge the two beams as well as support a small ball-style tripod head used to hold the camera. The exact placement of the platform's crosspieces will depend on how far back the camcorder has to be in order to focus through the telescope, but the four 1-inch-wide pieces should be close enough to one another so that the ball head can be moved if needed.

The ball-style tripod head need not be big and elaborate in order to hold the camera steadily. If the ball head is too tall, the camcorder will never properly line up with the eyepiece; too short, and it will require shims. Drill several holes in the crossbar platform area so that the position of the ball head can be changed to accommodate different eyepieces. Ward uses "eyepieces between 40 mm and 11 mm, depending upon the object in question. The holes are slightly offset so I can adjust the height of the camcorder by angling the ball. Of course, this means I am videotaping at an angle, but I have yet to have anyone notice. I did a lot of trial and error to make sure the camcorder would be centered for each particular eyepiece."

Admittedly, there are two obvious drawbacks to Ward's design. First, if you are using a fork-mounted instrument, the platform will inevitably collide with the mounting as the instrument is swung up toward the zenith. Another problem is balancing, but this is easily dealt with by adding counterweights onto the front of the telescope tube.

So, given these disadvantages, how does it work? As a novice astrovideographer myself, I can say that home camcorders can capture surprisingly good images of bright objects such as the Moon and planets. Ward writes, "With this setup, using a 19-mm wide-field eyepiece in my Celestron 8, I can easily record the Galilean satellites of Jupiter as well as the giant planet's bands. I have gotten better results when I manually set the camera's focus to infinity and adjusted the exposure to a darker image than the auto-exposure feature would give. I also learned that if I changed the white balance setting to the incandescent setting, it would add a blue filtering effect, which improves the image resolution on Jupiter." See you in the movies!

The "Egg-piece" Caddie

The saying "Don't put all your eggs in one basket" certainly doesn't apply to Bill Elison of Missoula, Montana. Shortly after purchasing his Celestron Star Hopper 8 Dobsonian, he realized that he needed a convenient place to hold his eyepieces when outside observing. Certainly a coat pocket would not do, and sometimes you want to hold a quick observing session without dragging out a table to hold accessories. To solve this problem shared by many amateurs, Elison devised the simple eyepiece caddie shown in Figure 7.10.

Simplicity is the key to the success of Elison's "egg-piece" caddie. The caddie is made from the bottom half of a cardboard egg carton held in an open-top box of 0.25-inch-thick plywood. Two metal hooks fasten the caddie onto the telescope's rocker box. Construction should take only an hour or so using the items listed in Table 7.8.

First, select the egg carton that you will be using. Because no industry-wide standard set of dimensions for egg cartons exists, cut the plywood sides and bottom based on the outside dimensions of your own egg carton. Make the box just a fraction smaller than the egg carton and then trim the cardboard carton a little with scissors so that it will fit snugly into the plywood box. Don't forget to allow 0.25 inch for the thickness of the plywood when figuring the

Figure 7.10   *Bill Elison's eyepiece caddie, attached to his Newtonian reflector. Photo courtesy of Bill Elison.*

Table 7.8   **Egg-piece Caddie Parts List**

Description	Quantity
Plywood, 0.25″ thick, dimensions to suit	2
Egg carton, bottom half only	1
Steel mending strip, 5″ long × 0.5″ wide	2

sizes of the individual pieces. Assemble the caddie using 4-penny (4d) finishing nails and wood glue. Set the box aside to dry.

Next, decide how long the hooks need to be in order to hold the caddie onto the front of your telescope's rocker box. As before, the dimensions shown in the above table reflect those used by Elison for his Celestron Star Hopper; they will need to be modified slightly for other telescopes. For instance, the hooks need to be longer to reach over the recessed front panels of Orion's Deep-Space Explorers.

Shape the hooks to fit over the top edge of the rocker box and then drill two holes in each as shown. Use two wood screws to attach the hooks to the back of the caddie box. Make sure that if the wood screws protrude into the box, they do so below the top of the egg carton to avoid scratching the eyepieces and your fingers. You should also cover the metal hooks with adhesive-backed felt to protect the rocker box from being scratched.

Elison notes that he always puts his eyepieces in the egg-carton slots closest to the rocker box and uses the adjacent slots for the eyepieces' dust caps. If the eyepieces are placed in order of focal length, it is easy to identify them in the dark without a flashlight.

The Binocular Viewing Chair

One of the most popular projects detailed in the first edition of *Star Ware* was the flexible-parallelogram binocular mount; I use mine all the time! But like most people, I usually stand when using the mounting, and this can grow tiring after a while. Sure, you can look through the binoculars while seated, but that also proves inconvenient when shifting between opposite ends of the sky. This binocular viewing chair, however, makes sweeping from horizon to horizon easy, quick, and convenient.

Randall McClelland, a retired medical doctor from Quincy, Illinois, designed his chair to be easy to make, simple to set up, and comfortable to use. The McClelland binocular chair (Figure 7.11) incorporates a canvas chair fashioned after folding lawn chairs, mounted on a base that swivels in azimuth atop a ball-bearing lazy Susan. The binoculars themselves are held on a pivoting platform whose altitude axis is coaxial with the observer's ears. By using a combination of foot and hand power, an observer can comfortably view any part of the sky from a seated position. Table 7.9 offers a parts list.

Figure 7.11 *Randall McClelland seated in his innovative binocular observing chair. Photo by Bruce Ketchum.*

The McClelland binocular viewing chair is made up of three main components: the rotating base, the chair itself, and the instrument platform (Figure 7.12). Begin construction of the base first and then work your way up. Cut two pieces of 0.75-inch plywood, one 16 inches by 16 inches, the other 24 inches by 24 inches. Drill a hole in the center of the smaller piece that is just large enough for a 2-inch-long, 0.375-inch bolt. Connect the two boards to one another with a 12-inch-diameter ball-bearing lazy Susan, available at most hardware stores and home repair centers. Make sure that the head of the bolt is held between the two boards. On the bottom of the bottom (ground) board, add three 2-inch-square blocks spaced in an equilateral triangle pattern.

The chair is constructed from plywood and several lengths of 1″ × 2″ hardwood lumber (poplar, maple, etc.) and is designed to fold for storage. (You might consider substituting 1″ × 3″ hardwood lumber if you weigh more than, say, 220 pounds; if so, some of the dimensions may have to be adjusted accordingly.) Construction is a little complicated but should go smoothly if you follow these steps in order. Begin by cutting each of the chair's pieces according to the

Table 7.9 **Binocular Chair Parts List**

Description	Quantity
1″ × 2″ stud, premium grade, 8′ long	3
Plywood, 0.75″ thick, 48″ × 48″	1
Cloth sling, canvas, 18″ wide	60″
Hardwood dowel, 0.5″ diameter, 18″ long	1
Lazy Susan (12″–16″ diameter)	1
Strap hinge, 1″ wide	2
Binocular tripod adapter	1
Miscellaneous hardware (0.25″ nuts, bolts, and washers recommended)	—

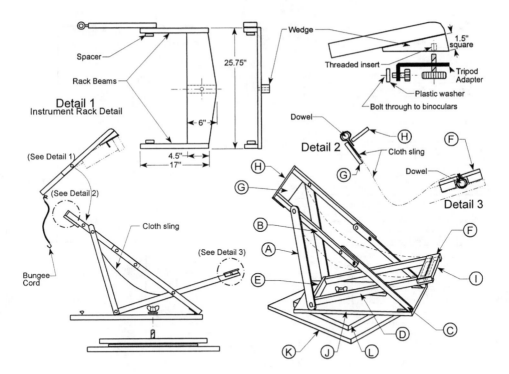

Figure 7.12 *Plans for the McClelland binocular chair.*

lengths shown in Figure 7.13, drilling clearance holes for 0.25-inch nuts, bolts, and washers where shown. Glue and screw the two lower side beams (labeled D in the drawing) 1.5 inches in from the ends of the lower crossbeam (E) and then glue and screw the foot board (F) to the opposite ends of the lower side beams (D) as shown. Drill pilot holes in the ends of the crossbeam (E) and screw in two 0.25-inch hanger bolts. Use these with two pairs of washers and wing nuts to hold the back beams (A) to the lower assembly. Finally, attach the lower front beams (C) to the lower side beams (D) with 0.25-inch T-nuts and thumb screws.

The upper half of the chair is made from a pair of 1″ × 2″ upper beams (B) attached to the head board (G and H). The back of the head board (G) is cut from 0.75-inch plywood, while the top (H) is a section of 1″ × 2″ hardwood. Leave a 0.25-inch gap between (G) and (H) for the cloth sling, which will be described shortly. Once assembled, the chair's upper half attaches to the back beam (A) and lower front beams (C) using 0.25-inch T-nuts and thumbscrews. Make sure you attach the T-nuts to the proper surfaces; at the A–B connection, the T-nut is nailed to the far side of B, while at the B–C joint, it is nailed to the far side of C. Remember, the head board is supposed to meet and fold over the foot board when the chair is collapsed for storage. To make sure that the chair will fold easily, McClelland recommends saving the marking and drilling of the holes that join parts A and B until last.

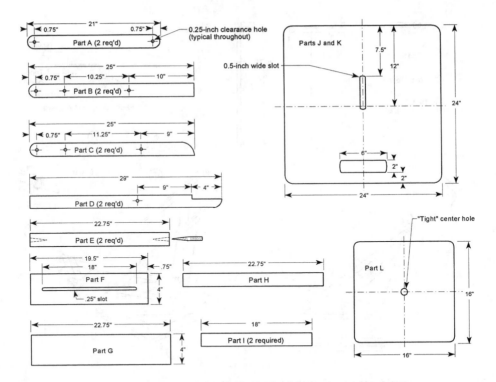

Figure 7.13 *Additional detailed drawings for the McClelland binocular chair.*

When the chair frame is completely assembled, attach it to the chair base using two strap hinges.

With the chair frame complete, it's time to make the cloth sling. Slip one end of the cloth between parts G and H, as shown in Detail 2 of Figure 7.12. Sew a small loop widthwise across the cloth that is just large enough for one of the hardwood dowels to pass through. Drape the cloth within the frame, inserting the other end through the slot in part F as shown in Detail 3. Cut the cloth to length, allowing enough droop for the observer's body, and sew another loop for the second hardwood dowel.

Finally, construct the instrument rack (Detail 1). Cut the pieces so that when assembled, the inner span of the instrument rack beams (I) matches the outer width of the chair's upper beams (B). Since the exact location of the rack will vary from one person to the next, the best solution is to sit in the chair first, positioning yourself for greatest comfort, and then mark where your ears line up with the chair's frame. That will become the pivot point for the instrument rack. To prevent binding, use a 1-inch-diameter hardwood spacer to separate the instrument rack from the chair frame. Mount everything together using recessed 0.25-inch T-nuts and thumbscrews, as shown. On one side of the chair, attach a 16-inch length of 1″ × 2″ hardwood to the instrument rack's beam. When the chair is set up, an elastic bungee cord attaches between this balance arm and the chair base to help counterbalance the binoculars.

Binoculars are mounted under the instrument rack's wedge using a standard 90° tripod adapter tilted sideways. If possible, slot the longer arm of the adapter so that the binoculars may be adjusted back and forth for different observers. Although you can use a thumbscrew for attaching the tripod adapter to the wedge, McClelland finds that using a wooden handle and a long 0.25-inch carriage bolt makes the binoculars easier to pivot in azimuth.

After setting up the chair for use, sit back, place a small pillow behind your head, and you are set to tour the universe through binoculars in style! Since his retirement, McClelland has made several versions of his binocular chair, including a two-seater that allows a pair of observers using two pairs of binoculars to view the same area of sky simultaneously. Another variation holds a pair of binoculars as well as a side-mounted 6-inch f/4 Newtonian reflector, which also serves as a counterweight!

The Denver Observer's Seat

Anyone who has ever spent more than, perhaps, half an hour at the eyepiece of a telescope knows that it can be a painful experience. Constantly bending over to look through the eyepiece can put a tremendous strain on an observer's neck, back, and legs. To help alleviate some of the discomfort, many amateur astronomers use commercially made observing chairs or stools.

The first edition of *Star Ware* detailed the construction of a folding observer's chair that was fashioned after a stepladder. Figure 7.14 shows a second,

Figure 7.14 *The Denver observing chair. Photo courtesy of Charles Carlson.*

somewhat simpler design developed by David Trott and Charles Carlson of the Denver Astronomical Society. The height of the seat is fully adjustable, from near ground level to more than three feet off the ground. When not in use, the chair will fold compactly for easy transport and storage.

Table 7.10 is an itemized list of materials for the chair.

Cut the 2″ × 4″ lumber into four pieces, two measuring 34 inches long, one measuring 24 inches long, and the remainder at a little less than 4 inches long. The two longest pieces will be used for the chair's upright beams, the shorter piece for the T-shaped cross beam, and the short leftover cut will be used in the seat assembly. Bevel one end of each of the upright beams at a 22.5° angle. If you do not have a table saw that permits such a precise cut, then judge the angle as best you can so that the beam ends will sit flat on the ground when they are tilted at a 45° angle.

Cut the two seat supports from the two pieces of 7″ × 14″ plywood and drill two 0.44-inch holes as shown at left in Figure 7.15. Carlson notes that these are the only pieces that must be cut and drilled *exactly* for proper fit. Now cut the seat itself from the 10″ × 13″ piece of 0.75-inch plywood. Bevel one of the long edges to match the 30° angle of the seat supports. Round and smooth the corners for a more refined look as well as to prevent splintering.

Center and glue the short piece of 2″ × 4″ onto the bottom of the seat, approximately 1.5 inches from the beveled (back) edge as shown. Use four equally spaced #10 flathead wood screws to fasten the block to the seat. Align the seat supports with the 2″ × 4″ block and attach them with 2-inch-long, #10 flathead wood screws, four on each side.

Before continuing assembly, seal all of the wood parts with several coats of a good-quality polyurethane varnish or similar waterproof finish. Allow all parts to dry fully before proceeding.

With the varnish dry, lay out the two upright beams so that their square edges touch each other and the beveled edges face up. Install the door hinge

Table 7.10 **Denver Observer's Seat Parts List**

Description	Quantity
2″ × 4″ stud, premium grade, 8′ long	1
Plywood, 0.75″ thick, 10″ × 13″	1
Plywood, 0.75″ thick, 7″ × 14″	2
Aluminum bar, 1″ × 0.125″, 28″ long	1
Door hinge, 3″ wide	1
Chest handle, 2.75″ wide	1
Safety stair tread, Scotch Tred (or equivalent) 4″ wide × 26″ long	1
Rubber automotive fuel-line tubing, 0.75″ diameter, 4″ long	1
Velcro strip, 12″ long	1
Thumbtacks	4
Miscellaneous screws, nuts, washers, nails	

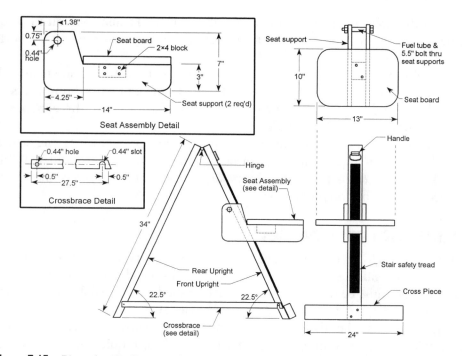

Figure 7.15 *Plans for the Denver observing chair.*

connecting the two square ends of the beams. Fold the beams over on each other and center the 24-inch-long cross beam flush with the beveled edge of one of the uprights. Secure the cross beam to the front upright with carpenter's glue and a pair of 2.5-inch-long, 0.375-inch-diameter lag bolts and 0.375-inch washers. Install the chest handle at the top of the front upright beam.

Fashion the crossbrace from a 28-inch-long piece of 1″ × 0.125″ aluminum bar. Bevel one end of the brace at a 22.5° angle as shown and drill two 0.44-inch holes. Use a saw to transform the hole nearest the bevel into a slot. Attach the brace to the seat by first opening the chair's beams to a 45° angle. Use a 0.375-inch-diameter, 2-inch-long lag bolt to attach the brace to one side of the rear upright, about two inches up from the bottom. Leave enough slack in the lag bolt so that the brace can move freely. Install another 0.375-inch-diameter, 2-inch-long lag bolt in the same location on the front upright, again allowing enough of the bolt to protrude so that the brace can snap down over it for a snug fit.

Trim the adhesive-backed stair safety tread to a width of 3 inches and attach it to the front upright, starting about three inches from the top. Use small finishing nails to hold the tread corners in place, being careful to hammer them flush to the wood.

Hold the seat assembly against the front upright and insert the fuel-line tubing in line with the two holes in the seat supports. Slide a 0.375″ × 5.5″ bolt through the holes in the seat supports and the tube. Fasten the bolt in place using a locknut, being careful not to tighten it so much that the seat supports warp and cause binding.

Finally, hold the seat closed when not in use with a strap made from Velcro. Attach the material onto the back of the rear upright with thumbtacks so that it loops and mates across the front beam. Allow for about three inches of material overlap for a secure hold.

Whether you choose to buy one or make your own, an observing seat is one of the most important accessories for the amateur astronomer. Being seated in a relaxed position is sure to make your time at the eyepiece more productive and much more enjoyable.

Hands-Off Focusing Device

Getting a sharp focus is often taken for granted by observers, but it is crucial for both high-magnification observations and through-the-telescope astrophotography. But you must be careful, because turning the focusing knob by hand may actually shift the target right out of the field of view. One of the latest innovations to help make this exacting task easier is the electric focusing device, outlined in the previous chapter. Although beneficial, these gadgets require power to run the device's motor and command prices that many amateurs find prohibitive.

Facing these same problems, Ken Florentino of Colorado Springs, Colorado, has come up with an inexpensive and easily made attachment that can be adapted to just about any rack-and-pinion or Crayford focusing mount for hands-off operation (Figure 7.16). His design is based around a pair of standard model airplane/car servos that are wired together. Although new servos can be expensive, scrap servos are frequently available from large hobby shops for very little money.

One servo, the "director," is held in the observer's hand, while the second, the "slave," is mounted on the telescope adjacent to the focuser. When the con-

Figure 7.16 *Steve Bygren's servo focusing device. Left: The slave servo attached to the telescope's rack-and-pinion focusing mount. Right: The handheld master servo with wooden handle attached to its spindle. Photos courtesy of Steve Bygren.*

trol arm of the director servo is turned by hand, the nylon gear train inside the servo spins a small DC motor rapidly, so rapidly that it acts as a tiny electric generator. Enough voltage and current are produced to travel down the connecting wires to the slave servo to turn its DC motor. The slave servo's output arm turns with enough *oomph* (a technical term) to turn the focusing knob.

The unit shown here was created and submitted by Steve Bygren, another member of the Colorado Springs Astronomical Society. Table 7.11 lists the basic parts that Bygren needed to adapt Florentino's concept to his Cassegrain reflector. Your requirements may differ slightly.

Begin the operation by performing some surgery on the servos. Since both need to be reworked in the same manner, it makes sense to do the same operation to each servo before moving on to the next step. First, remove each servo's cover that surrounds the internal motor, gear train, and electronics. Pay close attention to the configuration of the nylon gears as you remove them, because they must be reinstalled afterwards. Disconnect the small printed-circuit board from each and discard it. A small potentiometer may also have to be removed from the servo.

With the PC board removed, solder two wires between the pair of electrical connections on one servo motor directly to the two connections on the other servo motor. Keep in mind that the wire needs to be long enough to stretch between the focuser and the observer's hand when the assembly is completed. Bygren suggests that two feet should be enough for most applications, but figure out just how much you will need beforehand.

The next step can be a little tricky. Most servos have small tabs that limit the amount of travel of their output arms. These are intended to prevent the servo from spinning continuously in circles. In most cases, they are located on one of the nylon gears in the gear trains. Look for a small protrusion on one of the gears that runs into a stop after about $\pm 30°$ of rotation. Although they are necessary when used on radio-controlled planes and cars, they restrict motion when teamed with a focusing mount and so must be removed. Although a razor-style modeling knife can be used to trim the tab, Bygren quickly cautions that because the nylon gear can easily slip out of your fingers just as the knife is slicing through, a file is a safer alternative. Grind down the tab until it is flush with the gear.

Table 7.11 *Servo Focuser Parts List*

Description	Quantity
Servo, model airplane	2
Wire, same gauge that is used originally with servo	As needed to bridge the gap between the servos
O-ring (size depends on your application)	1
Plastic knob	1
L-bracket for holding servo to telescope	1

Finally, put the gear trains back into the servos just as they came out originally and close the plastic housings. To test the system out, turn one of the servo's spindle. If everything was done correctly, the other servo's shaft should also spin. Fashion and attach a wooden handle to the spindle of whichever servo you will be turning when focusing (that is, the director).

The challenge now is to connect the other servo (the slave) to your telescope's focusing mount. Different telescopes require different solutions, but typically a wheel is attached to the servo's spindle; then a drive belt, fashioned from a rubber O-ring, runs from the wheel to the focuser's shaft. Bygren notes, "The most important thing to keep in mind is that the output shaft of the slave servo must be parallel to the focuser's shaft or knob. I attached the slave servo to an L-shaped aluminum plate, and the plate to the back of the telescope below the focuser, using double-sided foam adhesive tape." Adjust the location of the servo so that the O-ring belt is just taut enough to turn the focuser without slipping.

An All-Sky Reflector for Astrophotographers

One of the most interesting types of astrophotographs is a horizon-to-horizon shot of the entire sky (Figure 7.17 shows a daytime shot). You've probably seen them, taken either at night or possibly during a total solar eclipse. So-called fish-eye photographs offer a perspective that is unattainable by the eye alone.

The most common way to take an all-sky photograph is with an ultra-wide-angle lens on your camera—in the case of a 35-mm camera, lenses with focal lengths of 16 mm or less. These fish-eye lenses produce wonderful results but may break the bank before you ever get a chance to use one. A recent advertisement in a photography magazine priced these lenses between $450 and $2,000. Happily, there are alternatives.

One of the alternate possibilities is shown in Figure 7.18. Built by Steve Edberg of the Jet Propulsion Laboratory in Pasadena, California, this all-sky accessory uses a normal 50-mm lens and a 35-mm single-lens reflex camera. In this case, a hemispherical mirror, convex toward the sky, reflects the sky into

Figure 7.17 *An all-sky shot of the daytime sky taken with the Edberg all-sky reflector. Photo courtesy of Steve Edberg.*

Figure 7.18 *Steve Edberg and his all-sky reflector, each ready to enjoy a clear sky at the Riverside Telescope Maker's Conference.*

the camera lens, which is aimed down toward the mirror and ground. A flat base holds the mirror in place, while a small, raised platform suspends the camera directly over the mirror. Three rods support the camera platform above the mirror, and a bracket holds the camera in place. Table 7.12 lists the major parts of the Edberg all-sky reflector.

The first and perhaps most difficult step is finding a suitable reflector. These might include silver Christmas ornaments, baby moon hubcaps, and even plastic egg-shaped packages from women's pantyhose. Only perfectly smooth surfaces yield good results. Edberg suggests looking in surplus optics houses, such as C&H Sales in Pasadena, California (2176 East Colorado Boulevard; Pasadena, California 91107; telephone [562] 681-4925). If a convex mirror is not available, you might also use a glass hemispherical dome, an optical dome, or a large condenser lens, although these will have to be sent to a coating

Table 7.12 **All-Sky Reflector Parts List**

Description	Quantity
Reflector (at least 4″ in diameter)	1
Plywood, minimum 0.38″ thick	Cut to suit
Threaded rods, 0.25-20	Length to suit
L-bracket	1
Hardwood wedge, if desired	1, size to suit
Thumbscrew, 0.25-20, 1.5″ long	1
Wing nut, 0.25-20	6
Fender washer, 0.25″ inside diameter	12
Stand-off, 0.25-20 (approximately 1″ long)	3

company to be aluminized. Keep in mind that you do not need a hemisphere to cover the entire sky. The reflector needs to span only 90°, because the reflection doubles the angle covered to the full 180° desired.

To determine if your reflector will cover the full sky, Edberg recommends laying a piece of soft string across the apex of the reflector to measure its "convex diameter" (don't use twine, or it might scratch the reflector's surface) and then measuring the reflector's "flat" diameter across its back surface. Divide the convex diameter by the flat diameter. If the answer is 1.11 or greater, the reflector will span 180° or more as seen by the camera.

The stand itself is made from commonly available hardware, pieces of 0.38-inch or thicker plywood, and a hardwood block. General dimensions are shown in Figure 7.19, but keep in mind that all of these can and will vary depending on the diameter and curvature of the reflector. Determine the proper dimensions for your system by first holding the camera over the reflector. Move the camera back and forth until the edge of the reflector just makes

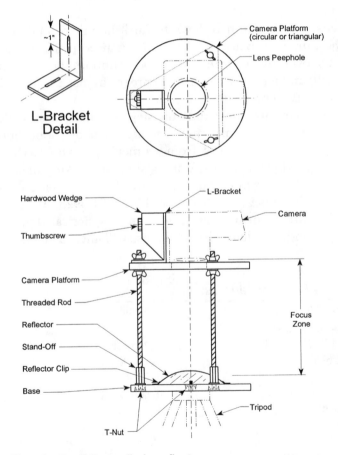

Figure 7.19 *Plans for the Edberg all-sky reflector.*

it into the camera's field of view and then focus the image. This can be tricky because, unlike normal photographs of the night sky, the reflected star images are *not* at infinity; in fact, they may be closer than a standard 50-mm camera lens can focus. To bring the sky into focus, an auxiliary close-up lens may have to be attached to the front of the camera lens, or a short extension tube might need to be placed between the camera lens and camera body. Both of these can be purchased at a camera store, but they may also add to the required height of the camera platform. That's why testing the distance between reflector and camera is so critical.

You will also find that because of the reflector's steep curvature, no one focus point makes all of the field sharp. Edberg finds that the best focus is about midway in the field of view (i.e., midway between horizon and zenith). Test the focus during the day by aiming toward distant mountains or clouds or at night by aiming toward the Moon. Once set, note the distance setting on the camera lens for future use. (Stopping down the camera lens to f/2.8 or more will mitigate the effect of the curved focal surface but will also necessitate using fast film for all-sky photography at night.)

Once the focus point is set, have an assistant measure the distance from the front of the camera lens to the reflector. That's the "focus zone" in the figure, which must be used to determine the stand's height.

Wait to cut the plywood base and camera platform until after you buy the reflector. Both pieces must be large enough to hold the reflector or camera as well as the three vertical rods used to support the camera platform. Typically, they should be the same diameter, although the camera platform may actually be trimmed to a truncated triangle to minimize its cross-section (as suggested by the dotted line in the figure). Edberg recommends making the base large enough to fill the camera's field of view completely. This way, photographs will show the reflector in the center of the frame surrounded only by the base rather than the ground.

Drill an oversized hole in the center of the base plate for a 0.25-20 T-nut, which will be used to attach the finished unit to a camera tripod. Hammer the T-nut into the hole so that its flange will be beneath the reflector when everything is assembled. Next, drill three equally spaced 0.25-inch holes around the edge of the base to hold the three support rods. Be sure not to locate them so close to the edge of the base that the wood fractures; allow at least 0.75 inch.

Cut the threaded rods to length, making certain to add an extra 1.5 inches to each for adjustment after assembly; you can always cut them shorter afterwards, if desired. Attach the rods to the base using threaded stand-offs and place fender washers between each stand-off and the base.

Finish the base by making three clips to hold the reflector in place. Edberg recommends using wood screws with washers or spring steel cable clamps, suitably bent. Place small pieces of cork or adhesive-backed felt between each clip and the reflector surface to prevent scratches. Paint the base and rods flat black.

With the base complete, transfer-drill the three holes used for the vertical support rods into the camera platform. Locate and drill a large hole with a hole saw, just a little smaller than the front housing of the camera lens, at the exact center of the platform. When set up, the camera's lens will peer through this hole, with the edge of the lens housing resting on the platform. When complete, paint the camera platform and base flat black.

An L-shaped bracket is used to hold the camera to the platform. You can use a binocular tripod adapter, but Edberg recommends making one from a piece of metal. Just make sure that the bracket you buy or make is tall enough to match the location of the tripod socket on the bottom of your camera.

Edberg concludes, "All-sky photography offers a new perspective on the sky. Large objects like the Milky Way and manifestations of the zodiacal light can be recorded in very illustrative ways."

8

Till Death Do You Part

If this were a book about how to run a business, then this chapter might be entitled "Standard Operating Procedures" because it details methods and strategies that although not part of the business's primary product or service, help to enhance the company's operational efficiency. This chapter contains lots of little tidbits to help you get the most out of your telescope. It addresses a wide variety of topics, ranging from care and maintenance to traveling with a telescope.

Love Thy Telescope As Thyself

Unlike so many other products in our throw-away society, in which planned obsolescence seems the rule, telescopes are designed to outlast their owners. They require very little care and attention, cost nothing to keep, and eat very little. With a little common sense on the part of its master, a telescope will return a lifetime of fascination and adventure. But if neglected or abused, a telescope may not make it to the next New Moon. Are you a telescope abuser?

Storing Your Telescope

Nothing affects a telescope's life span more than how and where it is stored when *not* in use. *How* to store a telescope will be addressed farther along in this chapter, but first let's consider *where* the best places are to keep an idle instrument. The choice should be based on a number of different factors that, at first glance, might appear unrelated. A good storage place should be dry, dust-free, secure, and large enough to get the telescope in and out easily. Ideally, a telescope should always be kept at or near the outside ambient air temperature; doing so reduces the cool-down time required whenever the telescope is first set up at night. The quicker the cool-down time, the sooner the telescope will be ready to use.

Without a doubt, the best place to keep a telescope is in an observatory. It offers a controlled environment and easy access to the night sky. Of course, not everyone can afford to build a dedicated observatory, and neither is an observatory always warranted. Clearly, an observatory is pointless if the nearest good observing site is an hour's drive away. In cases such as these, a few compromises must be struck.

If an observatory is not in the stars (so to speak), other good places to store telescopes include a vented, walled-off corner of an unheated garage or a wooden tool shed. I keep my telescopes in a corner of my garage that is completely walled off to protect the optics from the inevitable dust and dirt that accumulates in garages as well as from any automotive exhaust that could damage delicate optical surfaces. A pair of louvered vents were installed in the outside wall of the garage to let air move freely in and out of the telescope room, reducing the risk of mildew. Wooden tool or garden sheds share many of the advantages of observatories and garages for telescope storage, but again I recommend installing one or two louvered vents in the shed's walls for air circulation. Metal sheds are not as good, as they can build up a lot of interior heat on sunny days.

Many amateurs choose to store their equipment in basements, which are certainly secure enough and large enough to qualify. Furthermore, they offer easy access provided there is a door leading directly to the outside. Their cool temperatures also keep the optics closer to that of the outside air. While all of these considerations weigh in their favor, most basements fail when it comes to being dry and free of dust. If a basement is your only alternative, invest in a dehumidifier. Clothes closets, another favorite place to hide smaller telescopes, also fall short because clothes act as dust magnets. Remember, unless a spot meets all of the criteria, continue the search.

Regardless of where a telescope is kept, seal the optics from dust and other pollutants when it is not in use. Usually this is simply a matter of putting a dust cap over the front of the tube. Most manufacturers supply their telescopes with a custom-fit dust cap just for this purpose. Use it diligently. If the telescope did not come with a dust cap or if it has been lost over time, then a plastic shower cap makes a great substitute. If you are into a more sophisticated look, some of the companies listed in appendix C sell dust caps made of rubber in a wide variety of sizes. Further, if the telescope or binoculars came with a case, use it. Not only will a case add a second seal against dust, but it also will protect the instrument against any accidental knocks or bumps.

A dark, damp telescope tube is the perfect breeding ground for mold and mildew. To avoid the risk of turning your telescope into an expensive petri dish, be sure that all of its parts are dry before sealing it up for the night. Tilt the tube horizontally to prevent water from puddling on the objective lens, primary mirror, or corrector plate. No matter how careful you are, optics are bound to become contaminated with dust eventually. A moderate amount of dust has, surprisingly, little effect on a telescope's performance. But if there's a great deal of it, or if the optics have become coated with a film or mildew, the observer sees dimmer, hazier views that lack clarity.

Having a telescope that is a little dusty is cause not for panic but for cautious action. "Clean a telescope?!" you ask. "Isn't that a job that should be left to professionals?" Not at all. While to the uninitiated it might seem a formidable task, it's actually not, as you are about to discover.

An optic should be cleaned only when dust or stains are apparent to the eye; otherwise, leave well enough alone. *Never* clean a telescope lens or mirror just for the sake of cleaning it, because every time an optic is touched, there is always the risk of damaging it. Remember this rule: If it ain't broke, don't fix it.

The methods described here are for cleaning *outer* optical surfaces only. Unless you really know what you are doing, I strongly urge against dismantling sealed telescopes (such as refractors and catadioptrics), binoculars, and eyepieces. Dirt and dust will never enter a sealed tube if it is properly stored and protected. Nevertheless, if an interior lens or mirror surface in a sealed telescope becomes tainted by film or mildew, it should be disassembled and cleaned only by a qualified professional. Contact the instrument's manufacturer for recommendations on how to do so. If you don't, you may discover that the telescope is much easier to take apart than it is to put back together! I know someone who once decided to take apart his 4-inch achromatic objective lens to give it a thorough cleaning. All was going well until it came time to put the whole thing back together. Seems he tightened a retaining ring in the lens holder just a little too much and ... CRACK! The edges of the crown element fractured. Don't make the same mistake, y'hear?

Never start to clean a telescope if time is short. For instance, it is not time to decide that your telescope is absolutely filthy as the Sun is setting in a crystal-clear sky. Instead, check the optics well beforehand so there are no surprises. To help you along the way, I have divided the cleaning process into two parts, one for lenses and corrector plates and another for mirrors.

Cleaning Lenses and Corrector Plates

Begin the cleaning process by removing all abrasive particles that have found their way onto the lens or corrector plate. This does NOT mean blowing across the lens with your mouth; you'll only spit all over it. Instead, use either a soft camel's-hair brush (Figure 8.1a) or a can of compressed air, both available from photographic supply stores. Some brushes come with air bulbs to allow blowing and sweeping at the same time. If the brush is your choice, lightly whisk the surface of the lens in one direction only, flicking the brush free of any accumulated dust particles at the end of each stroke.

Many amateurs prefer to use a can of compressed air instead of a brush for dusting optical surfaces, as this way the lens is never physically touched. Hold the can perfectly upright, with the nozzle away from the lens *at least* as far as recommended by the manufacturer. If the can is too close or tilted, some of the spray propellant may strike the glass surface and stain it. Also, it is best to use several short spurts of air instead of one long gust.

With the dust removed, use a gentle cleaning solution for fingerprints, skin oils, stains, and other residue. Don't use a window glass–cleaning spray or other

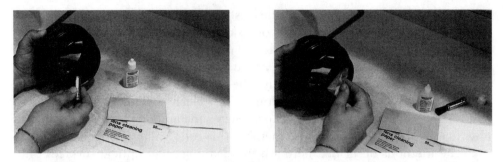

Figure 8.1 *The proper way to clean a lens or corrector plate: (a) Lightly brush the surface with a soft camel's-hair brush and then (b) gently wipe the lens with lens-cleaning solution.*

household cleaners. They could damage the lens' delicate coatings beyond repair. Photographic lens-cleaning fluid can be used but can occasionally leave a filmy residue behind. One of the best lens-cleaning solutions can be brewed right at home. In a clean container, mix three cups of distilled water with a half cup of pure-grade isopropyl alcohol. Add two or three drops of a mild liquid dishwashing soap (you know, the kind that claims it will not chap hands after repeated use).

Dampen a piece of sterile surgical cotton (not artificial so-called cotton balls) or lens tissue with the solution. Do not use bathroom tissue or facial tissue; they are impregnated with perfumes and dyes. Squeeze the cotton or lens tissue until it is only damp, not dripping, and gently blot the lens (Figure 8.1b). Avoid the urge to use a little elbow grease to get out a stubborn stain. The only pressure should be from the weight of the cotton ball or lens tissue. Once done, use a dry piece of lens tissue or cotton to blot up any moisture.

The operation for cleaning the corrector plate of a catadioptric telescope is pretty much the same as just detailed. The only difference is in the blotting direction; in this case, begin with the damp cotton or tissue at the secondary mirror holder in the center of the corrector and move out toward the edge. Follow a spokelike pattern around the plate, using a new piece of cotton or tissue with each pass. As you stroke the glass, turn the cotton or tissue in a backward-rolling motion to carry any grit up and away from the surface before it has a chance to be rubbed against the optical surface. Overlap the strokes until the entire surface is clean. Again, gently blot dry.

Mirror Cleaning

Cleaning a mirror is much like cleaning a lens in that special care must be exercised to ensure against damaging the fine optical surface. In fact, a mirror is even more susceptible to scratches than a lens. The mirror's thin aluminized coating is extremely soft, especially when compared to abrasive dirt, and so is easily gouged. This is not meant to scare you out of cleaning your mirror if it really needs it but only to heighten your awareness.

The operation for cleaning a telescope's primary or secondary mirror requires it to be removed from the telescope and the cell that holds it in place.

Consult your owner's manual for more details on mirror removal. With the naked mirror lying on a table, blow compressed air across its surface to rid it of any large dust and dirt particles. (Remember, keep the can of compressed air vertical.) Do not use a brush for this step, as even the softest bristles can damage a mirror's coating.

Next, inspect the mirror's coating for pinholes and scratches. A good aluminum coating should last at least ten years, even longer if the mirror has been well cared for. To check its condition, hold the mirror, its reflective side toward you, in front of a bright light. It is not unusual to see a faint bluish image of the light source through the mirror if the source is especially bright, but its image should appear the same across the entire mirror. If not, there may be thin, uneven spots in the coating. Any scratches or pinholes in the coating also will become immediately obvious as well. A few small scratches or pinholes, while not desirable, can be lived with. But if scratches or pinholes abound or if an uneven coating is detected, then the mirror should be sent out for re-aluminizing. Appendix C lists several companies that re-aluminize mirrors; consult any or all of them for prices and shipping details. It is also a good idea to have the secondary mirror re-aluminized at the same time, because it probably suffers from the same problems as the primary. In fact, most re-aluminizing companies will work on the secondary at no additional cost.

If the coating is acceptable, bring the mirror to a sink. Be sure to clean the sink first and lay a folded towel in the sink as a cushion, just in case—OOPS—the mirror slips. Run lukewarm tap water across the mirror's reflective surface. This should lift off any stubborn dirt particles that refused to dislodge themselves under the compressed air. End with a rinse of distilled water. Tilt the mirror on its side next to the sink on a soft, dry towel and let the water drain off the surface. Examine the mirror carefully. Is it clean? If so, quit.

If you want to go further, thoroughly clean the sink to remove any gritty particles that may not have washed down the drain. Next, fill the sink with enough tepid tap water to immerse the mirror fully and add to it a few drops of gentle liquid dish soap (the same as used in the lens-cleaning solution). As shown in Figure 8.2a, carefully lower the mirror into the soapy water and let it sit for a minute or two. With a big, clean wad of surgical cotton, sweep across the mirror's surface ever so gently with the same backward-rolling motion, being careful not to bear down. Now is not the time to act macho. After you've rolled the cotton a half-turn backward, discard it and use a new piece. If stains still exist after this step is completed, let the mirror soak in the water for five to ten minutes and repeat the sweeping with more new cotton.

With the surface cleaned to your satisfaction, drain the sink once again. Run tepid tap water on the mirror for a while to rinse away all soap. Then turn off the tap and pour room-temperature distilled water across the surface for a final rinse.

Finally, rest the mirror on edge on a towel where it may be left to air dry in safety. I usually rest it against a pillow on my bed (Figure 8.2b). Tilt the mirror at a fairly steep angle (greater than 45°), its edge resting on the soft towel, to let any remaining water droplets roll off without leaving spots. Close the

Figure 8.2 *The proper way to clean a mirror: (a) After using compressed air or a camel's-hair brush to remove any loose contaminants, wash the mirror in the sink using a gentle liquid dishwashing detergent. (b) After rinsing the mirror, tilt the mirror nearly vertically to prevent water from puddling.*

door behind you to prevent any nonastronomers from touching the mirror. When the mirror is completely dry, reassemble the telescope. Recollimate the optics using the procedure described later in this chapter, and you're done!

Other Tips

Other telescope parts require occasional attention as well. For instance, rack-and-pinion focusing mounts will sometimes bind if not lubricated occasionally. To prevent this from happening, spray a little silicone-based lubricant (such as WD-40) on the mount's small driving gear. To do this, remove the screws (typically two) that hold the small plate onto the side of the focuser housing, taking care not to lose anything along the way. With the plate out of the way, take a look inside at the small pinion gear that meshes with and drives the focusing mount's tube up and down. Squirt a little (*very* little) lubricant on the pinion teeth, reaffix the cover plate, and wipe off any drips as required.

If your telescope has a cardboard tube, make sure its ends do not fray and begin to unravel. Tilt the tube horizontally, remove or otherwise protect the optics, and brush on a thin layer of varnish to seal the ends. Checking the tube's condition once every now and then can add years to its life.

If a metal telescope mounting begins to move roughly or starts to bind, put a drop or two of lubricating oil on the axes' bearing points. This will keep the telescope moving freely and evenly, rather than binding and grabbing. Some manufacturers recommend this be done at specific intervals, while others make no mention of it at all. If nothing is said in the owner's manual, then do it once a year or so.

The typical wood-Formica-Teflon construction of Dobsonian mounts requires little in the way of maintenance. However, if your Dobsonian does not move freely in altitude or azimuth, take the mount apart and spray a little furniture polish on the contact surfaces. Buff the polish as you would a coffee table until it shines with a luster. Put the mounting back together and take it for a test spin. The difference should be immediately noticeable.

Some clock drives also need an occasional check to keep them happy. Carefully remove the drive's protective cover plate or housing and put a little thin grease between the two meshing gears. While the drive is open, put a drop or two of thin oil on the motor's shaft as well. Finally, reassemble the drive and turn it on. Listen for any noises. Most clock drives hum as they slowly turn. If unusually loud, angry, grinding noises are coming from it, turn the drive off immediately and contact the manufacturer for recommendations.

Get It Straight!

Have you ever gotten as frustrated with your telescope as the character depicted in Figure 8.3? (Hopefully you stopped short of whacking it with an axe.) There is nothing more vexing or disappointing to an amateur astronomer than to own a telescope that doesn't work as expected. "That lousy company," you think to yourself. "I spend over a thousand dollars on a telescope, and what do I get? A big, expensive, piece of junk!" All too often we are quick to blame the poor images produced by our telescopes on faulty optics and poor workmanship. Yet this is not necessarily the case. Although some telescopes are truly lacking in quality, most work acceptably *if they are in proper tune.* Even with the finest optics, a telescope will show nothing but blurry, ill-formed images if those optics are not in proper alignment. Correct optical alignment, or collimation, is a must if we expect our telescopes to work to perfection.

What exactly is meant by *collimation?* A telescope is said to be collimated if all of its optics are properly aligned to one another. Refractors, for instance, are collimated when the objective lens and eyepiece are perpendicular to a line connecting their centers (of course, this is assuming the lack of a star diagonal).

A Newtonian reflector is collimated when the optical axis of the primary mirror passes through the centers of the diagonal mirror and the eyepiece.

Figure 8.3 *From* Russell W. Porter *by Berton Willard (Bond Wheelwright Company, 1976). Reprinted with kind permission of the author.*

Furthermore, the eyepiece focusing mount must be at a right angle to the optical axis. For our purposes here, all Cassegrain reflectors and all Cassegrain-based catadioptric telescopes will be lumped together. To be precisely collimated, their primary and secondary mirrors must be parallel to one another, with the center of the secondary lying on the optical axis of the primary. The eyepiece holder must be perpendicular to the mirrors, with the telescope's optical axis passing through its center. Refer to Figure 2.4 for clarification if these descriptions are unfamiliar.

The optics in some types of telescopes are apt to go out of alignment more easily than others. For example, it is rare to find a refractor that is not properly collimated unless its tube is bent or warped, usually as the result of abuse and mishandling. On the other hand, most Schmidt-Cassegrain telescopes will experience some collimation difficulties in their lives, whereas Newtonian and Cassegrain reflectors are notoriously easy to knock out of alignment.

It's easy to tell if a telescope is collimated properly. On the next clear night, take your telescope outside and center it on a bright star using high magnification. Place the star slightly out of focus and take note of the diffraction rings that result. If the telescope is properly collimated, then the rings should appear concentric around the star's center. If, however, the rings appear oval or lopsided—and remain oriented in the same direction when you turn the focuser both inside and outside of best focus—then the instrument is in need of adjustment.

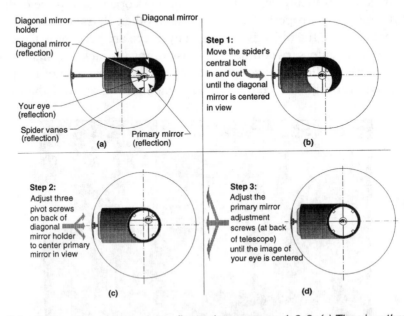

Figure 8.4 *Collimating a Newtonian reflector is as easy as 1-2-3. (a) The view through an uncollimated telescope. (b) Adjust the diagonal mirror's central post until the diagonal is centered under the eyepiece tube. (c) Turn the diagonal mirror's three adjustment screws behind the mirror until the reflection of the primary mirror is centered. (d) Finally, adjust the primary mirror's cell until the mirror image of your eye is centered in view.*

Collimating a Newtonian

In general, the faster the telescope (that is, the lower the f-number), the more critical its collimation. Though slower Newtonians may be collimated adequately by eye, I have chosen to lump them together with RFT scopes in the interest of brevity. Throughout the discussion, I will make it clear when a step is required for both and when it is for RFTs only.

Begin by purchasing or making a *sight tube*, which is essentially a long, empty tube with a pair of crosshairs at one end and a precisely centered pinhole at the other. The outer diameter of the sight tube must match the inside diameter of the telescope's eyepiece holder (typically, 1.25 inches). Tectron, the Telescope and Binocular Center, and AstroSystems make excellent sight tubes. Or you can make one yourself (see chapter 7).

As an aid for collimation, many Newtonians (especially those that are f/6 or less) come with small, black dots right in the middle of their primary mirrors. (Although at first it might appear that these dots will impede mirror performance, in reality they have no effect because they lie in the shadow of the diagonal mirror.) If your telescope mirror does not have a centered dot, now is the time to put one there. Remove the mirror from the telescope and place it on a large piece of paper. Trace its diameter with a pencil and then move the mirror to a safe place.

Now to find the *exact* center of the mirror tracing: Cut out the tracing precisely along its edge. Once you've done that, fold the paper circle in half and then in half again so that you end up with a quarter-circle, 90° wedge. Snip off the tip of the wedge with a pair of scissors and unfold the paper. The resulting hole marks the center of the mirror tracing.

Return with the mirror and align (gently, please) the paper so that the tracing matches the mirror's edge. Using the tracing as a template, make a tiny mark at the center of the mirror with a permanent felt-tip marker. Carefully check the spot's accuracy with the ruler; if acceptable, then enlarge the spot to about one-eighth of an inch (0.25 inch for mirrors of 12 inches and larger). Finally, stick an adhesive-backed hole reinforcer (the kind used to strengthen loose-leaf paper) on the mirror around the black dot.

Aim the telescope toward a bright scene (the daytime sky works well). Rack the focusing mount in as far as it will go, insert the sight tube, and take a look. The scene will probably look like one of those shown in Figure 8.4. Ideally, you should see the diagonal mirror centered in the sight tube. If it is, skip to the next step; if not, then the diagonal must be adjusted. Move to the front of the telescope tube and look at the back of the diagonal mirror holder. Most Newtonians use a four-vane spider mount to hold and position the diagonal in place. Spider mounts typically grasp the diagonal in a holder supported on a central bolt. Loosen the nut(s) holding the bolt in place and move the diagonal in and out along the optical axis until its outer diameter is centered in the sight tube. Before tightening the nut, check to make sure that the diagonal is not turned away from the eyepiece, because it can also rotate as it is moved. When done with this step, the view through the sight tube should look like Figure 8.4b.

Some less expensive telescopes use single rods attached to the eyepiece mounts to hold their diagonals in place. These are much more prone to being knocked out of adjustment and are unfortunately much harder to aim accurately. Loosen the setscrew that holds the rod in place, being certain to hold on to the diagonal or it will drop into the tube. By rotating the rod or moving it up and down, recenter the diagonal in the sight tube. Be careful not to bend the rod in the process. (Unless the rod is bent, the primary should appear centered in the diagonal at the end of this step. If so, skip the next step; if not, then repeat this process.)

With the diagonal centered in the sight tube, look through the tube at the reflection of the primary mirror. You ought to see at least part of the primary and the far end of the telescope tube. For the diagonal to be adjusted correctly, the primary must appear centered. To do this, most spider mounts have three equally spaced screws that, when turned, pivot the diagonal's angle. By alternately loosening and tightening the screws, move the reflection of the primary until it is centered in the sight tube. When properly aligned, the view after this step should look like Figure 8.4c.

Aiming the primary mirror will go much faster if you have an assistant. First, look at the back of the primary mirror's cell. You should see three (sometimes six) screws facing out. These adjust the tilt of the primary. Turn one or more of the adjustment screws until the diagonal's silhouette is centered in the sight tube.

With your helper at the adjustment screws, look through the sight tube. You should see the reflection of the primary centered in the crosshairs. In bright light, the primary's central spot should also be seen. Following your instructions, have your assistant turn one or more of the screws until the reflected image of your eye in the diagonal mirror appears centered as shown in Figure 8.4d. (Primary mirror mounts with six screws follow pretty much the same procedure. Three of the screws [usually the inner three] adjust the mirror, while the others prevent the mirror from rocking. If your telescope uses this type of mirror mount, the three outer screws must be loosened slightly before an adjustment can be made).

If you own an NFT Newtonian, the mirrors should now be collimated adequately. If, however, you own an RFT, then some fine-tuning will be needed. For this, it is strongly recommended that you use a Cheshire eyepiece. Here's how it works. Shine a flashlight beam into the eyepiece's side opening and look through the peephole. Centered in the dark silhouette of the diagonal will be a bright donut of light—the reflection of the eyepiece's mirrored surface. The dark center is actually the hole in that surface. Adjust the primary mirror until its black reference spot is centered in the Cheshire eyepiece's donut. That's it.

The tell-all test is the *star test*. Take the telescope outside and let the optics fully adjust to the outside air temperature. Aim toward a moderately bright star. Unless your telescope is polar-aligned and the clock drive is turned on, use Polaris for the star test. Unlike all other stars, Polaris has the distinct advantage of not moving (at least, not much) in the sky, which makes the test a little easier to perform. Using a medium-power eyepiece, place the star in the center of the field. Move it slightly out of focus, transforming it from a point into a tiny disk surrounded by bright and dark rings. If the mirrors are properly collimated, then the rings should be concentric, like a bull's-eye. If not, one or both of the mirrors may need a little fine-tweaking. Be sure to recenter the star in the field every time an adjustment is made. (Defects in the mirror's curve can also cause the rings to appear irregular, but more about this later.)

Collimating a Schmidt-Cassegrain

Unlike the Newtonian, in which both the primary and secondary mirrors can be readily accessed for collimation, commercially made Schmidt-Cassegrain telescopes have their primary mirrors set and sealed at the factory. As such, an owner cannot adjust the primary if misalignment ever occurs. Fortunately, SCTs are rugged enough to put up with the minor bumps that might occur during setup without affecting collimation.

This leaves only the secondary mirror to adjust, as shown in Figure 8.5. Take a look at the secondary-mirror mount centered in the front corrector plate. There you will see the heads of three adjustment screws spaced 120° apart. (On some models, a plastic disk covers the screw heads. If so, it must be removed—carefully—to expose the adjustment screws.) In addition, some telescopes have a fourth, large screw or nut in the center of the secondary holder. If so, DO NOT TOUCH IT! Loosening that central screw will release the secondary from its cell and drop it into the tube. Talk about a quick way to ruin your day!

Remove the star diagonal and insert a sight tube into the eyepiece holder. Take a look. Ideally, you should see your eye centered in the secondary mirror. If the view is more like Figure 8.5a, turn one or more of the adjusting screws until everything lines up as in Figure 8.5b.

To check your success, move outside on a clear night. Set up the telescope as you would for an observing session, giving it adequate time to cool to the night air. With the telescope acclimated to the outdoor temperature, remove the star diagonal, if so equipped, and insert a medium-power eyepiece into the telescope. Center the instrument on a very bright star. Turn the focusing knob until the star moves out of focus and its disk fills about a third of the eyepiece field. Take a look at the dark spot on the out-of-focus disk. That's the dark silhouette of the secondary mirror. For the telescope to be properly collimated, the secondary's image must be centered on the star, creating a donut-like illusion. If the donut is asymmetric, then the secondary must be adjusted.

Make a mental note of which direction the silhouette favors, go to the front of the telescope, and turn the adjustment screw that most closely coincides to that direction *ever so slightly*. Now, return to the eyepiece, recenter the star in view, and look at the dark spot again. Is it better or worse? If it's worse, turn the same screw the opposite way; if it's off in a different direction, turn one of the other screws and see what happens. Continue going back and forth between eyepiece and adjustment screws until the dark spot is perfectly centered in the star blob.

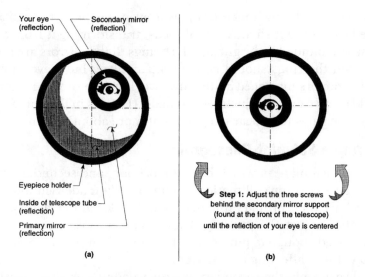

Figure 8.5   *Collimating a Schmidt-Cassegrain involves adjusting the secondary mirror (a) until its reflection appears centered in the primary (b).*

To double-check the adjustment, switch to a high-power eyepiece. Place the star back in the center of the field and defocus its image only slightly. If the secondary's dark outline still appears uniformly centered, then collimation was a success and the telescope is set to perform at its best. If not, repeat the procedure, but with finer adjustments.

If the image is still not correctly aligned even after repeated attempts, then there is a distinct possibility that the primary is not square to the secondary. Focus on a rich star field. If any coma (ellipticity) is evident around the stars at the center of view, then chances are good that the primary is angled incorrectly. In this case, your only alternative is to contact either the dealer from which the telescope was purchased or the manufacturer.

To learn more about the fine art of telescope collimation, consult the book *Perspectives on Collimation* by Vic Menard and Tippy D'Auria. The capsule review in chapter 6 gives further information.

Test Your Telescope

There is no way to guarantee that every telescope made, even those from the finest manufacturers, will work equally well. Although companies have quality-control measures in place to weed out the bad from the good, a lemon is bound to slip through every now and then. That is why you should always check an instrument right after it is purchased. Examine the telescope for any overt signs of damage. Next, look at the optics. They should be free of obvious dirt and scratches. The mounting should be solid and move smoothly. If any problems are detected, immediately contact the outlet from which the telescope was purchased so that the problem can be rectified.

It is easy to distinguish clean, scratch-free optics from those that are not, but a poor-quality lens or mirror may not be so obvious. Fortunately, an elaborate, fully equipped optical laboratory is not required to check the accuracy of a telescope's optics. Instead, all that you need are a well-collimated telescope, a moderate-power eyepiece, and a clear sky. With these ingredients, any amateur astronomer can perform one of the most sensitive and telling optical tests available: the star test. A fourth ingredient (this book) will help you interpret the star test's results.

But before you can run a valid star test, you must make certain that the optics are properly collimated. It is impossible to get accurate results from the star test if a telescope's optics are misaligned. Now might be a good time to review that procedure (found earlier in this chapter) if you are uncertain.

Once everything is in alignment, select a star with which to run the star test. Which star? That depends on your telescope. First, if your scope does not have a clock drive or an equatorial mount or is not polar aligned, it is best to aim toward a bright star near the celestial pole because these move more slowly in the sky than, say, stars near the celestial equator. Which individual star depends on the telescope's aperture. For 4-inch and smaller instruments, select a first- or second-magnitude star; 6- to 10-inch telescopes are best checked with a third- or fourth-magnitude star, and larger instruments may be best tested with a fifth-magnitude star. Use a moderately high power eyepiece for the test. Suiter[1] recommends a magnification equal to 25× per inch (10× per centimeter) of aperture. So for a 6-inch telescope, magnification should equal 150×, and so on.

Focus the star precisely and examine its image closely. You'll recall from chapter 1 that at higher magnifications, a star will look like a bull's-eye, that is, a bright central disk (the Airy disk) surrounded by a couple of faint concentric rings (diffraction rings). By definition, any telescope that claims to have "diffraction-limited" optics must show this pattern. Is that what you see?

Probably not, or at least not at first glance. Many factors affect the visibility of diffraction rings, such as telescope collimation, warm-air currents inside the telescope tube, the steadiness of the atmosphere, and the focal ratio of the telescope. Poor atmospheric steadiness, or "seeing" as it is commonly known, causes star images to "boil," making it impossible to detect fine detail. Diffraction rings can be seen only under steady seeing conditions (see the section in chapter 9 entitled "Evaluating Sky Conditions" for more on this). The aperture of the telescope also plays a big role in seeing diffraction rings. The larger the aperture, the smaller the diffraction pattern and, thus, the more difficult it is to see.

Slowly rack the image *slightly* out of focus. Just how far out to go is a matter of focal ratio. Most people foil the star test simply by being too aggressive when defocusing the star's image. Suiter recommends that for an f/10 instrument, the eyepiece should be moved no more than 0.20 inch off focus, while an f/4.5 telescope requires the eyepiece be moved a mere 0.06 inch! In either case,

1. Suiter, H.; *Star Testing Astronomical Telescopes;* Willmann-Bell, Inc., 1994.

the star's light should enlarge evenly in all directions, like ripples expanding after a small stone is tossed into a calm body of water.

Begin by moving the eyepiece inside of focus. Examine the out-of-focus star image. It should look like one of the patterns illustrated in Figure 8.6. Then reverse the focusing knob, bringing the star slightly outside of focus. Examine the ring pattern again. If the telescope has first-rate optics (that is, substantially better than merely diffraction-limited optics), both extrafocal images should appear identical.

What if they don't? Compare the exact shapes of the inside-focus and outside-focus patterns with those shown in Figure 8.7. Are the patterns at least circular? No? What geometric pattern do they resemble? An oval? If the oval shapes look the same both inside and outside of best focus, the optics are doubtless out of collimation. If the images are dancing wildly, then either the telescope optics are not yet acclimated to the outside air temperature or the atmosphere is too turbulent to perform the test. What if the images are either triangular or hexagonal? If the patterns have one or more sharp corners, then the optics are probably being pinched, or distorted, by their mounts. Pinched optics are especially common in Newtonian reflectors when the clips holding the primary in the mirror cell are too tight.

"My telescope never focuses stars sharply. I just did the star test, but the out-of-focus patterns appear oval. When the eyepiece is racked from one side of focus

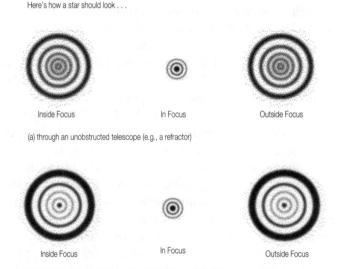

Here's how a star should look . . .

Inside Focus   In Focus   Outside Focus

(a) through an unobstructed telescope (e.g., a refractor)

Inside Focus   In Focus   Outside Focus

(b) through an obstructed telescope (e.g., a reflector or catadioptric)

Figure 8.6 *What should a star look like through the perfect telescope? Ideally all light should come to a common focus, causing a star to appear as a tiny disk (the Airy disk) surrounded by concentric diffraction rings. The star should expand to look like one of the illustrations here when seen out of focus, the image being identical on either side of focus: (a) through a refractor (unobstructed telescope) and (b) through a refractor on cata-dioptric (obstructed telescope).*

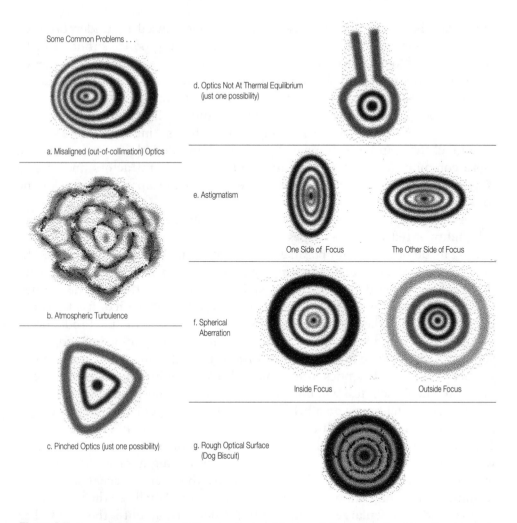

Some Common Problems . . .

a. Misaligned (out-of-collimation) Optics

b. Atmospheric Turbulence

c. Pinched Optics (just one possibility)

d. Optics Not At Thermal Equilibrium
(just one possibility)

e. Astigmatism

One Side of Focus   The Other Side of Focus

f. Spherical
Aberration

Inside Focus   Outside Focus

g. Rough Optical Surface
(Dog Biscuit)

Figure 8.7 *Something is amiss if the star test yields any of these results: (a) optics are out of collimation; (b) turbulent atmospheric conditions are present; (c) optics are being pinched or squeezed by their mounting; (d) optics have not acclimated to the outside air temperature; optics suffer from (e) astigmatism, (f) spherical aberration, and (g) rough optical surface.*

to the other, the ovals flip orientation 90°." If that is your telescope, then it suffers from astigmatism. Astigmatism may be caused by poorly figured optics, but it may also be caused by pinched optics, uncollimated optics, or by the cooling process after the scope is taken outdoors. In Newtonian reflectors, a slightly convex or concave secondary mirror is also a common cause of astigmatism. If the axis of the star oval is parallel to the telescope tube, suspect the secondary mirror.

The most common optical defect found in amateur telescopes is spherical aberration, which becomes evident when a mirror or lens has not been ground and polished to its required curvature. As a result, light from around the edge of the optic comes to a focus at a different distance than light from the center.

Spherical aberration comes in two varieties: one caused by undercorrected optics, and one caused by overcorrected optics. Both produce similar effects: On one side of focus, the outermost part of the bull's-eye pattern is brighter or sharper than on the other side of focus. In the case of a Newtonian or Cassegrain design, the shadow of the secondary mirror in the center of the star disk will look larger on one side of best focus and smaller on the other.

What if, even when a star is brought out of focus, rings cannot be seen at all and all that can be seen instead is a round, mottled blob? This condition, more common in reflectors and catadioptrics than in refractors, indicates a rough optical surface. Mirror makers have an especially appropriate nickname for this: *dog biscuit.*

Although not as accurate as the star test, another simple way to check optical quality of reflecting telescopes is the Ronchi test, which uses a diffraction grating made of many thin parallel lines printed on clear glass or plastic. The idea is to hold the diffraction grating up to the reflector's eyepiece holder, with the eyepiece removed, and examine a star's blurred image through the grating.

To perform a Ronchi test, you will need a 100-line-per-inch diffraction grating, available from the Telescope and Binocular Center, Schmidling Productions, and Edmund Scientific Company, among others. With the grating in hand, set up your telescope outside and allow it to cool to the ambient temperature. While you are waiting for the optics to acclimate, check and adjust the instrument's collimation. The Ronchi test will not work properly with poorly aligned optics.

Center the telescope on a bright star, remove the eyepiece (and star diagonal, if used), and hold the diffraction grating up to the empty eyepiece holder. Turn the focusing knob slowly in one direction until you see an enlarged disk with four superimposed black lines. Are the lines straight, or are they curved or crooked? If the latter, in what direction are they bent? Repeat the test by turning the focusing knob in the opposite direction. You'll see the star shrink in size and then enlarge again. Continue defocusing until the star's disk

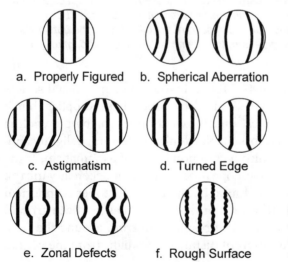

a. Properly Figured   b. Spherical Aberration

c. Astigmatism   d. Turned Edge

e. Zonal Defects   f. Rough Surface

Figure 8.8 *Some possible results from the Ronchi test and their causes. As explained in the text, ideally you should see straight lines during the test.*

expands and four dark lines are visible again. As before, examine their curvature. Note how this time the lines curve in the opposite direction.

Depending on how much and in what manner the lines curve, your optics may have one of several problems or no problem at all. Figure 8.8 interprets some of the common results. If the lines are straight, then your telescope *probably* enjoys a properly corrected mirror or lens (Even if it passes the Ronchi test, be sure to double-check the instrument with the more telling star test. Some instruments will pass the Ronchi test but fail the star test miserably.) If, however, the lines are bent or crooked, then the telescope suffers from one of the imperfections shown in the figure. Be forewarned that as in the star test, an improperly collimated or uncooled telescope may show the same result. Before you accuse your instrument of having poor optics, double-check their alignment, let the optics cool, and then repeat the test.

Both the star test and Ronchi test examine all of the optical components collectively; they can ascertain if there is a problem but cannot immediately distinguish *which* optical component is at fault. To point the finger of blame at a reflector, try rotating the primary mirror 90° and retesting. If the defect in the pattern turns with the mirror, then the primary is at fault. If it does not, then the secondary mirror is the culprit.

The moral to all this? Simple: There is no substitute for good optics. Pay close attention to manufacturers' claims and challenge those that appear too good to be true. If you are considering a telescope made by the XYZ Company, write or call to inquire about its optics. Who makes them? What tests are performed to evaluate optical quality? And finally, what kind of guarantee is offered? Remember, if you want absolute perfection, you will probably have to pay a premium for it.

Have Telescope, Will Travel

Not too long ago, the most traveling a telescope would ever do was the trip from the house to the backyard, but not anymore. With the increased popularity of large regional or national star parties, as well as the ever-worsening problem of light pollution, many amateur telescopes routinely travel tens or hundreds of miles in search of dark skies.

Whenever a telescope is picked up and moved, the risk of damage is present. That is why some hobbyists tend to shy away from driving around with such delicate instruments. With a little forethought and common sense, though, this risk can be minimized.

Let's first examine traveling with a telescope by car. Begin by examining the situation at hand. If you own, say, a subcompact sedan and an 18-inch Newtonian, you have a big problem with only one solution: Buy another car. (That is exactly what a friend of mine did after he purchased an NGT-18 from JMI a few years ago! He traded in his sports car for a minivan.)

If your telescope and automobile are already compatibly sized, then it is just a matter of storing the instrument safely for transport. Take apart as little of the telescope as necessary. If there is room in the car to lay down the entire instrument, mounting and all, then by all means do so. Take advantage of anything that can minimize the time required to set up and tear down.

If the telescope tube must be separated from its mounting, it is usually best to place the tube into the car first. Be sure to seal the optics from possible dust and dirt contamination, which usually means simply leaving the dust caps in place, just as when the telescope is stored at home. If the telescope does not come with a storage case, protect the tube from bumps by wrapping it in a clean blanket, quilt, or sleeping bag. Strategically placed pillows and pieces of foam rubber can also help minimize screw-loosening vibrations. If possible, strap the telescope in place using the car's seatbelt.

Next comes the mounting. Carefully place it into the car, making sure that it does not rub against anything that may damage it or that it may damage. Wrap everything with a clean blanket for added protection. Again, use either the car's seatbelts or elastic shock cords (bungee cords) to keep things from moving around during a sharp turn. Be sure to secure counterweights, which can become dangerous airborne projectiles if the car's brakes are hit hard.

Transporting telescopes by air presents many additional problems. Their large dimensions usually make it impossible to carry them on a plane and store them under the seat! Therefore, we must be especially careful when packing these delicate instruments.

Some owners of 8-inch and smaller Schmidt-Cassegrain telescopes prefer to wrap their instruments in foam rubber and place them in large canvas duffle bags and carry them onto the plane, placing the scopes into the overhead compartments. Although this method usually works (some overhead compartments are just not large enough), the chance of damaging the telescope is great. This practice, therefore, is *not* recommended. Neither should an SCT be shipped in its standard footlocker-type carrying case. They are not designed for that much (potential) abuse.

Scott Roberts of Meade Instruments recommends placing the telescope in its carrying case and then placing them in a double-walled cardboard shipping carton. Use the original padding that came with the telescope from the factory, if available. Celestron says that their deluxe hard cases (Celestron part number 302070), made by Penguin, well known in the photo industry for their excellent camera cases, are strong enough to take the jostling that air cargo can go through. These cases are packed with cubed high-density foam rubber so that they may be customized to hold any comparably sized instrument.

Even if you will be using one of these high-impact-resistant cases, take a few additional precautions before sending your baby off onto the loading ramp. Begin by completely enclosing the telescope in protective bubble wrap, available in larger post offices and stationery stores. Bubble wrap is a two-layer plastic sheet impregnated with air to form a multitude of cushioning bubbles. The bubbles are available in many different sizes—select the larger variety.

Owners of other types of telescopes face even greater challenges. First and foremost, the optics must be removed and placed into a suitable case to be carried on the plane. The empty tube may then be bubble-wrapped, surrounded by styrofoam pellets (both inside and outside the tube), and placed into a strong wooden crate. Make certain that the shipping carton can be used again on the return trip and bring along a roll of packing tape or duct tape just in case an emergency repair is needed.

Due to their weight, tripods and mountings pose special problems. I have traveled by air with a large tripod by first wrapping it in two thick sleeping bags and then packing everything in a large tent carrying case. Heavy equatorial mountings, on the other hand, must be packed professionally. Once again, seek a local crating company for help.

No industry-wide policy exists regarding the transport of telescopes by air. Some airlines permit telescopes to be checked as luggage provided they do not exceed size and weight restrictions. (The purchase of optional luggage insurance is strongly recommended.) Other airlines will not accept telescopes as check-in items at all. In these instances, you must ship the instrument separately ahead of time via an air cargo carrier. Owing to the amount of paperwork involved, especially on international shipments, air cargo services usually require that advance arrangements be made. Contact your airline well ahead of departure to find out exact details and damage insurance options.

Make sure that each piece of luggage has both a destination ticket and an identification tag and that both are clearly visible on the outside. Information on the ID tag should include your name, complete address, and telephone number. Although permanent plastic-faced identification tags are preferred, most check-in points provide paper tags that may be filled out on the spot. I always make it a habit to include a second identification label inside my luggage as well, just in case the outside tag is torn off.

Whatever you do, don't forget to bring along all the tools needed to reassemble the telescope once you arrive. It is best to keep the tools in a piece of checked baggage. While returning from Mexico after the July 1991 solar eclipse, a friend was stopped from boarding his flight because he was carrying a screwdriver. He ultimately had to check the screwdriver as a separate piece of luggage because all of his bags had already been boarded. It was quite a sight at the baggage claim area when his screwdriver came down the ramp among all these suitcases!

Finally, compile a thorough inventory of all equipment that you plan to bring. Include a complete description of each item, such as its dimensions, color, serial number, manufacturer, and approximate value. U.S. Customs requires owners to register cameras and accessories with them on a "Certificate of Registration for Personal Effects Taken Abroad" form before departure. Contact your nearest Customs office for further information. Keep a copy of the list with you at all times while traveling, just in case any item is lost or stolen. Carriers will be able to find the missing piece more quickly if they know what to look for.

9

A Few Tricks of the Trade

A telescope alone does not an astronomer make. Sure, telescopes, binoculars, eyepieces, and other assorted contraptions are all important ingredients for the successful amateur astronomer, but there is a lot more to it than that. If a stargazer lacks the knowledge and skills to use this equipment, then it is doomed to spend more time indoors gathering dust than outdoors gathering starlight. Here is a look at some techniques and tricks used by amateur astronomers when viewing the night sky.

Evaluating Sky Conditions

Clearly nothing affects our viewing pleasure more than the clarity of the night sky. However, just because the weather forecast calls for clear skies does not necessarily mean that it's time to get out the telescope. As amateur astronomers are quick to discover, "clear" is in the eye of the beholder. To most people a clear sky simply means an absence of obvious clouds, but to a stargazer it is much more.

To an astronomer, sky conditions may be broken down into two separate categories: *transparency* and *seeing*. Transparency is the measure of how clear the sky is, or in other words, how faint a star can be seen. Many different factors, such as clouds, haze, and humidity, contribute to the sky's transparency. The presence of air pollutants, both natural and otherwise, also adversely affects sky transparency. Artificial pollutants include smog and other particulate exhaust, whereas volcanic aerosols and smoke from large fires (for example, forest fires) are forms of natural air pollution. Still, the greatest threat to sky transparency comes not from nature but from ourselves. We are the enemy, and the weapon is uncontrolled, badly designed nighttime lighting.

Today's amateur astronomers live in a paradoxical world. On one hand, we are truly fortunate to live in a time when modern technology makes it possible for hobbyists to own advanced equipment once in the realm of the professional only. On the other, we hardly find ourselves in an astronomical Garden of Eden. Although technology continues to serve the astronomer, it is also proving to be a powerful adversary. The night sky is under attack by a force so powerful that unless drastic action is taken soon, our children may never know the joy and beauty of a supremely dark sky. No matter where you look, lights are everywhere: buildings, gigantic billboards, highway signs, roadways, parking lots, houses, shopping centers, and malls. Most of these lights are supposed to cast their light downward to illuminate their earthly surroundings. Unfortunately, many fixtures are so poorly designed that much of their light is directed horizontally and sky-ward. The result: light pollution, the bane of the modern astronomer.

Have you ever driven down a dark country road toward a big city? Long before you get to the city line, a distinctive glow emerges from over the hori-zon. Growing brighter and brighter with each passing mile, this monster slowly but surely devours the stars; first the faint ones surrender, but eventu-ally nearly all succumb. Several miles from the city itself, the sky has meta-morphosed from a jewel-bespangled wonderland to a milky, orange-gray barren desert. Although I know of no astronomer advocating the total and complete annihilation of all nighttime illumination, we must take a critical look at how it can be made less obtrusive.

Responsible lighting must take the place of haphazard lighting. But let's face it—not many people will be interested in light conservation if the point is debated from an astronomical perspective only. To win them over, they must be convinced that more efficient lighting is good for *them*. The public must be educated on how a well-designed fixture can provide the same amount of illu-mination over the target area as a poorly designed one but without extraneous light scattered toward the sky and with a lower operating cost. That latter phrase, *lower operating cost,* is the key to the argument. The cost of operating the light will be lower because all of its potential is specifically directed where it will do the most good. The wattage of the bulb may now be lowered for the same effect, resulting in a lower cost. Everybody wins—taxpayers, consumers, and, yes, even the astronomers!

Can one person effect a change? That was the dream of David Crawford, the person behind the International Dark-Sky Association. The nonprofit IDA has successfully spearheaded anti-light-pollution campaigns in Tucson, San Diego, and many other towns and cities. It provides essential facts, strategies, and resources to light-pollution activists worldwide. For more information on how you can join the fight against light pollution, contact the IDA at 3545 North Stewart, Tucson, Arizona 85716.

In midnorthern latitudes, the clearest nights usually take place immedi-ately after the crossing of an arctic cold front. After the front rushes through, cool, dry air usually dominates the weather for 24 to 48 hours, wiping the atmosphere clean of smog, haze, and pollutants. Such nights are characterized by crisp temperatures, high barometric pressure, and low relative humidity.

To help judge exactly how clear the night sky actually is, many amateurs living in the Northern Hemisphere use the stars of Ursa Minor (the Little Dipper) as a reference because they are visible every hour of every clear night in the year. Figure 9.1a shows the major stars of Ursa Minor, while Figure 9.1b shows those in the Southern Cross (Crux), a favorite check for observers in the Southern Hemisphere. The numbers next to several of the stars represent their visual magnitudes. Note that in each case the decimal point has been eliminated to avoid confusing it with another star. Therefore, the 20 next to Polaris indicates it to be magnitude 2.0, and so on. You may find that all of the Dipper stars are visible only on nights of good clarity, while other readers may be able to see them on nearly every night. Still others, observing from light-polluted environs, may never see them all.

The night sky is also judged in terms of seeing, which refers not to the clarity but rather the sharpness and steadiness of telescopic images. Fre-

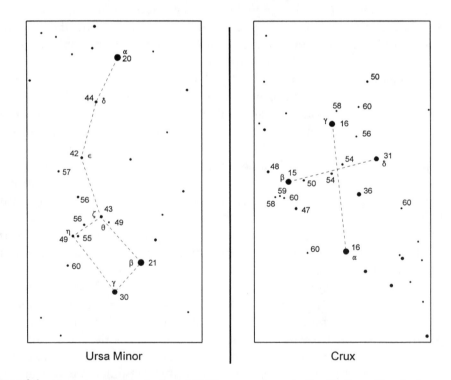

Ursa Minor | Crux

Figure 9.1 *How clear is the sky tonight? Many amateurs in the Northern Hemisphere use the visibility of the Little Dipper (Ursa Minor) as a gauge, while amateurs south of the equator use the Southern Cross (Crux). In both figures, decimals have been omitted to avoid confusing them with stars. Therefore, a magnitude 2.0 star is shown as "20" on the chart and so on.*

quently, on clear, dark nights of exceptional transparency, the twinkling of the stars almost seems to make the sky come alive in dance. To many, twinkling adds a certain romantic feeling to the heavens, but to astronomers, it only detracts from the resolving power of telescopes and binoculars.

The twinkling effect, called *scintillation,* is caused by turbulence in our atmosphere. Density differences between warm and cold layers refract or bend the light passing through, causing the stars to flicker. Ironically, the air seems steadiest when a slight haze is present. Although any cloudiness may make faint objects invisible, the presence of thin clouds can actually enhance subtle details in brighter celestial sights such as double stars and the planets.

Although the Earth's atmosphere greatly influences seeing conditions, image steadiness also can be adversely affected by conditions inside and immediately surrounding the telescope. If you are like most amateur astronomers, you probably store your telescope somewhere inside your house or apartment . . . your *warm* house or apartment. Moving the telescope from a heated room out into the cool night air immediately sets up swirling heat currents as the instrument and its optics begin the cooling process toward thermal equilibrium. Peering through the eyepiece of a warm telescope on a cool night is like looking through a kaleidoscope, with the stars writhing in strange ritualistic dances. As the telescope tube and its optics reach equilibrium with the outside air, the images will begin to settle down. The night's observing may then begin in earnest.

How long it takes for these heat currents to subside depends on the telescope's size and type as well as the local weather. Newtonian and Cassegrain reflectors seem to acclimate themselves the fastest. Still, these can require at least one hour in the spring, summer, and fall and up to two hours in the winter (the greater the temperature change, the longer it will take). Refractors and catadioptric instruments, because of their sealed tubes, need up to twice as long!

Several steps can be taken to minimize the time required for instrument cooling. For openers, find a cool and dry place to store the telescope when not in use. If a dedicated observatory is impractical, then good alternatives include a vented wooden garden shed or a sealed-off corner of an unheated garage. This way, the telescope's temperature will always be close to the outside temperature even before it is set up. If you must drive with your telescope to an observing site, try to travel without using the heater on the way there.

Tilting the open end of a reflector vertically like a smokestack can help speed air-temperature equalization because warm air rises. In the case of sealed tubes, where a lens or corrector prevents the smokestack effect, it is best to turn the instrument broadside to the wind. The cooling breeze will help wick away heat as it is radiated by the telescope tube.

If the telescope is of your own making, select a material other than metal for the tube. Wood, cardboard, and fiberglass have a lower heat capacity than metal and, therefore, will have less heat to radiate once set outside. Unfortunately, nonmetal tubes can also retain the warm air inside the tube, but in practice the benefit outweighs the disadvantage. Metallic tubes also dew over more quickly.

A variation on this theme is not to use a solid tube at all but rather an open-truss framework to support the optics. Such an idea has been popular among observatory-installed telescopes for a century or more, as it speeds thermal equalization while cutting down the instrument's weight. Many Dobsonian-style reflectors utilize this same principle for the same reasons, as well as the resulting ability to break down a large instrument into a relatively small package for transport. Unfortunately, while the speed of temperature adaptation is increased, so are the chances of stray light intruding into the final image. Open tubes are also more susceptible to interference from the observer's body heat, turbulence caused by crosswinds, and possible dewing of the optics. A sleeve made from black cloth and wrapped around the truss acts to slow these interferences but will not eliminate them.

As noted in chapter 4, some manufacturers build cooling vents into their instruments' designs—a good idea. Some, such as JMI's NGT instruments, Obsession and Starsplitter Newtonians, and Meade's Maksutovs, even offer small, flat, so-called muffin fans to help rid their telescopes of trapped pockets of warm air. Amateur telescope makers would do well to consider including some type of venting system in their instruments as well, for they go a long way in speeding up a telescope's reaching temperature equilibrium. Just be sure to seal the vents against dust infiltration when the telescope is not in use.

A telescope is not the only piece of equipment that needs to adjust to the change in temperature after being brought outdoors for the first time at night; the Earth must as well. The ground you place the instrument on, having been exposed to the comparatively warm temperatures of daytime, also has to adapt to the cool of night. Different surfaces absorb heat better than others. Concrete and blacktop are the worst offenders because they readily absorb and retain heat. Grass, though also requiring a cool-down period, is better because it does not retain as much heat.

Your Observing Site

This brings up another hot topic among amateur astronomers today: where to view the sky. Choosing a good observing site is becoming increasingly difficult. The ideal location should be far from all sources of light pollution and civilization in general. In addition, it should be as high above sea level as possible to avoid low-lying haze and fog, it must be safe from social ne'er-do-wells and other possibly harmful trespassers, and it should allow for an obstacle-free view of the horizon in all directions. Wouldn't it be nice if this was a description of your backyard? We can all dream of finding such a Shangri-la, but a few compromises usually must be made.

Over the years I have used several different observing sites with varying degrees of success. Many national, state, county, and local parks and beaches offer excellent areas, but their accessibility may be limited to daytime hours only or by residency. Ask your local park office if special access is available. The local authority that oversees the state parks near my home offers a stargazing permit that allows after-hours access to more than half a dozen parks for a small annual fee. The parks not only have much better horizons than most observing sites but also offer the added benefit of round-the-clock security patrols, an important consideration in these times. (Unfortunately, the patrol cars are outfitted with more lights than a small city, but I guess you have to take the good with the bad.)

Other good alternatives include both private and public golf courses. They have wide open expanses but may also suffer from restrictions and excessive security lighting around the clubhouse. If the owner of the course is apprehensive at first, why not offer to run a free observing session for club members in return for nighttime access? Although they may not be bona fide amateur astronomers, most people jump at the chance to see celestial wonders such as the rings of Saturn and the Moon. Flat farmland can also provide a secluded view, but unless the land is your own, be sure to secure permission from the owner beforehand. The last thing you want is to be chased by a gun-wielding farmer at two in the morning!

Where are the best observing sites? Here's my top-ten list:

10. A beach (watch out for fog and salt spray!)
 9. A flat rooftop (given a while to cool down after sunset)
 8. A town, county, state, or national park
 7. An open field or farmland
 6. A club observatory
 5. Your yard
 4. A *daytime-only* airport or landing strip
 3. The desert
 2. A golf course
 1. A hill or mountain

No matter where it is, a good observing site must be easy to reach. I know many urban and suburban amateurs who never see starlight during the week; instead, they restrict their observing time to weekends only, because the closest dark-sky site is more than an hour's drive away. Isn't that a pity? First, these amateurs may spend more time commuting to the stars than they do actually looking at them. Second, odds are they are forsaking many clear nights each month just because they believe the sky conditions closer to home are unusable. You really have to ask yourself if the local sky conditions are truly that bad. Remember, a telescope will show the Moon, Sun, and the five naked-eye planets as well from the center of a large city as it will from the darkest spots on Earth. Hundreds, if not thousands, of double and variable stars are also observable through even the most dismal of sky conditions. True, there is something extraspecial about observing under a star-filled sky, but never forgo a clear night just because the ambience is less than ideal. As the old saying goes, where there's a will, there's a way.

Star Parties and Astronomy Conventions

Perhaps as a reflection of the increasing interference of light pollution, the past few decades have seen a tremendous growth in regional, national, and international star parties and astronomy conventions (Figure 9.2). At these, hundreds, even thousands, of amateur astronomers travel to remote spots to hear top-notch speakers, set up their homemade and commercially purchased telescopes, and enjoy dark-sky conditions far superior to the skies back home. Table 9.1 lists some of the world's largest and oldest.

Dozens of smaller conventions are held across the country and around the world throughout the year. To help spread the word about when and where they will occur, astronomy magazines contain monthly listings giving information and addresses for further information. Take at look at a current issue to see if there is an upcoming event near you. If so, by all means try to attend. Astronomy conventions and star parties are great ways to meet new friends, learn a lot about your hobby and science, and get a great view of the night sky.

Finding Your Way

Once a site is selected, it is time to depart on your personal tour of the sky. Although most novice enthusiasts begin with the Moon and brighter planets, most objects of interest in the sky are not visible to the unaided eye or even through a side-mounted finderscope. How can a telescope be aimed their way if the observer cannot see the target in the first place? That is where observing

Figure 9.2 *The granddaddy of all amateur astronomy conventions, Stellafane hosts thousands of hobbyists each summer atop Breezy Hill in Springfield, Vermont.*

Table 9.1   ***Annual Astronomy Conventions and Star Parties***

Alberta Star Party (Caroline, Alberta; August)
  4612 17th Ave. NW, Calgary, AB T3B OP3, Canada
  Email: weisc©cadvision.com
  Web page: http://www.syz.com/rasc/
Apollo Rendezvous (Dayton, Ohio; June)
  Miami Valley Astronomical Society, 2600 Deweese Pkwy., Dayton, OH 45414
  Email: starsrus@infinet.com
  Web page: http://www.mvas.org/
Astrofest (Kankakee, Illinois; September)
  Chicago Astronomical Society, P.O. Box 30287, Chicago, IL 60630
Enchanted Skies Star Party (Socorro, New Mexico; October)
  P.O. Box 743-I, Socorro, NM 87801
  Web page: http://www.nmt.edu/~astro
Eta Aquilae Star Party (Rucava, Latvia; August)
  Latvian Astronomical Society, Blvd. Rainis 19, Riga LV-1586, Latvia
  Email: astro@acad.latnet.lv
  Web page: http://www.lanet.lv/members/LU/astro/ast_erg.html
Grand Canyon Star Party (Grand Canyon National Park, Arizona; June)
  Dean Ketelsen, 1122 E. Greenlee Place, Tucson, AZ 85719
  Email: ketelsen@as.arizona.edu
  Web page: http://www.kaibab.org/stars/gc_stars.htm
Great Plains Star Party (Osawatomie, Kansas; September or October)
  15506 Beverly Ct., Overland Park, KS, 66223
  Email: robinson@sky.net
  Web page: http://www.icstars.com
Mason-Dixon Star Party (York, Pennsylvania; May)
  Jeri Jones, York County Parks, 400 Mundis Race Road, York, PA 17402
  Email: jlj276@aol.com or ullery@act.org
Mount Kobau Star Party (Osoyoos, British Columbia; August)
  P.O. Box 20119 TCM, Kelowna, BC V1Y 9H2, Canada
  Web page: http://www.bcinternet.com/~mksp/
Nebraska Star Party (Valentine, Nebraska; August)
  640 S. 30th St., Lincoln, NE 68510
  Email: NSP@4w.com
  Web page: http://www.4w.com/nsp/
Northeast Astronomy Forum (Suffern, New York; April)
  Rockland Astronomy Club, 73 Haring Street, Closter, NJ 07624-1709
  Email: donaldurbanl@juno.com
Nova East Star Party (Fundy National Park, Nova Scotia; September) 108 Aspen Crescent,
  Lower Sackville, NS B4C 1E1, Canada
  Email: pgray@hercules.stmarys.ca
  Web page: http://halifax.rasc.ca/ne/
Okie-Tex Star Party (Ardmore, Oklahoma; October)
  Rte. 1, Box 96, Union City, OK 73090
  Email: kcarr@icon.net
  Web page: http://www.icon.net/ ~seadkins/okie-tex.html

Table 9.1   continued

Oregon Star Party (Ochoco National Forest, Oregon; August)
   P.O. Box 91416, Portland, OR 97291
   Email: ospinc@teleport.com
Peach Tree Star Gaze (Jackson, Georgia; May)
   Ken Poshedly, 3440 Everson Bay Court, Snellville, GA 30278
   Email: k.poshedly@mci2000.com
Queensland Astrofest (Linville, Queensland; August)
   145 Ardoyne Road, Oxley Qld 4075, Australia
   Email: scharfpb@eis.net.au
   Web page: http://www.ozemail.com.au/~mhorn2/afest.html
Riverside Telescope Maker's Conference (Big Bear Lake, California; May)
   9045 Haven Avenue, Suite 109, Rancho Cucamonga, CA 91730
   Email: robert_stephens@eee.org
   Web page: http://home.sprynet.com/sprynet/hrmeyer/rtmchome.htm
Rocky Mountain Star Stare (near Colorado Springs, Colorado; June/July)
   Colorado Springs Astronomical Society, 8125 Avens Circle,
   Colorado Springs, CO 80920
   Email: dwilder@kktv.com or bygrens@aol.com
   Web page: http://members.aol.com/bygrens/rmss97.html
Starfest (Mount Forest, Ontario; August)
   North York Astronomical Association
   26 Chryessa Ave., Toronto, ON M6N 4T5
   Email: tward@visionol.net
   Web page: http://www.unicom/~nyaa/nyaa0.htm
Stellafane (Springfield, Vermont; August)
   P.O. Box 50, Belmont, MA 02178
   Web page: http://www.stellafane.com
Swiss Star Party (Gurnigel; August)
   Peter Kocher, ufem Berg 23, CH-1734 Tentlingen/FR, Switzerland
   Email: peter.kocher@profora.ch
   Web page: http://www.ezinfo.ethz.ch/astro/stp/
Table Mountain Star Party (Ellensburg, Washington; August)
   P.O. Box 785, Puyallup, WA 98371
   Email: TMSP@triax.com
Texas Star Party (Fort Davis, Texas; May)
   TSP Registrar, 1326 Mistywood Lane, Allen, TX 75002
   Email: kastro@aol.com
   Web page: http://www.metronet.com/~tsp/
Winter Star Party (West Summerland Key, Florida; February)
   Bob and Sharon Grant, 5401 S.W. 110th Avenue, Miami, FL 33165
   Web page: http://www.scas.org

technique comes into play. To locate these heavenly bodies, one of two different methods must be used. But before any of these systems can be discussed, it is best to become fluent in the way astronomers specify the location of objects in the sky.

Celestial Coordinates

Like the Earth's spherical surface, the celestial sphere has been divided up by a coordinate system. On Earth, the location of every spot can be pinpointed by its unique longitude and latitude coordinates. Likewise, the position of every star in the sky may be defined by *right-ascension* and *declination* coordinates.

Let's look at declination first. Just as latitude is the measure of angular distance north or south of the Earth's equator, declination (abbreviated *Dec.*) specifies the angular distance north or south of the celestial equator. The celestial equator is the projection of the Earth's equator up into the sky. If we were positioned at 0° latitude on Earth, we would see 0° declination pass directly through the zenith, while 90° north declination (the North Celestial Pole) would be overhead from the Earth's North Pole. From our South Pole, 90° south declination (the South Celestial Pole) is at the zenith.

As with any angular measurement, the accuracy of a star's declination position may be increased by expressing it to within a small fraction of a degree. We know there are 360° in a circle. Each of those degrees may be broken into 60 equal parts called *minutes of arc*. Further, every minute of arc may be broken up to 60 equal *seconds of arc*. In other words:

$$\text{one degree } (1°) = 60 \text{ minutes of arc } (60')$$
$$= 3{,}600 \text{ seconds of arc } (3{,}600'')$$

When minutes of arc and seconds of arc are spoken of, an angular measurement is being referred to, not the passage of time.

Right ascension (abbreviated *R.A.*) is the sky's equivalent of longitude. The big difference is that while longitude is expressed in degrees, right ascension divides the sky into 24 equal east–west slices called *hours*. Quite arbitrarily, astronomers chose as the beginning, or zero-mark, of right ascension the point in the sky where the Sun crosses the celestial equator on the first day of the Northern Hemisphere's spring. A line drawn from the North Celestial Pole through this point (the Vernal Equinox) to the South Celestial Pole represents 0 hours right ascension. Therefore, any star that falls exactly on that line has a right ascension coordinate of 0 hours. Values of right ascension increase toward the east by one hour for every 15° of sky crossed at the celestial equator.

To increase precision, each hour of right ascension may be subdivided into 60 minutes and each minute into 60 seconds. A second equality statement summarizes this:

$$\text{one hour R.A. } (1 \text{ h}) = 60 \text{ minutes R.A. } (60 \text{ m})$$
$$= 3{,}600 \text{ seconds R.A. } (3{,}600 \text{ s})$$

Unlike declination, where a minute of arc does not equal a minute of time, a minute of R.A. does.

The stars' coordinates do not remain fixed. Due to a 26,000-year wobble of the Earth's axis called *precession,* the celestial poles actually trace circles on the sky. Right now, the North Celestial Pole happens to be aimed almost exactly at Polaris, the North Star. But in 13,000 years it will have shifted away from Polaris and will instead point toward Vega in the constellation Lyra. The passage of another 13,000 years will find the pole aligned with Polaris once again.

Throughout the cycle, the entire sky shifts behind the celestial coordinate grid. Although this shifting is insignificant from one year to the next, astronomers find it necessary to update the stars' positions every 50 years or so. That is why you will notice that the right ascension and declination coordinates are referred to as "epoch 2000.0" in this book and most other contemporary volumes. These indicate their exact locations at the beginning of the year 2000 but are accurate enough for most purposes for several decades on either side.

Star-Hopping

The simplest method for finding faint sky objects, and also the one preferred by most amateur astronomers, is called *star-hopping.* Star-hopping is a great way to learn your way around the sky while developing your skills as an observer. Before a telescope can be used to star-hop to a desired target, its finderscope must be aligned with the main instrument. Take a look at the telescope's finder. Chances are it is held by six thumbscrews in a set of mounting rings. Begin the process by aiming the telescope toward a distant identifiable object. Though the Moon, a bright star, or a bright planet may be used, I suggest using instead a terrestrial object such as a distant light pole or mailbox. The reason for this is quite simple: They don't move. Celestial objects appear to move because of the Earth's rotation, making constant realignment of the telescope necessary just to keep up. Center the target in the telescope's field and lock the mounting's axes. To alter the finder's aim, adjust the front three thumbscrews until the finder's tube is centered in the front mounting ring. Now loosen the back three screws. Move the finder by hand until the target is centered in the finder's crosshairs. Once set, tighten all of the adjustment screws and check to see that the finder or telescope did not shift in the process.

Some finderscopes are held by only three adjustment screws and are adjusted in much the same way (though they are much more prone to misalignment). Aim at a terrestrial target as previously suggested and lock the telescope's axes. Look through the finder to see which direction is out of alignment. Loosen the two opposite screws and move the finder by hand until the target is centered. Tighten all thumbscrews, making a final check to see that the finder is aimed correctly.

With the finder correctly aligned to the telescope, the fun can begin. After deciding on which object you want to find, pinpoint its position on a detailed star atlas, such as one of the charts found in the next chapter. Scan the atlas page for a star near the target that is bright enough to be seen with the naked eye. Once a suitable star is located on the atlas, turn your telescope (or binoculars) toward it in the night sky. Looking back at the atlas, try to find little geometric patterns among the fainter suns that lie between the naked-eye star and the target. You might see a small triangle, an arc, or perhaps a parallelogram. Move the telescope to this pattern, center it in the finderscope, and return to the atlas. By switching back and forth between the finderscope and the atlas, hop from one star (or star pattern) to the next across the gap toward the intended target. Repeat this process as many times as it takes to get to the area of your destination. Finally, make a geometric pattern among the stars and the object itself. For instance, you might say to yourself, "My object lies halfway between and just south of a line connecting star A and star B." Then locate star A and star B in the finder, shift the view to the point between and south of those two stars, and your target should be in (or at least near) the telescope's field of view. Don't worry if you get lost along the way; breathe a deep sigh and return to the starting point.

The charts found in chapter 10 are drawn with the star-hopper in mind. Each has at least one naked-eye star shown, and many display large portions of prominent constellations. In addition, the written description of each featured object includes a short passage describing how to star-hop to its position.

Let's imagine that we want to find M31, the Andromeda Galaxy. (M31 is this galaxy's catalog number in the famous Messier listing of deep-sky objects. If you don't know what the Messier catalog is, or for that matter, what a deep-sky object is, read the next chapter.) Begin by finding it in Figure 9.3. M31's celestial address is right ascension 0 hours 42.7 minutes (written $00^h 42^m.7$), declination $+41°16'$. Notice that it is located northwest of the naked-eye stars Beta (β), Mu (μ), and Nu (ν) Andromedae. Find these stars in the sky and center your telescope's aim on them. Hop from Beta to Mu and then on to Nu. M31 lies a little over a degree to the west-northwest of this last star and should be visible through the finderscope (indeed, it is visible to the naked eye under moderately dark skies).

Easy, right? Now let's tackle something a little more challenging. Many observers consider M33 (seen back in Figure 5.3), the Great Spiral Galaxy in Triangulum, to be one of the most difficult objects in the sky to find. Most references note it as magnitude 5.9, which means that if the galaxy could somehow be squeezed down to a point, that would be its magnitude. However, M33 measures a full degree across, so its *surface brightness,* or brightness per unit area, is very low.

M33 and its home constellation are also plotted on Figure 9.3. How can we star-hop to this point? One way is to locate the naked-eye stars Alpha Trianguli and our old friend Beta Andromedae. Aim at the former with the finderscope and move slightly toward the latter. About a quarter of the way between Alpha Trianguli and Beta Andromedae and a little to the south lies a

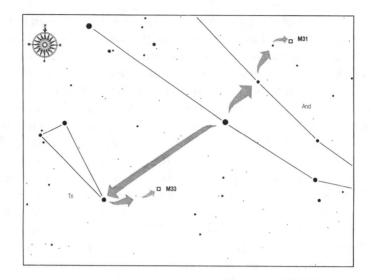

Figure 9.3 *Star-hopping is the only way to find deep-sky objects, according to some observers. Here are suggested plans of attack for locating M31, the Andromeda Galaxy, and M33, the Great Spiral in Triangulum. Both are in the autumn sky.*

6th-magnitude star. Spot it in your finderscope. Connect an imaginary line between it and Alpha and then extend that line an equal and opposite distance from the 6th-magnitude star. There you should see the galaxy's large, hazy glow. Nevertheless, when you peer through your telescope at that point, nothing is there. Recheck the map. No, that's not it; the telescope is aimed in the right direction. It's got to be there! What could be wrong?

At frustrating times such as this, it is best to pull back from your search and take a breather. Then go back, but this time use the averted-vision technique mentioned later in the section of this chapter entitled "Eye Spy." Look a little to one side of where the galaxy should be. By glancing at it with peripheral vision, its feeble light will fall on a more sensitive area of the eye's retina. Suddenly, there it is! You will wonder how you ever missed it before.

Setting Circles

This brings us to the second method that amateur astronomers use to find faint objects in the sky. It sounds easiest in theory, because you don't have to know the sky. But in fact it's no shortcut—as the next few pages of (complicated) instructions should make clear.

Nowadays, most telescope mountings can be retrofitted with digital setting circles, which are, in effect, tiny computers that help the observer aim a telescope toward a preselected target. These were examined at some length in chapter 6, but the discussion here deals with old-fashioned, mechanical-style setting circles. Mechanical setting circles—round, graduated scales—are only found on equatorial mounts, one on each axis (Figure 9.4). The circle on the

Figure 9.4 *Setting circles allow an observer to take aim at a sky object by knowing its celestial coordinates.*

polar axis is divided into 24 equal segments, with each segment equal to one hour of right ascension. The setting circle attached to the declination axis is divided into degrees of declination, from 90° North to 90° South. With an equatorial mounting accurately aligned to the celestial pole (or any mounting with the properly encoded alt-azimuth digital setting circles), a sky object may be located by dialing in its pair of celestial coordinates.

Is that all there is to it? Unfortunately, no. First, to use setting circles, the mounting's polar axis must be aligned to the celestial pole. Begin by leveling the telescope. This may be done by adjusting the length of each tripod leg (if so equipped) or by placing some sort of a block (a piece of wood, a brick, etc.) under one or more corner footpads of the mount. Don't spend a lot of time trying to make the mount perfectly level; close is good enough. In fact, to be completely correct, it is not necessary to level the mount at all. The only thing that matters is that the polar axis be aimed at the celestial pole. In practice, however, it is easier to polar-align a mount that is level than one that is not, so take the time to do so. As an aid, many instruments come outfitted with bubble levels on their mountings.

Next, check to make sure that your finderscope is aligned to the telescope and that the entire instrument is parallel to its polar axis. This latter step is usually accomplished simply by swinging the telescope around until the declination circle reads 90°. Most declination circles are preset at the factory, although some are adjustable; others may have slipped over the years. If you believe that the telescope is not parallel to the polar axis when the declination circle reads 90°, consult your telescope manual for advice on correcting the reading. If your manual says nothing, or worse yet you cannot find it (it must be here somewhere, you think), try this test. Align the telescope to the polar axis as best you can by eye and lock the declination axis. Using only the horizontal and vertical motion (azimuth and altitude, respectively) of the mounting, center some distant object, such as a treetop or a star, in your finderscope's view. Now, rotate the telescope about the polar axis *only*. If the telescope/finderscope combination is parallel to the polar axis, then the object will remain fixed in the center of the view; in fact, the entire field will appear to pivot around it.

If, however, the object moved, then the finder and the polar axis are not parallel to one another. Try it again, but this time pay close attention to the direction in which the object shifted. If it moved side to side, shift the entire mounting in azimuth (horizontally) exactly half the *horizontal* distance it moved. If it moved up and down, then the mounting's altitude (vertical) pivot is not set at the correct angle. Loosen the pivot and move the entire instrument one-half the *vertical* distance that the object moved. Because, in all likelihood, it shifted diagonally, this will turn into a two-step procedure. Take it one step at a time, first eliminating its horizontal motion, then the vertical.

With the polar axis and telescope now parallel, it is time to set the polar axis parallel to the Earth's axis. Some equatorial mounts come with a polar-alignment finderscope built right into the polar axis. These come with special clear reticles surrounding the celestial pole. Consult the telescope's manual for specific instructions.

Polar-alignment finderscopes are certainly handy, but for those of us without such luxuries, the following method should work quite well. With the telescope level and the declination axis locked at +90°, point the right-ascension axis by eye approximately toward Polaris. Release the locks on both axes and swing the instrument toward a star near the celestial equator. Once aimed at this star, spin the right-ascension circle (taking care not to touch the declination circle) until it reads the star's right ascension. Table 9.2 suggests several suitable stars for this activity. Note that the stars' positions are given at five-year intervals. Choose the pair closest to your actual date. (Although this slight shift is not of much concern when aligning a telescope to use setting circles, it is of great consequence for long-exposure through-the-telescope astrophotography.)

Swing the telescope back toward the celestial pole, stopping when the setting circles read the position of Polaris (also given in Table 9.2). If the telescope is properly aligned with the pole, then Polaris should be centered in view. If not, lock the axes and shift the entire mounting horizontally and vertically until Polaris is in view. Repeat the procedure again until Polaris is in view when the circles are set at its coordinates. Owing to the coarse scale of most circles supplied on amateur telescopes, a polar alignment within roughly 0.5 to 1° of the celestial pole is usually the best you can get. (Digital setting circles are much more accurate, but still the alignment need not be overly precise to be useful.)

Once the mounting is adjusted to the pole and the right-ascension circle (also known as the *hour circle*) is calibrated with the coordinates of a known star, the mount's clock drive must be turned on. Many equatorial mounts have a direct link that turns the hour circle in time with the telescope. This way, as the sky and its coordinate system shift relative to the horizon, the hour circle will move along with them. Some less-sophisticated equatorial mounts, however, do not have driven hour circles. In these instances, the hour circle must be recalibrated not just once a night but before each use—inconvenient, to say

Table 9.2 **Suitable Stars for Setting-Circle Calibration**

Star	Epoch	Right Ascension h	Right Ascension m	Declination °	Declination '
Alpheratz (Alpha Andromedae)	(2000)	00	08.5	+29	05
	(2005)	00	08.7	+29	07
	(2010)	00	09.0	+29	09
Hamal (Alpha Arietis)	(2000)	02	07.5	+25	28
	(2005)	02	07.9	+25	29
	(2010)	02	08.0	+25	31
Aldebaran (Alpha Tauri)	(2000)	04	35.9	+16	30
	(2005)	04	36.2	+16	31
	(2010)	04	36.4	+16	31
Procyon (Alpha Canis Minoris)	(2000)	07	39.3	+05	14
	(2005)	07	39.6	+05	13
	(2010)	07	39.9	+05	12
Regulus (Alpha Leonis)	(2000)	10	08.3	+11	58
	(2005)	10	08.6	+11	57
	(2010)	10	08.9	+11	55
Arcturus (Alpha Boötis)	(2000)	14	15.7	+19	11
	(2005)	14	15.9	+19	10
	(2010)	14	16.1	+19	08
Altair (Alpha Aquilae)	(2000)	19	50.8	+08	53
	(2005)	19	51.0	+08	53
	(2010)	19	51.3	+08	54
Polaris (Alpha Ursae Minoris)	(2000)	02	31.8	+89	16
	(2005)	02	37.6	+89	17
	(2010)	02	43.7	+89	18
South Pole Star (Sigma Octantis)	(2000)	21	08.6	−88	57
	(2005)	21	13.1	−88	56
	(2010)	21	17.2	−88	55

the least. The easiest way to do this is to reset it on a reference star immediately before swinging the telescope toward a new target.

Incidentally, though the previous paragraphs describe the procedure for aligning to the North Celestial Pole, the actions are the same for observers in the Southern Hemisphere, except the mounting must be aligned to the South

Celestial Pole. Everywhere you see "Polaris" or "North Star," substitute in "Sigma Octantis" or "South Star." Sigma Octantis, a 5.5-magnitude sun, is located about 1° from the South Celestial Pole. Its celestial coordinates are also given in Table 9.2. Although not as obvious as its northern counterpart, Sigma works well for the task at hand.

With the mounting polar-aligned, the setting circles calibrated, and the clock drive switched on, it is now a relatively simple matter to swing the telescope around until the circles read the coordinates of the desired target. If all was done correctly beforehand, the target should appear in, or at least very near, the eyepiece's field of view. Be sure to use your eyepiece with the widest field when first looking for a target's field before moving up to a higher-power ocular.

What if it's not there? Don't give up the ship immediately. Instead, try recalibrating the setting circles on a known star near the intended target's location. This technique, called *offsetting*, is a lot more accurate than trusting the setting circles to give the correct reading by themselves.

Now that the technique for using setting circles is familiar, here is why they should *not* be used, especially by beginners who are unfamiliar with the sky. To my way of thinking (and you are free to disagree), using setting circles or a computer to help aim a telescope reduces the observer to little more than a couch-potato sports spectator flipping television channels between football games on a Sunday afternoon. Where is the challenge in that? Observational astronomy is not meant to be a spectator's sport; it is an activity that is best appreciated by doing. There is something very satisfying in knowing the sky well enough to be able to pick out an object such as a faint galaxy or an attractive double star using just a telescope, a finder, and a star chart. Even if you get your setting circles perfectly aligned and master their use, you will be missing out on the thrill of the hunt. Stalking an elusive sky object is much like searching for buried treasure: you never know what else you are going to uncover along the way. As you set your attention toward one target, you might turn up other objects that you have never seen before. Of course, there is also that slight possibility of striking astronomical pay dirt by stumbling onto an undiscovered comet. Imagine that thrill!

All right, so maybe I have been hard on setting circles, maybe even a bit unjustifiably. Make no mistake; setting circles are very useful tools *when used for the right reasons*. There is no question that they are required to aim the large, cumbersome telescopes in professional observatories where viewing time is at a premium. They also serve a very real purpose for advanced amateurs who are involved in sophisticated research programs, such as searching for supernovae in distant galaxies or estimating the brightness of variable stars. Both of these activities involve rapid, repetitive checking of a specific list of objects. Setting circles can also come in very handy when light pollution makes star-hopping difficult. However, they should never be used as a crutch. Most amateur astronomers will do much better by looking up and learning how to read the night sky rather than looking down at the setting circles.

Eye Spy

Although we may think of our binoculars and telescopes as wonderful optical instruments, there is no optical device as marvelous or versatile as the human eye. Experts estimate that 90% of the information processed by our brains is received by our eyes. There is no denying it; we live in a visual cosmos. To better understand how we perceive our surroundings, both earthly and heavenly, let us pause a moment to ponder the workings of our eyes.

The human eye (Figure 9.5) measures about an inch in diameter and is surrounded by a two-part protective layer: the transparent, colorless *cornea* and the white, opaque *sclera*. (Remember, don't shoot until you see the whites of their scleras.) The cornea acts as a window to the eye and lies in front of a pocket of clear fluid called the *aqueous humor* and the *iris*. Besides giving the eye its characteristic color (be it blue, brown, or any of a wide variety of other hues) the iris regulates the amount of light entering the eye and, more importantly, varies its focal ratio. Under low-light conditions, the iris relaxes, dilating the *pupil* (the circular opening in the center of the iris) to about 7 mm. Under the brightest lighting, the iris contracts the pupil to 2.5 mm or so across, increasing the focal ratio and masking lens aberrations to produce sharper views. From the pupil, light passes through the eye's *lens* and across the eyeball's interior, the latter being filled with fluid called the *vitreous humor*. Both the lens and cornea act to focus the image onto the *retina*, which is composed of ten layers of nerve cells, including photosensitive receptors called *rods* and *cones*. Cones are concerned with brightly lit scenes, color vision, and resolution; rods are low-level light receptors that cannot distinguish color. There are more cones toward the *fovea centralis* (the center of the retina and our perceived view), while rods are more numerous toward the edges. There are neither rods nor cones at the junction with the optic nerve (the eye's *blind spot*).

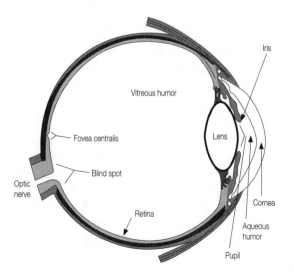

Figure 9.5 *The human eye, the astronomer's favorite tool.*

From an astronomical point of view, we are most interested in the eye's performance under dimly lit conditions, our so-called night vision. Try this test on the next clear, moonless night. Go outdoors from a well-lit room and look toward the sky. Chances are you will be able to pick out a few of the brightest stars shining in what looks like an ink-black background. Face away from the sky, wait three or four minutes, and then look up again. You will immediately notice that many more stars seem to have appeared in that short time, an indication that the eye has begun to adapt to its new, darker environment.

Complete the exercise by turning away from the sky a second time, making certain to block your eyes from any stray lights. Wait another 15 to 20 minutes and then look skyward once again. This time there will be even more stars than before. Over the ensuing time, the eye has become fully adapted to the dark. Not only has the pupil dilated, but a shift in the eye's chemical balance has also occurred. The buildup of a chemical substance called *rhodopsin* (also known as *visual purple*) has increased the sensitivity of the rods. Most people's eyes become adjusted to the dark in 20 to 30 minutes, though some require as little as 10 minutes or as long as one hour. This is enough to begin a night's observing, but *complete* dark adaptation takes up to another hour to occur.

Although the eye's sensitivity to dim lighting increases dramatically during the dark-adaptation process, it loses most of its sensitivity to color. As a result, visual observers can never hope to see the wide range of hues and tints in sky objects that appear so vivid in astronomical photographs. Nebulae are good examples of this. Unlike stars, which shine across a broad spectrum, nebulae are only visible at specific, narrow wavelengths. Different types of nebulae shine at different wavelengths. For instance, emission nebulae, those clouds that are excited into fluorescence by the energy from young stars buried deep within them, shine with a characteristic reddish color. Red, toward the long end of the visible spectrum, is all but invisible to the human eye under dim illumination. As a result, emission nebulae are among the toughest objects for the visual astronomer to spy.

The eye is best at perceiving color among the brighter planets, such as Mars, Jupiter, and Saturn, as well as some double stars where the color contrast between the suns can be striking. Most extended deep-sky objects display little color, apart from the greenish and bluish tints of some brighter planetary nebulae and the golden and reddish-orange tinges of some star clusters.

While the eye's blind spot does not adversely affect our night vision, the fact that the fovea centralis is populated only by cones does. What this means, quite simply, is that the center of our view is *not* the most sensitive area of the eye to dim light, especially when it comes to diffuse targets such as comets and most deep-sky objects. Instead, to aid in the detection of targets at the threshold of visibility, astronomers use a technique called *averted vision*. Rather than staring directly toward a faint object, look a little to one side or the other of where the object lies. By averting your vision in this manner, you direct the target's dim light away from the cone-rich fovea centralis and onto the peripheral area of the retina, where the light-sensitive rods stand the best chance of revealing the faint target.

Another way to detect difficult objects is to tap the side of the telescope tube very lightly. Your peripheral vision is very sensitive to motion, so a slight back-and-forth motion to the field of view will frequently cause faint objects to reveal themselves. I know it sounds a bit strange, but try it. It works, but be gentle.

Here's an aside for all readers who wear eyeglasses. If you suffer from either nearsightedness or farsightedness, it is best to remove your glasses before looking through the eyepiece of a telescope or binoculars, refocusing the image until everything is sharp and clear. On the other hand, if you suffer from astigmatism or require eyeglasses with thick, curved lenses, then it is best to leave the glasses on. Of course, if you use contact lenses, leave them in place when observing as you would doing any other activity.

Frequently, localized light pollution will also mask a faint object. The distraction caused by glare seen out of the corners of your eyes from nearby porch lights, streetlights, and so on can be enough to cause a faint celestial object to be missed. To help shield our eyes from extraneous light, many eyepieces and binoculars sold today come with built-in rubber eyecups. Although they prove adequate under most conditions, eyecups may not block out all peripheral light. Here are a couple of tricks to try if eyecups alone prove inadequate.

This first idea was already mentioned in my book *Touring the Universe through Binoculars*, but I think it merits repeating here. Buy a cheap pair of ordinary rubber underwater goggles, the kind you can find at just about any toy or sporting goods store. Cut out half of the goggles' front window. If you prefer to use one eye over the other for looking through your telescope, make certain to cut out the correct side. Of course, both windows would need to be cut out if used with binoculars. Spray the goggles with flat black paint, and they're done. To test your creation, put the goggles on (after the paint has dried, please!) and go out under the stars. The blackened goggles should provide enough added baffling to keep stray light from creeping around the eyepiece's edge and into your eyes.

Here another approach to the same situation that works well for viewing through telescopes but is not really applicable to binocular use. This involves wearing a dark turtleneck shirt or sweater, but not in the way you are used to. Instead of slipping it on from bottom to top, stick your head into the shirt through the neck opening. Let the shirt rest on your shoulders. Whenever you look through the eyepiece, simply pull the shirt up and over your head to act as a cloak against surrounding lights. This idea was first used by photographers a century ago, and it still works today. The only drawback, apart from looking a little strange to civilians, is that the eyepiece may dew over more quickly because of trapped body heat. Still, I can sometimes see up to a half-magnitude fainter just by using this cloaking device. Give it a try.

Record Keeping and Sketching

One of the best habits an amateur astronomer can develop is keeping a logbook of everything he or she sees in the sky. Recording observations serves the dual purpose of both chronicling what you have seen as well as how you have developed as an observer. It's also a great way to relive past triumphs on cloudy nights.

Although you are free to develop your own system, I prefer to record each object on a separate sheet of paper, including a few descriptive notes and a drawing. See Figure 9.6 for an example of a generic observation form. Most of the entries should be self-explanatory. *Transparency* rates sky clarity on a scale of 0, indicating complete overcast, to 10, which is perfect. *Seeing* refers to the steadiness of the atmosphere, from 0 (rampant scintillation) to 10 (very steady, with no twinkling even at high power). Because both of these are subjective judgments, as previously noted, I like to include the magnitude of the faintest

Observation Record

Object: _____ Constellation: _____

Date: _____ Time: _____

Observing Site: _____

Sky Trans'y: _____ Naked-eye Limit: _____ Seeing: _____

Telescope: _____

Eyepiece: _____ Filter: _____

Notes: _____

_____

_____

_____

_____

Figure 9.6 *Suggested observation form for recording celestial sights.*

star in Ursa Minor visible to the naked eye. This helps put the other two values in perspective. (The only exception to this is if the sky near the object under observation is noticeably different than that near Ursa Minor. In this case, record the faintest star visible near the object itself.)

It is difficult, if not impossible, to convey the visual impact of subtle heavenly sights with words alone, which is why including a drawing with all written observations is so important. The drawing should convey the perceived image as accurately as possible. Right now, some readers might be thinking, "I can't even draw a straight line." That's all right, because few objects in space are straight. Actually, sketching celestial objects is not as difficult as might be thought initially. It just takes a little practice.

Although astrophotographers require elaborate, expensive equipment to practice their trade, the astro-artist can enjoy his or her craft with minimal apparati. Besides a telescope or tripod-mounted binoculars, all you need to begin are a clipboard, a pad of paper, and a few pencils. All of these supplies are available from almost any art or stationery supply store for about twenty dollars. Here are a few specific things to look for. The delicate textures of celestial bodies are best rendered using an artist's pencil with soft lead such as H or HB or with sketching charcoal. (Just be careful with the charcoal, as it tends to get away from those unfamiliar with it.) As for surface media, most astro-artists prefer smooth white paper as opposed to rag bond typing paper. The grain of rag paper tends to overwhelm the fine shading of most astronomical sketches. Most sketching pads have a fine surface grain ideal for the activity, but even computer printer paper will suffice. Lastly, the paper must be kept from blowing away while the artist moves between eyepiece and sketch. The simplest approach is a clipboard with a dim red flashlight clipped on or otherwise held in place.

Sketching a sky object is a multistep process, one that requires patience and close attention to detail. First, examine the target at a wide range of magnifications. Select the one that gives the best overall view as the basis for the sketch. Begin the drawing by lightly marking the positions of the brighter field stars surrounding the target object. Next, go back to each star and change its appearance to match its perceived brightness. Convention has it that brighter stars are drawn larger than fainter ones. Once the field is drawn accurately, lightly depict the location and shape of the target itself. The last step is to shade in the target to match its visual impression. Using your finger, a smudging tool, or a soft eraser to smudge the lead, recreate the delicate shadows and brighter regions. Remember, the drawing is, in effect, a negative image of the object. As such, the brightest areas should appear the darkest on the sketch, and vice versa. Finally, examine the target again with several different eyepieces, penciling in any detail that previously went unseen.

Here are some final tips for beginning astro-artists. First, never try to hurry a drawing, even if it is bitter cold and your hands are turning blue. Better to put the pencil down, take a five- or ten-minute warm-up break, and then go back. Second, do not try to create a Rembrandt at the telescope. Instead, make only a rough (but accurate) sketch at the telescope itself; the final drawing may be made later indoors at your leisure. Last, avoid the urge to add a little stylistic license, such as putting spikes around the brighter stars or drawing in detail that isn't quite visible but that you think is there from looking at photographs. Remember, a good astro-artist is an impartial reporter.

Once filled out, the observation record may be filed in a large loose-leaf notebook. It is handiest to separate observations by category. Individual headings might include the Moon, planets, variable stars, deep-sky objects, and so on. Interstellar objects may be further broken down first by type and then by increasing right ascension beginning at 0 hours. Members of the Solar System might be separated by object and then filed chronologically.

The eye of the experienced observer can detect much more tenuous detail in sky objects than a beginner can spot. Does this mean that the veteran has better eyesight? Probably not. Like most things in life, talent is not inborn. It has to be nurtured and developed with time. That's why it is important to keep notes. You will be amazed at how far your observing skills have come when you look back at your early entries a few years later.

Observing versus Peeking (A Commentary)

Are you an astronomical observer or an astronomical peeker? There is a difference . . . a big one! A peeker flits from one object to the next, barely looking at each before . . . *swish* . . . it's off to another. He or she never writes down what was seen, let alone makes a simple drawing. Whenever asked what he or she saw during an observing session, the peeker will only say, "Oh, I don't know, just some stuff." The fact of the matter is that peekers usually cannot remember what they have seen and what they have not.

An observer takes a slow, methodical approach to the study of the night sky. Most compile long lists of objects they want to see before going outside in an effort to use each moment under the stars as effectively as possible. Unlike our friend the peeker, the true observer is not out to break any land-speed records. He or she prefers to take it a little more slowly, savoring each photon that reaches the eye.

Perhaps it is a sign of the hectic times in which we live, but more and more amateur astronomers seem to be peekers. If you are one of them, I have to ask you this: What's your rush? Take a deep breath and relax. Resist the urge to race impulsively across the sky. By taking a slower, more deliberate tour, you will see the heavens in a new and exciting light. Become an astronomical observer, a connoisseur of the universe.

10

It's Time to Solo!

Although this book is intended as a guide to equipment and not as a guide to observing the sky, I would be remiss not to devote some space to what can be seen with the telescopes, binoculars, and accessories of which I speak. With that in mind, this chapter will take you on a quick trip across the heavens. If this is your maiden voyage, you are about to witness sights that few people are aware even exist. Many will defy description. If you are a seasoned traveler, then you know what I mean. Regardless of how many years you look toward the sky, there will always be something new and wondrous to observe.

The Solar System
The Moon

The Moon is a fascinating place to visit through any telescope, regardless of aperture, magnification, or degree of sophistication. Even the smallest instruments (yes, including binoculars) will display the stark lunar terrain in all its "magnificent desolation," as Apollo 11's Edwin Aldrin put it. The surface of the Moon today appears much as it did in the past. Rugged mountain ranges, expansive flat plains, deep valleys, and innumerable craters all dating back to the beginnings of our Solar System await the amateur explorer.

Each lunar phase holds something exciting for the sightseeing astronomer. Figures 10.1 and 10.2 show the Moon at its first-quarter and last-quarter phases, respectively. To help guide your way across the barren lunar surface, each identifies several of the most interesting and spectacular features the Moon has to offer. The crescent phases after New Moon through first quarter

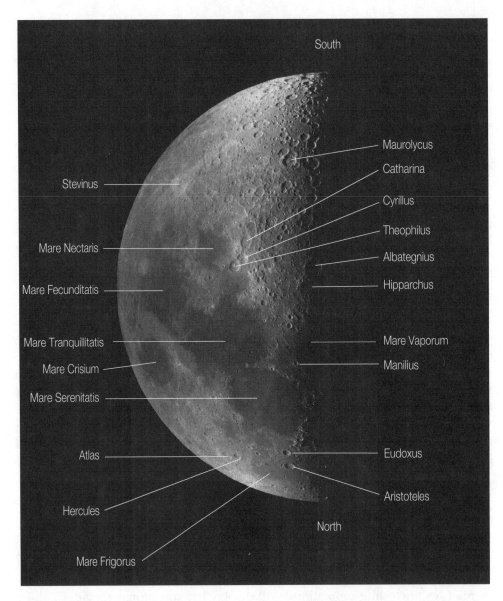

South

Maurolycus

Catharina

Stevinus

Cyrillus

Theophilus

Mare Nectaris

Albategnius

Hipparchus

Mare Fecunditatis

Mare Tranquillitatis

Mare Vaporum

Mare Crisium

Manilius

Mare Serenitatis

Atlas

Eudoxus

Hercules

Aristoteles

North

Mare Frigorus

Figure 10.1   *First-quarter Moon. Photograph by Richard Sanderson (6-inch f/12 Astrophysics refractor, T-Max 100 film, 1/30 second at f/12). South is up.*

display a tremendous variety of lunar terrain. Dominating the equatorial zone are the vast expanses of the lunar seas: (or *maria;* the singular form is *mare*) Mare Crisium, Mare Fecunditatis, Mare Tranquilitatis, and Mare Serenitatis. To their north are many scattered large craters, while the south polar region is awe-inspiring in its coarse beauty. Of special interest are the craters Clavius and Tycho. Both premier the night after first quarter.

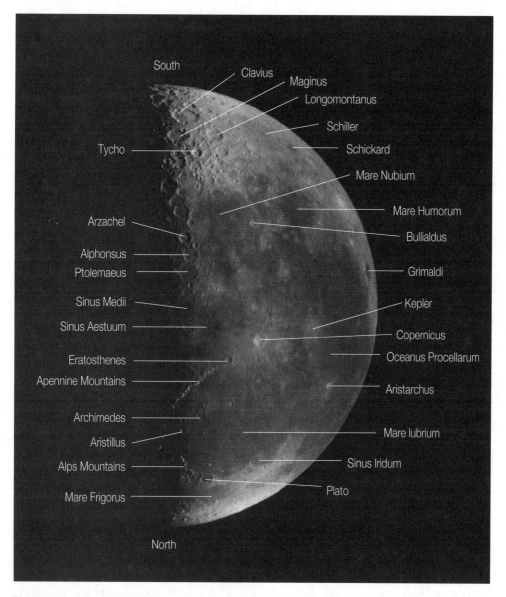

Figure 10.2 *Last-quarter Moon. Photograph by Richard Sanderson (6-inch f/12 Astrophysics rafractor, T-Max 100 film, 1/30 second at f/12). South is up.*

After first quarter, sunlight slowly pours into Oceanus Procellarum, the Ocean of Storms, which is the largest of all lunar maria. The Ocean of Storms holds many wondrous sights, including the craters Copernicus and Kepler. To the north, the crater remnant Sinus Iridum (the Bay of Rainbows) and the unusual dark-floored crater Plato receive first light. To the south, Tycho seems ablaze with fire as its magnificent system of bright rays scatters nearly to the Moon's equator. Once Full Moon passes, the sequence reverses, with shadowing and lighting effects adding a different perspective to our neighbor.

Many amateurs, especially those with larger instruments, shy away from looking at the Moon because of its often-overwhelming glare. But this is no excuse, because the Moon's brightness can be easily controlled. For about fifteen dollars, a Moon filter will reduce the intensity to a comfortable level. Though there can be benefits to viewing the Moon through color filters (green is especially useful for increasing subtle contrasts), most people prefer either neutral-density or variable-polarizing filters to maintain proper color balance. Another way to cut down on the glare is to stop down your telescope with a mask. Stopping down a telescope to about three inches will make the moonlight more manageable, although it will bring a loss of detail at high power.

For example, a 4-inch f/7 refractor (focal length = 28 inches) can be turned into an f/10 instrument by reducing its aperture to 2.8 inches. Simply cut a circle 2.8 inches across in the cardboard and secure the mask centrally in front of the objective lens.

A telescope with a central obstruction (that is, a secondary mirror) requires a slightly different approach. Measure the distance between the outer edge of that obstruction and the inside edge of the telescope tube. Cut a circle of this diameter in the cardboard. Make sure to offset the opening in the mask from the telescope's optical axis to ensure that no part of the secondary-mirror support blocks any portion of the view when the mask is attached to the telescope. For instance, a 14-inch f/5 Newtonian (focal length = 70 inches) requires an offset 7-inch mask to change the focal ratio to f/10. But because the distance between the diagonal mirror and the tube will be less than this, a smaller circle must be cut in the cardboard accordingly.

There is no one magnification that is best for viewing the Moon. Instead, try experimenting. Scan the surface with a low power to take the whole thing in. Then, when you spot an interesting area, switch to a medium or high power for detailed study. Under steady seeing conditions, it is often possible to exceed the 60×-per-inch rule when studying the Moon. (This is best attempted with a clock-driven telescope; viewing at extremely high power without one is like trying to play tag with a speeding train! Use the lunar drive rate, if your telescope is so equipped.)

Often as it orbits Earth, the Moon will pass in front of, or *occult*, stars and, on infrequent occasions, even a planet. Watching as the Moon slowly crosses in front of another body can be a lot of fun. For advanced amateurs, precisely timing the occurrence of an occultation is also a useful scientific activity. By knowing the exact moment of disappearance (called *ingress*) or reappearance (*egress*) of a star or planet, scientists may detect slight fluctuations in the Moon's orbital distance and speed. It is also possible to determine if the occulted star is single or an unresolved double.

Occultation predictions are computed and published annually by the International Lunar Occultation Center (ILOC) in Japan. ILOC predictions for the occultations of brighter stars are also available in the annual Royal Astronomical Society of Canada's *Observer's Handbook* as well as in both *Sky & Telescope* and *Astronomy* magazines.

Also listed in the same sources are predictions for total, partial, and penumbral lunar eclipses. A lunar eclipse occurs when the Full Moon passes through Earth's shadow. Because the Moon's orbit about Earth is not quite parallel to Earth's path around the Sun but rather tilted 5°, eclipses do not occur every Full Moon. Instead, the Moon misses Earth's shadow, passing either above or below. No more than three times a year, the Moon will cross Earth's orbit at Full Moon (at a point called a *node*), causing a lunar eclipse.

The Planets

Each of the nine planets that circle our Sun has a unique appearance through amateur telescopes. Saturn has its rings, Jupiter has its belted atmosphere and four Galilean moons, Mars displays its polar caps and mottled orange surface, and Venus drifts through different Moonlike phases. Each of these traits, and many more, can be seen through even modest telescopes.

Because detail on the planets is usually quite subtle, observers want as much contrast as possible. Three factors affect the visibility and clarity of planetary features: optical quality, telescope design, and seeing conditions. All other things being equal, the degree of contrast visible depends on the size of a telescope's central obstruction (for example, from a secondary mirror), if any—the smaller the central obstruction, the greater the contrast. Refractors are ideal because they do not have any central obstruction; that is, they have a clear aperture. Long-focus Newtonian reflectors have a very small central obstruction (the longer the focal length, the smaller the secondary mirror) and, therefore, good contrast. Short-focal-length Newtonians, Cassegrains, and Schmidt-Cassegrains have large central obstructions, ergo lower contrast for a given aperture.

If seeing conditions (or the optics) are poor, try stopping down its aperture to about three or four inches by using a cardboard mask. This is, at best, a compromise, for as aperture is decreased, so is resolution. With this in mind, many amateurs prefer using color filters to heighten contrast between different-colored areas on planets.

From the discussion in the previous chapter, you'll recall that *seeing* refers to the steadiness of the night sky. A turbulent night (that is, a night when the stars appear to twinkle with great fervor) is a poor night for planet-watching. For the best views, choose instead a night when the stars shine steadily (often when a slight haze or cloudiness is present). Also, the brightness of the naked-eye planets may dazzle the eye when viewed through larger instruments. Filters can help lessen the glare. To learn more about the art of observing the other members of our Solar System, consider joining the Association of Lunar and Planetary Observers (P.O. Box 16131, San Francisco, CA 94116).

Appendix F lists the locations for the four easily seen naked-eye planets—Venus, Mars, Jupiter, and Saturn—through the year 2004. For further details, as well as the locations of Mercury, Uranus, Neptune, and Pluto, consult one of the monthly astronomy magazines listed in chapter 6.

Mercury. Timing is everything when it comes to finding Mercury. Because of its proximity to the Sun, it is seen only near the horizon in very heavy twilight, no more than an hour after sunset or before sunrise. Therefore, you must have a clear view in the direction of sunset or sunrise, free of any low-lying fog, haze, or clouds. Good places for Mercury-spotting include a beach; a large, flat field; or the top of a hill.

Certain times of the year are better for spotting Mercury than others. The best chance for observers in the Northern Hemisphere to see Mercury in the evening is when the planet is at greatest eastern elongation in the spring or greatest western elongation in the early predawn morning hours of autumn. It is during these seasons that the ecliptic makes its largest angle with the horizon, carrying Mercury higher above the horizon than at any other time of the year.

To find Mercury, scan the horizon slowly and carefully near where the Sun just set or is about to rise for a bright starlike object. Once spotted, Mercury will show only a tiny disk. The only distinctive characteristic of Mercury is that because it is located between Earth and Sun, it goes through phases like the Moon. Trying to determine the phase can be difficult because of glare from the bright surrounding sky. To help improve contrast and reduce the effect of atmospheric turbulence, try using a Wratten #21 (orange), #23A (red), or #25 (deep red) color filter. A light blue Wratten #80A filter may help bring out some hazy surface mottlings, as may a polarizing filter.

Though Mercury will usually pass above or below the Sun as seen from Earth, every now and then it will appear to slide directly across the solar disk. Such an event is called a *transit*. By using at least 50× and following proper safety precautions (outlined later in this chapter), astronomers can watch as tiny Mercury slowly moves across the Sun's disk.

The last transit of Mercury took place in November 1993, although it was not visible from North America. The next transits are set for 1999 November 15; 2003 May 6–7; and 2006 November 8. The 1999 transit will be visible only from Antarctica and southernmost Australia. The 2003 event will be visible from Europe, Asia, Africa, and Australia, whereas the 2006 transit will be seen from North and South America and the Pacific. As the dates draw closer, consult the astronomical periodicals and annuals for exact times and circumstances of the transits.

Venus. Like Mercury, Venus's orbit lies inside of Earth's, therefore holding it hostage to either the western sky after sunset or the eastern sky before sunrise. However, although Venus is visible for no more than about three hours after sunset or before sunrise, the planet's outstanding brightness makes it an outstanding sight. At times reaching a brilliance greater than magnitude −4, Venus outshines all other objects in the sky except the Sun and Moon. For all its eminence, however, Venus hides its secrets well. The planet's atmosphere proves impenetrable even with the largest telescopes on Earth. Although its surface may be hidden from view, Venus can be observed going through a series of phases similar to the Moon's with even the smallest telescopes. Watching the planet's phases over the course of time can be great fun to track and record.

By knowing the planet's current phase, it is also possible to know where Venus is in its orbit. Venus displays a large, thin crescent phase (Figure 10.3) when it is closest to Earth and about to pass between us and the Sun (a point in its orbit called *inferior conjunction*). When Venus is farther from Earth, its globe will look noticeably smaller, even though more of its disk will appear lit by the Sun. At these times, Venus looks more like a quarter or gibbous Moon. As it rounds the far side of the Sun (referred to as *superior conjunction*), nearly the entire lighted side of the planet will face Earth.

On rare occasions, Venus may be seen to transit the face of the sun at inferior conjunction. Transits of Venus occur in pairs separated by eight years, with over one hundred years between successive pairs. The last transits of Venus were seen in 1874 and 1882. The next will take place on 2004 June 8 and 2012 June 6. At least part of the 2004 event will be visible from Europe, Africa, Asia, and Australia. The 2012 transit will be seen from western North America, the Pacific region, eastern Asia and Australia. The Midnight Sun will allow the few astronomers north of about latitude 70° (inside the Arctic Circle and parallel to northern Alaska) to see both events in their entirety. Once again, check the periodicals and annuals for exact times.

Figure 10.3 *Crescent Venus. Photograph taken at high noon (!) by George Viscome (12.5-inch f/10 Cassegrain, Tech Pan 2415 film, with #25A [red] filter).*

Mars. No planet has quite the allure for the planetary observer as does Mars (Figure 10.4). Although the canals and tales of Martians of a century ago are gone, something still fascinates us about this distant world.

Part of that fascination might be the fact that Mars is well-placed for observation only once every 26 months, when our two planets are closest. This is referred to as *opposition*. At opposition, Mars lies directly opposite the Sun in our sky, rising in the East at sunset and remaining visible all night long.

Some oppositions are better for observing Mars than others. At a poor opposition, Mars will grow to only about 14 seconds of arc (abbreviated 14″) across. Even at its best, Mars will never appear larger than 25 arc-seconds (25″). This is slightly larger than a ringless Saturn but much smaller than either Venus or Jupiter. Therefore, at least a high-powered 3-inch refractor or 6-inch reflector and steady seeing are required to see any of Mars's famed features, such as the polar caps or larger dark markings. Be sure to use your best-quality eyepieces when searching for elusive Martian detail.

Mars also benefits greatly from the use of color filters. Orange (Wratten #21) and red (#23A or #25) filters increase the contrast of the dark surface markings against the bright orange desert regions. Blue (#38A or #80A) and green (#58) filters are best for showing the polar caps as well as any haze, fog, or clouds in the planet's thin atmosphere.

Regardless of telescope, eyepiece, or filters, Mars is not an easy planet to study. Even with the planet near opposition, the best instrumentation will reveal little if the observer doesn't possess two ingredients that money can't buy: determination and patience. Long periods may lapse when Mars appears as little more than a quivering mass swimming in the turbulence of our atmo-

Figure 10.4 *Mars. Photograph by Gregory Terrance (16-inch f/5 Newtonian reflector, Lynxx PC CCD camera, exposure 1 second at f/45).*

sphere. But on the rare occasion of atmospheric tranquility, the real Mars will come through to reward the observer.

From Earth, the most prominent features of Mars are its two polar caps. During an opposition, small telescopes reveal the caps as small but unmistakably white. During the oppositions when Mars comes comparatively close to Earth (next occurring in 2003); the southern cap will be tilted toward Earth. At distant oppositions (such as those in the mid- and late-1990s), the northern cap is exposed toward Earth.

Scattered across the Martian surface are *albedo features*, the famed dark markings on Mars. Although most of the Martian surface appears bright orange, these regions appear noticeably darker by contrast. Albedo features are permanent markings, but they do appear to change slightly in size and shape. This metamorphosis is believed to be caused by the swirling Martian winds. Although the planet's atmosphere is much thinner than Earth's, its winds will occasionally reach hurricanelike force, shifting sands that alternately cover and uncover different portions of the rocky surface.

The most easily spotted albedo feature on Mars is Syrtis Major, a triangular wedge extending from north to south across the planet's equator. Almost equally dark but quite a bit smaller is Meridiani Sinus, a dark patch along the equator about 90° to the west of Syrtis Major.

Jupiter. No doubt about it, Jupiter is one of the most impressive members of the Solar System to observe. What makes it so special? For openers, Jupiter is one of the brightest objects in the sky, surpassed only by the Sun, Moon, Venus, and rarely Mars. Jupiter is also huge: 88,000 miles in diameter at the equator. Even at more than three hundred and ninety-one million miles from Earth, this translates to an average apparent equatorial diameter of 47 arc-seconds, larger than any other planet except Venus when it is near inferior conjunction. All this adds up to a spectacular view.

Even with the smallest astronomical telescopes, observers can spot the distinctive equatorial bulge of Jupiter. Although Jupiter measures about 47 arc-seconds, across its equator, it spans only about 44 arc-seconds from pole to pole. This *polar flattening* is the result of Jupiter's rapid rotation—a day on Jupiter is approximately 9 hours and 55 minutes long.

Telescopes both large and small can also divide the planet's impenetrable clouds into bright zones and dark belts, as seen in Figure 10.5. Especially prominent are Jupiter's broad, bright Equatorial Zone and the North and South Equatorial Belts. Larger telescopes and medium to high powers will show swirls, ovals, and festoons within the turbulent atmosphere. The Great Red Spot, a huge oval cyclonic storm found in Jupiter's South Tropical Zone, may also be seen. Don't expect to see the Red Spot immediately. There's only a 50–50 chance of it being on the Earthward side of Jupiter in the first place, and even then its coloration will be very subtle. Rarely a bright red, the Red Spot varies in color from a pale pink or orange to very pale tan-white.

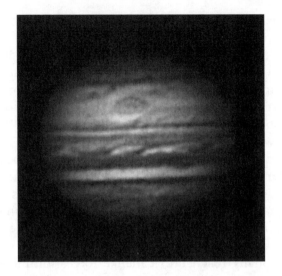

Figure 10.5 *Jupiter. Photograph by Gregory Terrance (16-inch f/5 Newtonian reflector, Lynxx PC CCD camera, exposure 1/3 second at f/30).*

Filters can help highlight details in the Jovian cloud bands. Try using a yellow-green (Wratten #11) or orange (#21) filter to pick out subtle features in the darker belts. A green (#56 or #58) or blue (#38A or #80A) filter helps accentuate the Red Spot's visibility, although it still may be difficult to pick out.

Jupiter owns at least 16 moons by modern reckoning, although only the 4 brightest are visible in most amateur telescopes. These are, moving in order from the planet outward, Io, Europa, Ganymede, and Callisto. All were discovered by Galileo when he turned his first crude telescope toward Jupiter, and so today they are called the *Galilean satellites.* All slowly but constantly change their positions relative to the planet and each other. Io moves the quickest, circling Jupiter in a little less than two days, while distant Callisto takes about two weeks. Watch and record their changing positions and see if you can tell which moon is which.

In addition to appearing dutifully by the side of Jupiter, the Jovian satellites also perform great sleight-of-hand tricks with their master world. Frequently, one or more of the satellites will be eclipsed or occulted behind Jupiter, while at other times they will transit in front of the planet.

Transits are two-part events. As the satellite crosses in front of Jupiter, its tiny, round shadow may also be seen projected onto the top of the Jovian atmosphere. The entire affair may last several hours.

The location of the shadow relative to the satellite depends on where Jupiter is in our sky relative to the Sun. Before Jupiter reaches opposition, the satellite's shadow precedes the satellite itself, allowing both to be seen prior to the transit's beginning. After opposition, the reverse is true, with the shadow now following the satellite. At opposition (and for several nights either side) the shadow is cast almost exactly behind the satellite.

Although the shadow of a satellite is easy to spot during a transit, seeing the satellite itself is not. Usually, the tiny whitish disk of the satellite is overpowered by the brighter clouds of Jupiter, rendering it invisible. Your best chance at spying the satellite in front of Jupiter will come by monitoring the scene before the event starts and then tracking the satellite as it begins its trans-Jovian trek.

An eclipse and occultation both occur every time a Jovian satellite hides behind Jupiter. The only difference between these two events is in what's doing the hiding. When the satellite moves into Jupiter's shadow, it is said to be in eclipse, while an occultation takes place when a satellite disappears behind Jupiter itself.

Saturn. Who can forget the rush of excitement at his or her first view of the planet Saturn? There, before the observer's eyes, stands a golden globe encircled by a bright, white ring. It's a heavenly sight surpassed by few others.

Saturn (Figure 10.6) is surrounded by a thick atmosphere that is similar in nature and composition to that of Jupiter. However, Jupiter's atmosphere is rich in detail; Saturn's is usually dull and nearly featureless. Most outstanding is the planet's whitish equatorial region and bordering northern and southern temperate zones. Try a deep yellow (Wratten #15) or orange (#21) to highlight some of the subtle features in the Saturnian atmosphere.

Every now and then, Saturn will surprise observers, such as back in September 1990. A bright white region grew to encircle fully one-quarter of the planet's equatorial zone. The so-called white spot of Saturn then slowly faded from view, disappearing almost two months after it first formed. Similar spots occurred in 1903, 1933, and 1960, approximately the same span of time it takes Saturn to complete one 29.5-year orbit of the Sun.

Figure 10.6   *Saturn. Photograph by Gregory Terrance (16-inch f/5 Newtonian reflector, Lynxx PC CCD camera, exposure 1 second at f/30).*

At first glance, Saturn might appear encircled by a single, solid ring. Closer inspection, however, shows that the ring is actually divided into separate zones in much the same way as the selections on a phonograph record (remember them?) are divided by blank spaces. In the case of Saturn's rings, most amateur telescopes will reveal two major divisions. The outermost part of the ring is called the *A ring* and appears a bright gray. The broader, inner ring is called the *B ring* and is colored pure white. Separating the two is a pencil-thin dark gap called *Cassini's Division*. Under exceptional seeing conditions, all three may be distinguished through telescopes as small as 2 inches in aperture. Larger instruments will also show a fourth component to the ring system. The *C* (or *Crepe*) *ring* lies inside the B ring and appears dark gray. Due to its darkness, the C ring is most easily seen against the planet's bright disk. Trained observers can see great detail in the rings through large telescopes in perfect seeing, whereas the eyes of passing spacecraft Voyagers 1 and 2 back in the 1980s revealed hundreds of intricate subdivisions within the Saturnian family of rings.

The amount of detail visible in the rings depends strongly on their tilt to our line of sight. Because the rings are tilted approximately 27° to the ecliptic, they present different faces at different points in Saturn's 29-year solar orbit. For approximately 13 years we see one side of the rings, and then for the next 16 we see the other. In between, the rings are oriented edge-on, and they disappear from view. The rings of Saturn last appeared edge-on in 1996, while maximum presentation of their southern face will occur in 2003. After that, the rings will again appear edge-on in 2010, with their northern face seen for the next several years thereafter.

Saturn has at least 18 satellites orbiting it. The brightest is named Titan and shines at 8th magnitude. With a sharp eye and a little luck, an observer using just a pair of 7× binoculars can spot Titan. Small telescopes will show Titan easily as it circles Saturn once every 16 days. At least five others—Enceladus, Tethys, Dione, Rhea, and Iapetus—are within range of 8- to 10-inch telescopes, although they may prove difficult to tell apart from faint background stars. Only when the Saturnian system is tilted edge-on to our view will the satellites perform disappearing acts, such as occultations, eclipses, and transits, like Jupiter's.

Uranus, Neptune, and Pluto. Uranus shines between magnitude 5.5 and 6, making it barely perceptible with the unaided eye under superb skies. From most of our observing sites, however, it will probably require a telescope or binoculars to be seen. Through small instruments, it looks like an ordinary star, standing out from the crowd only by its barest hint of pale green. A 6- or 8-inch telescope will display a tiny, featureless disk, while 10-inch and larger instruments can also reveal the brightest of its 15 satellites.

Neptune's challenge to amateur astronomers is simply in finding it. Again, its pale but peculiar blue-green color helps it to stand out from the crowd of background stars. Measuring a few seconds of arc in diameter, Neptune appears so small that it remains a point when viewed with less than about 100×. Higher magnifications will show its tiny disk, though there is little hope of seeing any detail in its cloudy atmosphere.

We finally arrive at distant, icy Pluto. Even when it lies at its minimum distance of 2.76 billion miles from the Sun (which last occurred in 1989), Pluto is still an extremely faint object. Appearing no brighter than 13th magnitude, it usually requires at least an 8-inch telescope to be seen (although Pluto has been spotted in telescopes as small as 6-inchers under exceptional conditions). The only way to tell Pluto apart from the myriad of other faint stars is by using a detailed finder chart. Once the chart's field is within the telescope's view, look very carefully for a faint star-like object. That will be Pluto. Plot its position on the chart and then revisit the spot over several nights to confirm your sighting by watching the planet creep slowly against the background stars.

Annual finder charts for Uranus, Neptune, and Pluto are published in the January issues of *Sky & Telescope* and *Astronomy* magazines. Maps are also found in the Royal Astronomical Society of Canada's annual *Observer's Handbook* and Guy Ottewell's yearly *Astronomical Calendar*. These star maps are absolutely essential when looking for these elusive planets.

Asteroids. Lying between the orbits of Mars and Jupiter are thousands of small, irregularly shaped chunks of rocky debris. These are the asteroids or minor planets, leftovers from the formation of the Solar System over four and a half billion years ago. Many asteroids are bright enough to be seen through amateur telescopes, though not even the largest, 600-mile-wide Ceres, will show up as anything more than a point of light.

Whenever a minor planet of about 8th magnitude or brighter is due to be visible, astronomical periodicals will publish a finder chart with its path plotted against the stars. To find the asteroid, employ the same technique used for locating Pluto as outlined above. Once your telescope or binoculars is centered on the area of sky where the asteroid is located, glance back and forth between the eyepiece field and the asteroid's finder chart. The asteroid will appear as a point of light near the predicted position on the chart.

If asteroiding interests you, then consider subscribing to the *Minor Planet Observer,* a monthly periodical written and published by Brian Warner. Each issue of the *MPO* features several asteroids that will be visible in the evening sky for the month of publication. To inquire about subscriptions, write BDW Publishing, Box 818, Florissant, CO 80816.

Comets. These are the nomads of our Solar System. Traveling in highly elliptical orbits, these frozen chunks of dirty ice arrive from beyond the orbit of Neptune, swing close by the Sun, and then return to the farthest reaches of the Solar System from where they came. As a comet swings by the Sun, solar energy sublimates frozen gases and dust on its surface to create the comet's *coma,* or head. Radiation pressure from the Sun and the solar wind combine to press against the coma's gas and dust, thus creating the comet's long tail.

Each year, a dozen or more comets either are discovered (about half by amateur astronomers) or reappear on schedule. Most are quite faint, but several may eventually become bright enough to be seen in amateur telescopes and binoculars. Each appears as a round, amorphous glow (the coma) highlighted by a brighter center. Careful telescopic study may reveal intricate filamentary structures in the coma, the result of gaseous jets erupting from the comet's solid nucleus.

Brighter comets (that is, those that become visible to the naked eye) may also display ghostly tails extending behind the coma and opposite the Sun. The brightest comets have dual tails (one called the *ion tail,* the other called the *dust tail*) that can extend 30° or more in length. These, the so-called great comets, are among the most spectacular sights ever to grace the night sky. The most recent great comets were Comet Hyakutake in 1996 and Comet Hale-Bopp in 1997 (and in Figure 10.7). Unfortunately, astronomers cannot predict when the next great comet will appear in our sky. Maybe one will be discovered tonight, or maybe not for another 50 years or more!

It is always fascinating to observe and record how a comet changes in brightness and appearance as it moves across the sky and around the Sun. Use a low-power, wide-field eyepiece to view the comet at first and then increase magnifica-

Figure 10.7 *Comet Hale-Bopp. (Photograph by author using a 400-mm lens at f/6.3, Fujichrome 400 slide film, and a 10-minute exposure).*

tion to check areas of interest. Be especially watchful of any fine detail in the comet's coma. Note its overall magnitude by comparing the coma's in-focus image to the out-of-focus glows of nearby stars of known brightness. A star atlas such as the *AAVSO Star Atlas* will be a great help in locating stars for comparison.

Next, estimate the apparent diameter of the coma by comparing its size to the true field of the eyepiece. Also check for any coloration in the coma or tail. Brighter comets frequently exhibit a bluish tinge. Lastly, estimate the length and shape of the comet's tail. Try using a Wratten #80A (light blue) filter to increase the contrast of the comet's tail against the background sky.

The most readily available source of information on comets that are currently visible may be found on the Internet at the Jet Propulsion Laboratory's Comet Observation Home Page, http://encke.jpl.nasa.gov/. If you would prefer receiving written notices of all new comet discoveries and recoveries, consider subscribing to the International Astronomical Union Circulars, produced by the Central Bureau for Astronomical Telegrams. For information, write to the CBAT, Mail Stop 18, Smithsonian Astrophysical Observatory, 60 Garden Street, Cambridge, MA 02138. Finally, facts about brighter comets may be heard on a recorded telephone message called "Skyline," sponsored by *Sky & Telescope* magazine. Skyline is updated every Friday with news of the latest astronomical discoveries and events and may be heard by calling (617) 497-4168 (toll rates to Cambridge, Massachusetts, apply).

The Sun. Before discussing *what* to look for, it is critically important to emphasize *how* to look at the Sun. Quite simply: **ALWAYS PRACTICE SAFE SUN!** The Sun's ultraviolet rays, the same rays that cause sunburn, will burn the retinas of your eyes almost instantaneously when concentrated by a telescope or binoculars. Moments later, the retina will be further cooked by the focused heat of visible and infrared light. Without taking proper safety precautions, permanent eye damage, even blindness, will result.

Happily, there are simple methods of looking at the Sun in complete safety. The safest is to use your telescope as a projector to cast an image of the Sun onto a white projection screen. The biggest drawback to solar projection is that without proper baffling the Sun's image will be washed out by the bright surroundings. To overcome this, many dedicated solar amateurs have constructed projection boxes that attach to their instruments.

If you wish to view the sun directly through a telescope or binoculars, special solar filters are required. DO NOT use photographic neutral-density filters, smoked glass, or overexposed photographic film, as suggested by some people. They can all lead tragically to blindness. The proper filters are commonly made of aluminized mylar film or glass coated with a nickel-chromium alloy and were discussed in detail in chapter 6. Regardless of type, the filter must be securely mounted IN FRONT OF the telescope or binoculars before looking through the eyepiece. This way, the dangerously intense light and heat from the Sun are reduced to safe levels before entering the optics and your eyes.

Never place a sun filter between the eyepiece and your eyes (even though some inexpensive telescopes come with such filters). Telescopes magnify not only the Sun's light but also its intense heat, which can quickly crack glass filters or burn through thin mylar film, allowing undiminished sunlight to burst into your unprotected eyes and cause permanent blindness.

Solar filters must be treated with great care, or they will quickly become damaged and unsafe to use. Regularly inspect the filters (especially the mylar type) for pinholes and irregularities in the coating by holding the filter up to a bright light. A small hole can be sealed with a tiny dot of flat black paint without causing the image to suffer. Dab just a bit of paint over the hole using a toothpick. If, however, more serious damage is detected, then the filters must be replaced immediately.

Whether you use a filter or projection, never look through the telescope's finderscope to aim the instrument at the Sun. (Be sure to keep the front dust cap on the finder; otherwise, the Sun's focused heat may melt the finder's crosshairs!) To aim at the Sun, take a look at the telescope's shadow cast on the ground. Move the telescope back and forth, up and down, until its shadow is at its shortest. The Sun should then be in the field of view.

One of the most interesting observations to make is to monitor the fluctuating number of sunspots across the Sun's visible surface, or *photosphere* (Figure 10.8). Sunspots appear at disturbed areas in the Sun's powerful, complicated magnetic field. Each consists of a black central portion, the *umbra*, and a surrounding grayish area called the *penumbra*. They may range in size from hundreds to thousands of miles in diameter.

Try to draw the Sun's disk every second or third day over a span of a month or more. With each successive sketch, sunspots will be seen to come and go, changing size and shape as they travel across the solar disk.

As described in chapter 6, hydrogen-alpha solar filters display the Sun at one precise wavelength—656 nanometers. Viewing at that wavelength reveals a turbulence within our star that goes unsuspected through white-light sun filters. Huge, flamelike solar prominences; filaments; and other fascinating sights, normally visible only during a total solar eclipse, can be watched and monitored daily with a hydrogen-alpha filter.

As often as three times a year, the New Moon will pass in front of part or all of the Sun, eclipsing it as seen from somewhere on the Earth's surface. The occurrence of a solar eclipse over a populated area always generates wide publicity in the news media and excites the curiosity of the general public.

Solar eclipses come in three varieties: partial, annular, and total. The same precautions must be exercised when viewing either a partial or annular (when a bright ring, or annulus, of sunlight remains even at maximum eclipse) solar eclipse as for viewing the uneclipsed Sun. Only when the Moon completely covers the Sun during the total phase of an eclipse is it safe to view without protection.

Figure 10.8 *The Sun, Photograph by Brian Kennedy (8-inch f/10 Celestron 8 SCT, f/5 telecompressor, T-Max 100 film, 1/250 second).*

Deep-Sky Objects

Beyond the edge of our Solar System lies the rest of the universe, a most fascinating place to visit. Out there lie buried treasures that our eyes alone cannot see. But with the aid of a telescope or binoculars, exciting sights unequalled by any here on Earth await the amateur astronomer. Thousands upon thousands of stars, either single, multiple, or set in vast clusters, dot the night sky. Huge clouds of gas and dust—the nebulae—signal where new stars are being born or mark the graves of stars that once shone mightily. All of these lie within our galaxy, the Milky Way. Beyond it lie thousands of other galaxies, each a separate island in the endless ocean of the universe.

Be forewarned that most deep-sky objects appear disappointing at first. After being inundated with magnificent photographs of spectacular star clusters, colorful nebulae, and swirling galaxies, you get a telescope, aim it toward one of these distant masterpieces, and look through the eyepiece. What do you see? No color, no swirls, no magnificence; only a faint, gray smudge. Sad but true, the human eye cannot possibly perceive the level of detail or color that can be recorded on photographic film.

Disappointed? You'll probably be a little let down at first, but think about it. Here you are looking at sights hundreds, thousands, even millions of light years away with your own telescope. Also consider *what* you are looking at. As your friends watch reruns on television, you are seeing things that far less than 1% of humanity has ever seen before! There's a certain satisfaction in knowing that. Without even being aware of it, you might be looking toward unknown worlds scattered throughout gigantic systems of stars. Perhaps, on one of those distant worlds, another creature is looking through his or her (or its) equivalent of a telescope toward us, pondering the universe and your existence as well! Astronomy, it turns out, can be very cerebral. That can be its greatest attraction of all.

Here is an inventory of some of the wonders of the universe that await your inquisitive eye.

Double and Multiple Stars

Nearly half of the stars we see at night are double or multiple stars. Though not all doubles can be resolved visually through telescopes, of those that can, no two pairs appear exactly the same. Some are separated by great distances and others seem to touch each other. Of course, there are also the impostors—two stars that just happen to lie along the same line of sight from Earth. These are referred to as *optical doubles*.

Many double stars shine pure white, while others flicker with distinctive colors. Most appear in pairs, though in some cases as many as six or seven stars belong to a single system.

Variable Stars

Most stars shine consistently at the same brightness, day after day, year after year. Others, called *variable stars,* appear to fluctuate in brightness. Some variables oscillate rhythmically over a predictable period of time; others burst unpredictably, flaring up to six magnitudes in brilliance in a matter of minutes or hours.

There are three major classes of variables: *eclipsing binaries, pulsating stars,* and *eruptive variables.* Eclipsing binaries are pairs of stars that are strictly by chance, seen nearly edge-on from our earthbound vantage point. The stars alternately pass in front of each other, causing temporary diminishings of their combined light. Pulsating stars actually expand and contract in size, causing them to alternately fade and brighten. The most common type of pulsating variables are the long-period variables. These can take weeks or months to complete brightness cycles. Eruptive variables change in brightness very quickly.

Examples of all of these are visible through amateur telescopes and binoculars. To learn more about observing variable stars and how you can make a valuable contribution to their study, contact the American Association of Variable Star Observers at 25 Birch Street, Cambridge, MA 02178.

Star Clusters

These come in two varieties. Randomly shaped swarms of mostly young blue and white stars are called *open* or *galactic clusters*. Each may contain from a dozen to several hundred individual points of light. Some are loose groupings with little or no apparent central concentration, whereas others are squashed into stellar traffic jams. Most open clusters are found within the spiral arms of our galaxy.

The second type of star clusters, called *globular clusters,* encircle the Milky Way like moths around a flame. Globulars are huge, spherical conglomerations of hundreds of thousands or even millions of stars. Unlike the young, brash stars in open clusters, the stars in globular clusters are generally among the oldest known. Even at the vast distances at which they lie, globulars appear surprisingly bright. Of the more than one hundred known within our galaxy, most are visible in 3-inch telescopes. Typically, at least a 6-inch telescope is needed to resolve some of the stellar constituents into separate points of light.

Nebulae

Diffuse nebulae are enormous clouds of dust and gas—primarily hydrogen—and are found chiefly in the spiral arms of the Milky Way and in other spiral galaxies. Diffuse nebulae that shine brightly (relatively speaking) are called *bright nebulae* and come in two different varieties. If the gases in a nebula are excited into luminescence by the energy from young, hot stars within, the nebula itself will glow, much like the gas in a neon sign. These are called *emission nebulae.* Other nebulae are illuminated only by reflecting light from the stars they surround and are therefore called *reflection nebulae.* Emission nebulae appear reddish in photographs, while reflection nebulae appear bluish.

Dark nebulae are similar in constitution to their brighter cousins above but do not emit or reflect light. Instead, they obscure light from all objects behind, appearing as starless holes silhouetted against a star-filled field.

Although diffuse nebulae mark stellar birth, *planetary nebulae* chronicle stellar death. Planetary nebulae are small, spherical shells of gas expelled by less massive stars during their death throes. These spheres are characterized by the glow of ionized oxygen, causing most to appear green or turquoise. The star that expelled the gaseous shell remains buried in the middle of the nebula and is referred to as the *central star.*

When the most massive stars go through their death throes, they go out in a big way! During its final moments, much of the star's mass is expelled in a tremendous supernova explosion. When the explosion clears, all that remains of the star is its dense core and an expanding cloud of gaseous debris called a *supernova remnant*.

Galaxies

All of the objects spoken of until now are within the confines of our galaxy, the Milky Way. Beyond lie other galaxies, each made up of tens of millions to hundreds of billions of individual stars.

Galaxies are divided into three major classifications. Those that resemble huge pinwheels, with two or more long, curved arms extending away from a central core, are called *spiral galaxies*. The Milky Way is considered to be a spiral galaxy. The second type, *elliptical galaxies,* show no hint of spiral structure. Instead, ellipticals appear as huge, oval spheres with no internal organization of any sort. Finally, if a galaxy has no distinctive shape, then it is categorized as an *irregular galaxy*. Some irregulars seem to exhibit hints of ill-defined spiral arms, although any evolutionary tie remains unknown.

A Celestial Inventory

To inventory the universe, astronomers have devised several different ways of labeling stars and other celestial objects. Only the brightest stars, such as Vega and Betelgeuse, bear names from antiquity that remain familiar today. While many others also have exotic names, few modern-day astronomers take the time to remember them.

Only a few short years before the invention of the telescope, astronomer Johannes Bayer created that era's most detailed atlas of the night sky. He chose to identify the brightest stars in each constellation by lowercase letters from the Greek alphabet. He usually labeled a constellation's brightest star *alpha* and then, working his way through the traditional constellation figure from head to toe, labeled succeeding stars *beta, gamma,* and so on. Once completed, he repeated the head-to-toe sequence for any fainter stars that remained, sometimes until all 24 letters of the Greek alphabet were used up. There are many exceptions to this pattern, but it holds true for the most part. Bayer's classification stuck and is still in use today. Table 10.1 lists the Greek alphabet by name and corresponding letter.

Therefore, the bright star Vega is called Alpha Lyrae, while nearby Albireo is Beta Cygni (*Lyrae* and *Cygni* being the genitive forms of the constellation names Lyra and Cygnus, respectively).

In order to extend the Greek alphabet system, British astronomer John Flamsteed assigned numbers to all stars of about fifth magnitude and brighter in each constellation. These Flamsteed numbers begin at 1 in each constellation and increase from west to east. Fainter stars have subsequently been cat-

Table 10.1 **The Greek Alphabet**

alpha	α
beta	β
gamma	γ
delta	δ
epsilon	ε
zeta	ζ
eta	η
theta	θ
iota	ι
kappa	κ
lambda	λ
mu	μ
nu	ν
xi	ξ
omicron	o
pi	π
rho	ρ
sigma	σ
tau	τ
upsilon	υ
phi	φ
chi	χ
psi	ψ
omega	ω

aloged in many lists such as the SAO (Smithsonian Astrophysical Observatory), HD (Henry Draper), and other directories.

Nonstellar deep-sky objects are also cataloged individually. Some of the more spectacular examples have unofficial nicknames (such as the Orion Nebula or the Andromeda Galaxy), but most do not; they do, however, have catalog numbers assigned to them. For instance, in this and other astronomy books, the Orion Nebula is referred to as M42, and the Andromeda Galaxy is listed as M31. These are the entries assigned to them in the famous Messier catalog of deep-sky objects. The Messier catalog, listing 109 of the finest nonstellar objects in the sky, was largely created in the eighteenth century by Charles Messier, a French comet hunter. (The list actually goes up to M110, but it is now generally agreed that M102 was a mistaken repeat observation of M101.)

Another important catalog of deep-sky objects is the New General Catalog, or NGC, compiled at the end of the nineteenth century by an astronomer named John Dreyer. The NGC lists more than seven thousand eight hundred clusters, nebulae, and galaxies, including all but a few of the Messier objects. For example, M42 is NGC 1976, while M31 is NGC 224. An extension of the NGC is the Index Catalog (IC).

Selected Deep-Sky Targets

Searching for and observing deep-sky objects has become one of the most popular pastimes for today's amateur astronomer. It's a great way to learn the night sky and hone your observational talents all at the same time.

The following is a brief selection of my favorite deep-sky objects. Each is accompanied by a description of its visual appearance through a variety of telescopes as well as a finder chart and a photograph. All (except open clusters) are also accompanied by a drawing made through either an 8-inch or 13.1-inch telescope, as noted. By relating the descriptions and drawings to the photos and then to what you see through your own telescope, subtle details will become more readily visible. Other data listed for each object include its constellation, epoch 2000 right-ascension (R.A.) and declination (Dec.) coordinates, visual magnitude, apparent diameter, and tips on how to find it. Refer to the all-sky chart (Figure 10.9) to find out which of these celestial masterpieces will be visible tonight.

These showpiece objects are just a beginning. Once you have conquered these, move on to appendix G, which lists the entire Messier catalog as well as the finest non-Messier objects. There is a lot out there to see.

Spring

M44   (NGC 2632)   Open Cluster in Cancer
R.A. 08^{h}40^m.1   Dec. +19°59′   Mag. = 3.1   Diameter = 95′
Where to look: Find the keystone-shaped body of Cancer the Crab about halfway between the twin stars Castor and Pollux of Gemini and Regulus in Leo. Under moderately dark skies, M44 is visible as a soft glow inside the Crab's body. (See Figure 10.10a.)
Description: Not all observers will immediately appreciate the grandeur of M44, the Beehive Cluster (Figure 10.10b). Due to its incredible 95′ span, M44 is difficult to fit into the field of view of ordinary eyepieces above 30× to 40× magnification. As a result, the clustering effect becomes lost, and the observer may feel less than impressed. However, through binoculars, finders, and rich-field instruments, the bees in this celestial hive come alive in sparkling style! Close to six dozen points of light are seen across a field littered by the combined glow of another 130 fainter stars.

M81   (NGC 3031)   Spiral Galaxy in Ursa Major
R.A. 09^{h}55^m.6   Dec. +69°04′   Mag. = 7.0   Diameter = 26′ × 14′
M82   (NGC 3034)   Irregular Galaxy in Ursa Major
R.A. 09^{h}55^m.8   Dec. +69°41′   Mag. = 8.4   Diameter = 11′ × 5′
Where to look: Draw a diagonal line between Phecda (Gamma Ursae Majoris, the southeast star in the Big Dipper's bowl) and Dubhe (Alpha Ursae Majoris, the northwest bowl star) and extend that line an equal distance toward the northwest. Just west of that point is an isosceles triangle made up of the 5th-magnitude stars Rho, Sigma[1] and Sigma[2] Ursae Majoris. Follow the

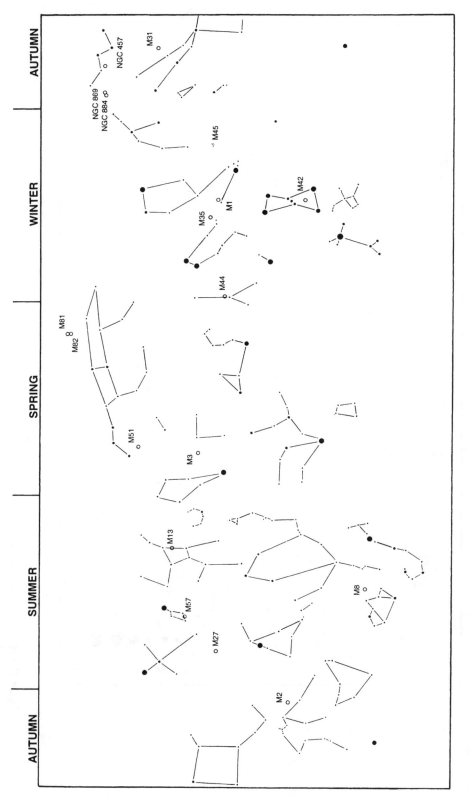

Figure 10.9 *Locations of the 16 selected deep-sky objects to come.*

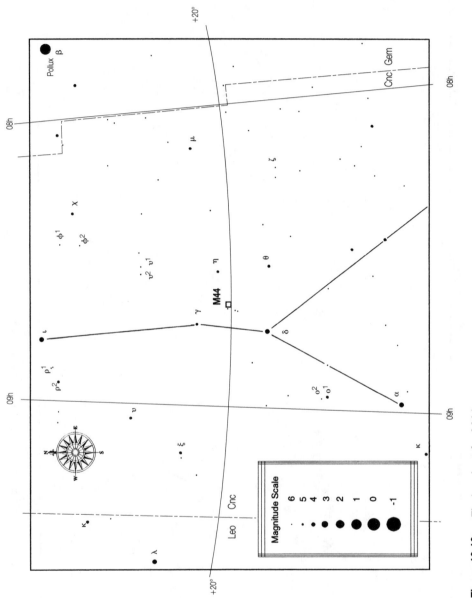

Figure 10.10a *Finder chart for M44.*

Figure 10.10b *M44, the Beehive Cluster. Photograph by George Viscome (135-mm tele-photo lens at f/3.5, hypered Tech Pan film, 20-minute exposure). South is up.*

base of the triangle from Sigma[1] to Sigma[2] and extend it about 3.5° to the northeast, where lies 5th-magnitude 24 Ursae Majoris and, just over a degree farther east, M81 and M82. (See Figure 10.11a.)

Description: M81 and M82 (Figures 10.11b and 10.11c) combine to create one of the finest close-set pair of galaxies found anywhere in the sky! M81 is an outstanding example of a type-Sb spiral galaxy. Given sharp eyes and moderately-dark skies, observers can spy its distinctive oval shape even in binoculars. Small- and medium-aperture amateur telescopes show the bright core surrounded by the fainter spiral halo, while 12-inch and larger scopes begin to suggest the galaxy's curled structure.

Irregular galaxy M82 reveals a long, slender form in all amateur telescopes. Six- to 8-inch instruments hint at the unusual dark lane that cuts through the galaxy's core, while 10-inch and larger scopes add many mottled bright and dark patches across the length of the object.

M51   (NGC 5194)   Spiral Galaxy in Canes Venatici
R.A. 13h29m.9   Dec. +47°12′   Mag. = 8.4   Diameter = 11′ × 8′

Where to look: Begin at Alkaid (Eta Ursae Majoris) at the end of the handle of the Big Dipper. Viewing through your finder, slide 2° to the west of Eta and pause at 6th-magnitude 24 Canum Venaticorum. This star forms a slender isosceles triangle with Eta and M51. Recreate that triangle in your finderscope

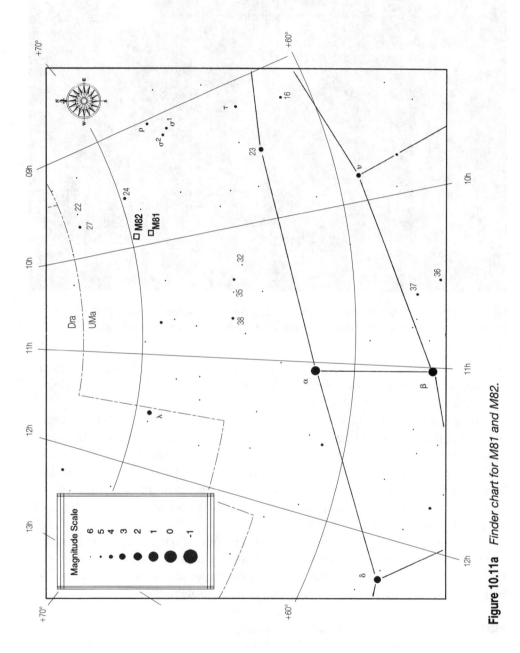

Figure 10.11a *Finder chart for M81 and M82.*

Figure 10.11b *M81 and M82. Photograph by George Viscome (14.5-inch f/6 Newtonian, hypered Tech Pan 2415 film, 20-minute exposure). South is up.*

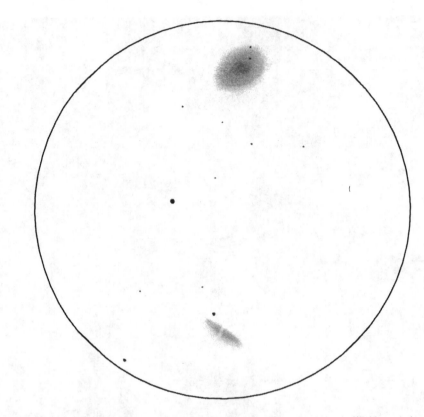

Figure 10.11c *M81 and M82. Drawing by the author using an 8-inch f/7 Newtonian and 24-mm Tele Vue Wide Field eyeplece (59×).*

and then look through your telescope. You should be close enough to the object to find it with a low-power eyepiece. (Though not shown in Figure 10.12a, a small parallelogram of four 7th-magnitude stars surrounds M51. Hop from Eta to 24 Canum Venaticorum and then into this parallelogram. Once these four stars are spotted, look for M51 just inside the northeastern corner.)

Description: One of the finest examples of an Sc spiral galaxy found anywhere in our skies. M51 and its close companion galaxy NGC 5195 (Figures 10.12b and 10.12c) are both visible in 10 × 70 binoculars (maybe even smaller) as a pair of faint fuzzies situated in a nice star field. Their apparent sizes expand as telescope aperture grows, but not until the 8-inch aperture class is used will they take on a new and exciting appearance. In these scopes (and all larger), M51 begins to show some of its remarkable spiral structure. An 8-inch will hint at the spirality only on clear, dark nights, while my 13.1-inch f/4.5 Newtonian will show the arms regularly from suburban skies. And the view I had of M51 through a 24-inch f/5 Newtonian a few years ago took my breath away! It looked just like a photograph.

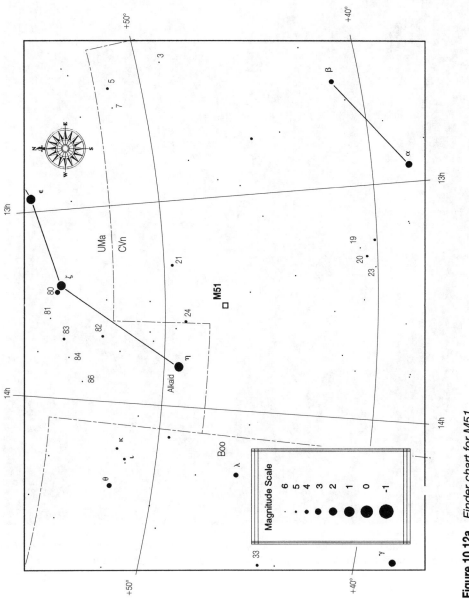

Figure 10.12a *Finder chart for M51.*

Figure 10.12b *M51, the Whirlpool Galaxy. Photograph by George Viscome (14.5-inch f/6 Newtonian, hypered Tech Pan 2415 film, 77-minute exposure). South is up.*

M3 (NGC 5272)   Globular Cluster in Canes Venatici
R.A. 13h42m.2   Dec. +28°23′   Mag. = 6.4   Diameter = 16′
Where to look: From Beta Comae Berenices, scan with your finderscope about 6° due east, keeping an eye out for a lone 6th-magnitude star. Right next door, and about equally bright, will be M3. (Expanding the view beyond what

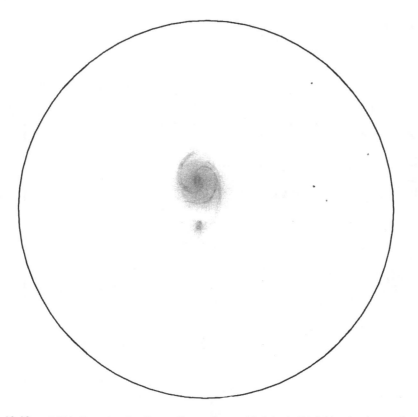

Figure 10.12c   *M51. Drawing by the author using a 13.1-inch f/4.5 Newtonian and a 12-mm Tele Vue Nagler eyepiece (125×).*

is shown in Figure 10.13a, M3 lies a little less than halfway between Arcturus in Boötes to Cor Caroli in neighboring Canes Venatici.)

Description: M3 (Figures 10.13b and 10.13c) is a beautiful globular cluster for all amateur telescopes. Although small instruments will show only a nebulous patch of interstellar cotton, 6-inch and larger scopes begin to resolve the group into a myriad of faint suns. Try the highest power that optics and sky conditions permit for the best view.

Summer

M13   (NGC 6205)   Globular Cluster in Hercules
R.A. 16h41m.7   Dec. +36°28′   Mag. = 5.9   Diameter = 16′
Where to look: Begin at the Hercules Keystone quadrangle, found about two-thirds of the way from the bright star Arcturus to brilliant Vega. Aim at Eta Herculis, the Keystone's northwestern star. Move your telescope along the

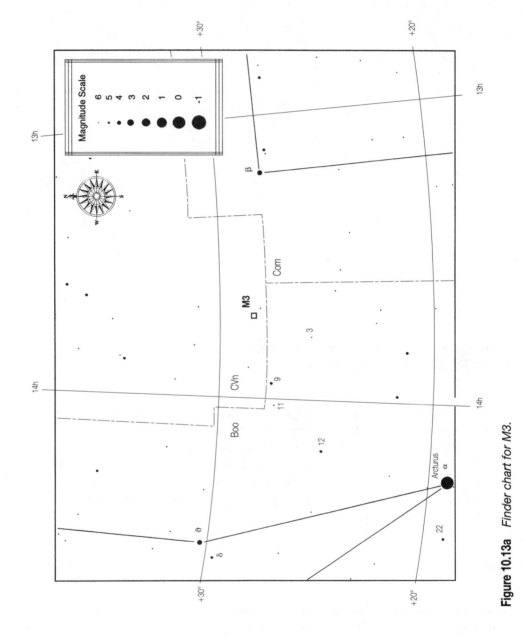

Figure 10.13a *Finder chart for M3.*

Figure 10.13b   *M3. Photograph by George Viscome (14.5-inch f/6 Newtonian, Tri-X film, 20-minute exposure). South is up.*

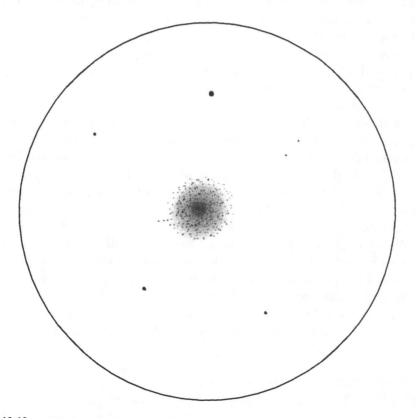

Figure 10.13c   *M3. Drawing by the author using an 8-inch f/7 Newtonian and a 12-mm Tele Vue Nagler eyepiece (119×).*

western edge of the Keystone. M13 lies about one-third of the way from Eta to Zeta Herculis at the southwest corner. (See Figure 10.14a.)

Description: This is one of the finest globular clusters visible from mid-northern latitudes. Like M3 in the Spring sky, M13 (Figures 10.14b and 10.14c) requires a 6-inch telescope for partial resolution of its estimated 100,000 stars. Again, medium- and high-power oculars should produce the best results, but avoid the temptation to overpower your telescope and spoil the image.

As you gain experience, be sure to revisit M13 often to look for some fine structural detail. Many eighteenth- and nineteenth-century astronomers commented on how many of the cluster's stars form chains or rows; they remind me of the legs of a spider. This effect is quite apparent in 8-inch and larger telescopes but not as clear in smaller instruments (though some have recorded them in telescopes as small as 6 inches). At the same time, look for three peculiar dark lanes forming what looks like a propeller set off-center in the cluster. Once again, they grow more obvious as aperture increases. Can you see either of these unusual features?

M8   (NGC 6523)   Bright Nebula in Sagittarius
R.A. 18^{h}03^m.8   Dec. −24°23′   Mag. = 5.8   Diameter = 90′ × 40′

Where to look: With your eyes alone, draw an imaginary line between Sigma Sagittarii and Lambda Sagittarii, stars in the handle and lid of the teapot, respectively. Again, using your eyes alone, extend that line an equal distance toward the west-northwest and aim your finderscope toward that imaginary point. The finder should reveal the pair of 5th-magnitude stars 4 and 7 Sagittarii. Aiming at the latter will put the western half of M8 in your telescope's view. (See Figure 10.15a.)

Description: M8 (Figures 10.15b and 10.15c) appears as a great cloud through amateur telescopes. Telescopes up to 6 inches in aperture unveil a soft glow of remarkable complexity sliced in half by a dark lane, described once as a "lagoon." Though most modern-day observers fail to see the similarity to a lagoon (it strikes me more like a canal or channel), the name stuck, dubbing this the Lagoon Nebula. Eight- to 10-inch instruments uncover some of the fainter portions of the nebula, while larger telescopes unleash a tumultuous cloud. Always begin a visit to M8 with a low-power, wide-field ocular. This way, most of its full 90′ × 40′ girth will squeeze inside a single field of view. Then switch to a higher power to zero in on areas that appear particularly interesting.

Engulfed in the clouds of M8 is the open cluster NGC 6530, an attractive congregation of about two dozen stars ranging from 7th to 9th magnitude and scattered across 15 arc-minutes. This isn't a bad deal: two objects for the price of one!

M57   (NGC 6720)   Planetary Nebula in Lyra
R.A. 18^{h}53^m.6   Dec. +33°02′   Mag. = 9.7   Diameter = 70″ × 150″

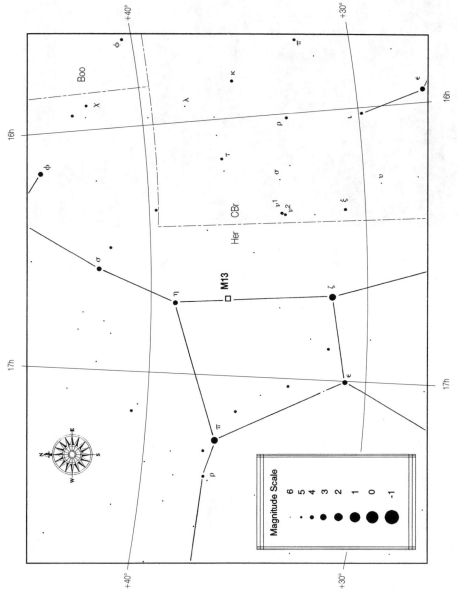

Figure 10.14a *Finder chart for M13.*

Figure 10.14b *M13, the Great Globular Cluster in Hercules. Note the 12th-magnitude galaxy NGC 6207 in the lower-right corner. Photograph by George Viscome (14.5-inch f/6 Newtonian, Tri-X film in cold camera, 30-minute exposure). South is up.*

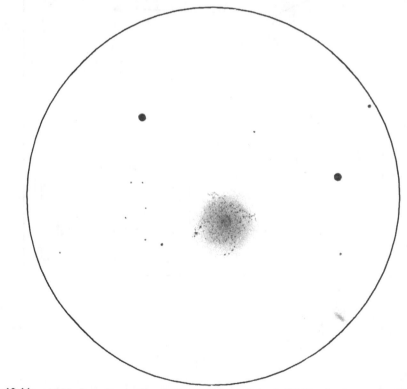

Figure 10.14c *M13. Drawing by the author using an 8-inch f/7 Newtonian and a 26-mm Tele Vue Piössi eyepiece (55×).*

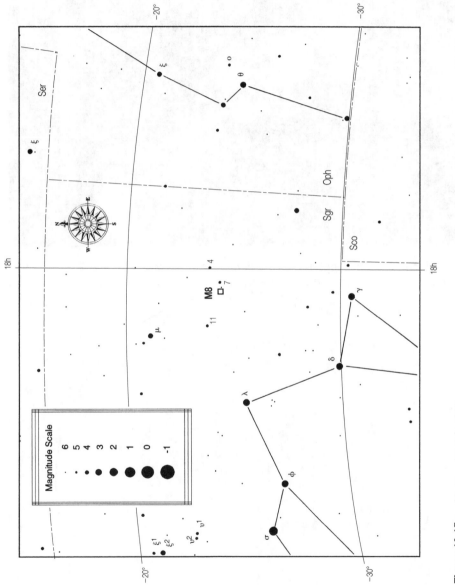

Figure 10.15a *Finder chart for M8.*

Figure 10.15b   *M8, the Lagoon Nebula. Photograph by George Viscome (8-inch f/5.6 Newtonian, hypered Tech Pan 2415, 46-minute exposure). South is up.*

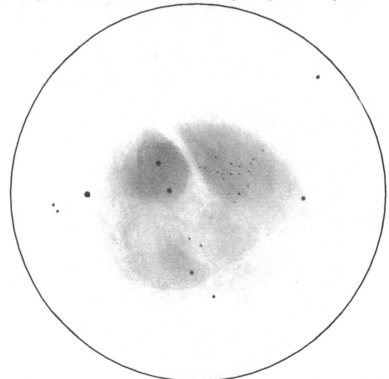

Figure 10.15c   *M8. Drawing by the author using a 13.1-inch f/4.5 Newtonian, a 24-mm Tele Vue Wide Field eyepiece and a DayStar LPR filter (62×).*

Where to look: M57 lies almost exactly halfway between Beta Lyrae (Sheliak) and Gamma Lyrae (Sulafat) along the southern edge of Lyra's parallelogram. (See Figure 10.16a.)

Description: This is the famous Ring Nebula, an outstanding little smokering of stellar debris. Visible in binoculars as a faint point of light, M57 (Figures 10.16b and 10.16c) displays its annular shape in telescopes as small as 2 inches in aperture. A 6-inch instrument adds to the prominence of the ring effect, while 8-inch and larger scopes begin to show a subtle fraying along the edge of the nebula's east-west axis. Conditions permitting, try an eyepiece yielding at least 100× for the best view.

Can you spy the Ring's elusive 15th-magnitude central star? The star's visibility is greatly hampered by its inherent dimness against the intrinsic brightness of the surrounding nebulosity. Though some observers claim to have seen it in 10-inch telescopes, I have never seen it through my 13.1-inch Newtonian. Perhaps I'm not trying hard enough?

M27   (NGC 6853)   Planetary Nebula in Vulpecula
R.A. 19h59m.6   Dec. +22°43′   Mag. = 8.1   Diameter = 480″ × 240″

Where to look: Center your view on Gamma Sagittae, the easternmost star in the Arrow, and then turn north. About 3.5° later, watch for 6th-magnitude 14 Vulpeculae. M27 lies just to its southeast, near a 9th-magnitude sun. (See Figure 10.17a.)

Description: M27 (Figures 10.17b and 10.17c), nicknamed the Dumbbell Nebula, has one of the highest surface brightnesses of just about any planetary nebula in the northern sky. Even binoculars will show its fuzzy disk settled among a rich portion of the summer Milky Way. Its nickname comes from its resemblance to an old-fashioned dumbbell, but it strikes me as more reminiscent of an hourglass (indeed, another pet name for M27 is the Hourglass Nebula). Viewing through an 8-inch telescope reveals some blue-green coloring to the Dumbbell and begins to fill in the hourglass curve with fainter nebulous extensions, especially when viewed with a narrow-band LPR filter in place. These extensions grow in brightness as aperture increases, impairing the hourglass analogy but adding to the overall majesty. Large amateur scopes will also reveal the nebula's faint central star.

Autumn

M2   (NGC 7089)   Globular Cluster in Aquarius
R.A. 21h33m.5   Dec. −00°49′   Mag. = 6.5   Diameter = 13′

Where to look: Begin at the star Alpha Aquarii (Sadalmelik). Orient yourself to the finder's field by identifying the fainter stars 28, 32, and Omicron Aquarii. Next, extend a line from Alpha to 28 and then beyond. Four degrees to the northwest of 28 lies 11 Pegasi (identified on some star charts as 27 Aquarii). This star forms the tip of an equilateral triangle with 25 and 26 Aquarii. Center this triangle in your finder and then head southwest 2° toward a lone 6th-magnitude star.

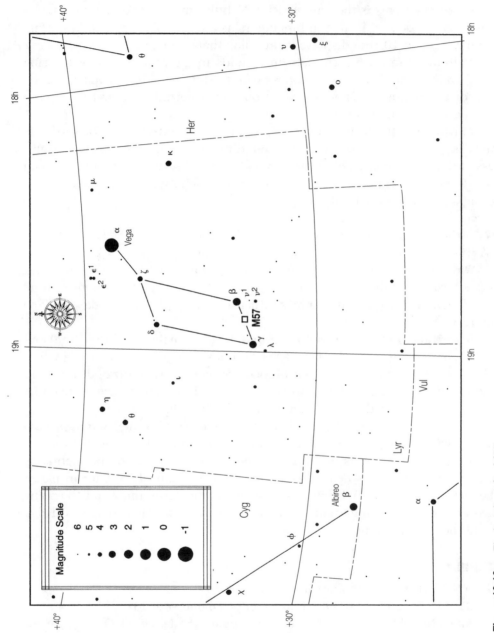

Figure 10.16a *Finder chart for M57.*

Figure 10.16b   *M57, the Ring Nebula. Photograph by George Viscome (14.5-inch f/6 Newtonian, hypered Tech Pan 2415 film, 15-minute exposure). South is up.*

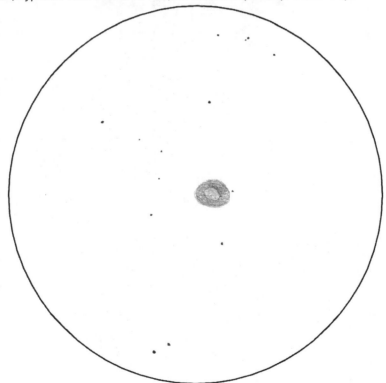

Figure 10.16c   *M57. Drawing by the author using a 13.1-inch f/4.5 Newtonian, a 7-mm Tele Vue Nagler eyepiece, and a Daystar LPR filter (214×).*

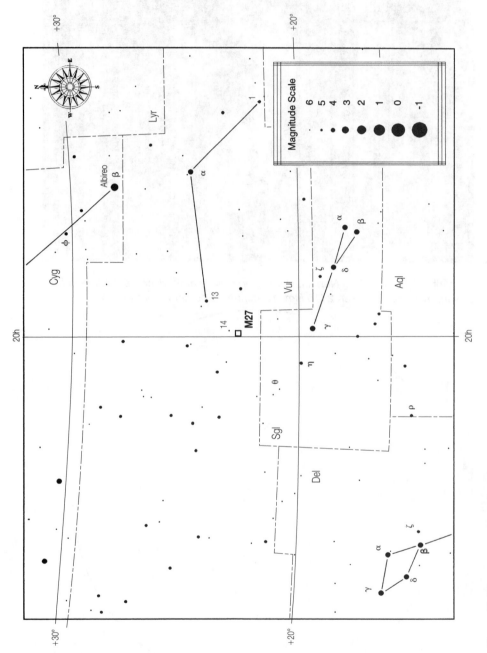

Figure 10.17a *Finder chart for M27.*

Figure 10.17b   *M27, the Dumbbell Nebula. Photograph by George Viscome (14.5-inch f/6 Newtonian, hypered Tech Pan 2415 film, 26-minute exposure). South is up.*

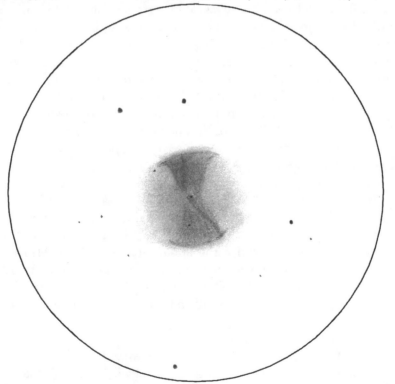

Figure 10.17c   *M27. Drawing by the author using a 13.1-inch f/4.5 Newtonian, a 12-mm Tele Vue Nagler eyepiece, and a Daystar LPR filter (125×).*

From here, M2 lies about 1° farther west-southwest and should be visible in finderscopes as a faint smudge. (See Figure 10.18a.)

Description: M2 (Figures 10.18b and 10.18c) is one of autumn's finest globular clusters. Small telescopes reveal it as a round puff of cotton marked by a bright center and diffuse edges. Partial resolution of M2 is possible in a 6-inch telescope, with some of the myriad of 13th-magnitude cluster stars visible around its outer edge. Twelve-inch and larger scopes smash the core of M2 by revealing uncountable stars teeming throughout.

M31   (NGC 224)   Galaxy in Andromeda
R.A. 00^{h}42^m.7   Dec. +41°16'   Mag. = 3.5   Diameter = 160° × 40°

Where to look: As outlined in chapter 9, begin your quest for M31 at Beta Andromedae (Mirach). Head 4° northwest to the naked-eye star Mu Andromedae and then northwest again for 3° to Nu Andromedae. You might have some trouble spotting these latter two stars, especially Nu, without optical aid if you are a captive of light pollution, but your finder will show them easily. From Nu, nudge your telescope due east for 1.33° for M31 which should be visible in the finder as a knockwurst-shaped smudge (See Figure 10.19a.)

Description: M31, the Andromeda Galaxy (Figures 10.19b and 10.19c), is a delight to behold in all instruments regardless of size. Under dark skies, low-power binoculars, and even the eyes alone, reveal the galaxy's huge span. Because of its size, most telescopes can only squeeze portions of the galaxy into single fields of view. A well-planned tour of M31 begins at the galaxy's center. The bright, oval galactic core reaches a visual crescendo as it draws toward an intense stellar nucleus. Stretching out toward the northeast and southwest from the core is the comparatively faint glow of the spiral arms. A 6-inch telescope will reveal one, possibly two, bands of dark nebulosity along the northwest side of the great oval. These dark lanes grow more obvious as aperture increases. Also coming into view with larger telescopes is NGC 206, a rectangular star cloud found about 0.66° southwest of the core.

Two smaller companion galaxies stay close by the side of M31. The brighter of the pair is M32 (NGC 221), an elliptical galaxy found 24' due south of M31's core. Telescopes reveal M32 as an oval glow also punctuated by a bright stellar nucleus. The third member of this galactic family is M110, also known as NGC 205. Though twice the apparent size of M32, M110 appears much fainter because of its lower surface brightness. Look for its nondescript oval glow about 35' northwest of the center of M31. In place of the bright nucleus found in M32, M110 reveals only a homogeneous glow across its face.

NGC 457   Open Cluster in Cassiopeia
R.A. 01^{h}19^m.1   Dec. +58°20'   Mag. = 6.4   Diameter = 13'

Where to look: Extend the line along the eastern leg of the W shape of Cassiopeia from Epsilon to Delta Cassiopeiae and then, looking through your finder, about one-third farther to 5th-magnitude Phi Cassiopeiae. Center Phi in

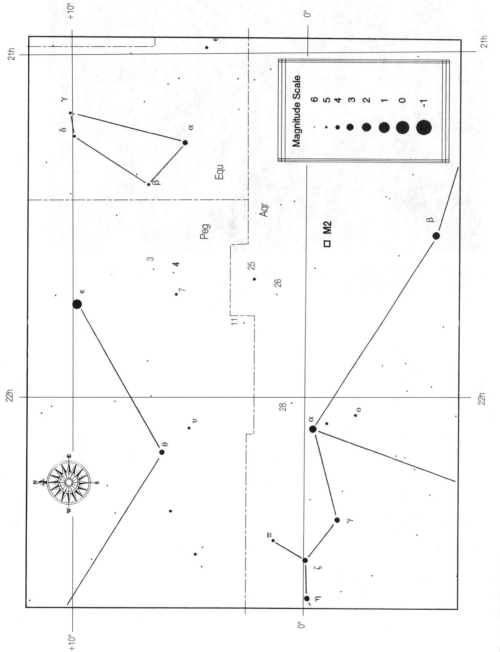

Figure 10.18a *Finder chart for M2.*

Figure 10.18b   *M2. Photograph by George Viscome (14.5-inch f/6 Newtonian, Tri-X film, 60-minute exposure). South is up.*

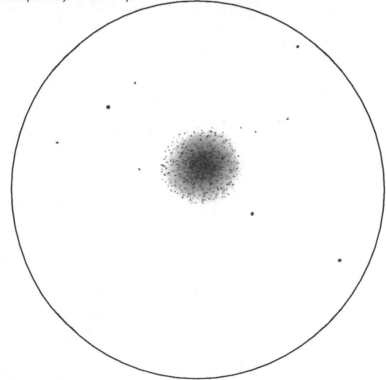

Figure 10.18c   *M2. Drawing by the author using a 13.1-inch f/4.5 Newtonian and a 12-mm Tele Vue Nagler eyepiece (125×).*

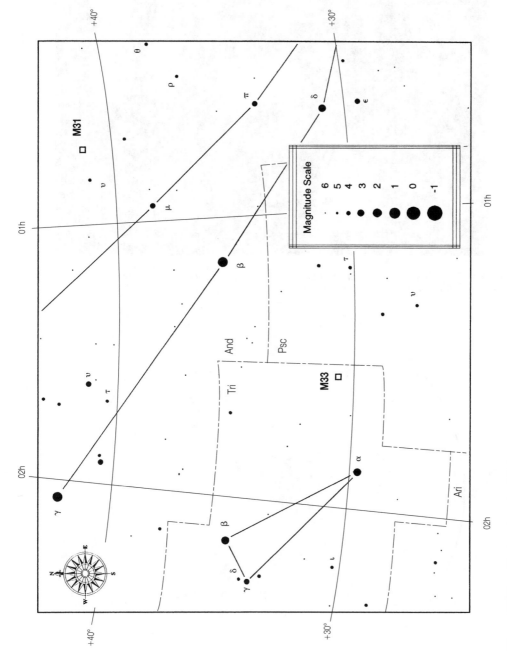

Figure 10.19a *Finder chart for M31.*

Figure 10.19b   *M31, the Andromeda Galaxy. Photograph by George Viscome (3-inch f/6.6 refractor, hypered Tech Pan 2415 film, 86-minute exposure). South is up.*

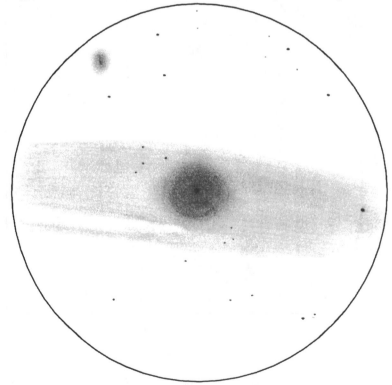

Figure 10.19c   *M31. Drawing by the author using an 8-inch f/7 Newtonian and a 24-mm Tele Vue Wide Field eyepiece (59×).*

the field and switch to your telescope to see NGC 457 bordering Phi to the north. (See Figure 10.20a.)

Description: Although Phi appears to lie within the cluster's boundaries, some studies indicate that it is only a foreground star that happens to lie along the same line of sight as the cluster. True cluster member or not, Phi combines with a second foreground star, 7th-magnitude HD 7902, to create a pair of eyes staring back at us (Figure 10.20b). The cluster's true members shine between 8th and 11th magnitude and fall into a pattern that resembles a bird in flight. The bird's body is created from about a dozen stars of magnitudes 9 to 11. A pair of 10th-magnitude suns marks its tail feathers. The wings are comprised of about a half-dozen stars each, set in long, graceful arcs. The east wing is highlighted by a distinctive 8th-magnitude orange star, which is the cluster's brightest star. Overall, the pattern resembles an owl, and so NGC 457 became known as the Owl Cluster.

NGC 869   Open Clusters in Perseus
R.A. 02^{h}19^m.0   Dec. +57°09'   Mag. = 4.3   Diameter = 30'
NGC 884
R.A. 02^{h}22^m.4   Dec. +57°07'   Mag. = 4.4   Diameter = 30'

Where to look: These open clusters, known collectively as the Double Cluster, are also easily found by using the W of Cassiopeia. Referring ahead to Figure 10.20a, extend the line from Gamma to Delta Cassiopeiae toward the stars of neighboring Perseus to the southeast. Center your attention about halfway between Delta Cassiopeiae and Gamma Persei. Even from suburbia, most observers can spot a dim smudge of light at this point. Aim your finderscope there to see two tight clumps of stars scattered in a rich starry field. Those clumps are NGC 869 and NGC 884.

Description: The Double Cluster (Figure 10.21) creates one of the most striking views to be had through a telescope or binoculars. Though attractive in all instruments, they are best seen through low-power short-focal-length telescopes and giant binoculars. These display a field of stardust of unparalleled beauty. Countless stars are strewn across the view, clustering together into two tight knots. Look toward the center of each group to see several tiny triangles and other geometric patterns created by the cluster stars. Most of the suns appear blue-white and white, although with some effort a few shining with subtle hues of yellow and red can be detected.

Winter

M45   Open Cluster in Taurus
R.A. 03'47^h Dec. +24°07'   Mag. = 1.2   Diameter = 110'

Where to look: M45, better known as the Pleiades or Seven Sisters, is visible to the naked eye as a tiny dipper-shaped pattern of stars riding on the back of Taurus the Bull, west of the Bull's head and its bright star Aldebaran. (See Figure 10.22a.)

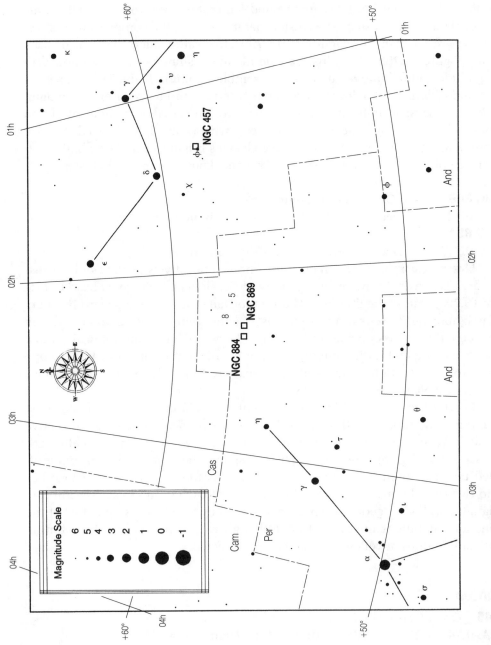

Figure 10.20a *Finder chart for NGC 457, NGC 869, and NGC 884.*

Figure 10.20b   *NGC 457, the Owl Cluster. Photograph by George Viscome (14.5-inch f/6 Newtorian, hypered Tech Pan 2415 film, 45-minute exposure). South is up.*

Figure 10.21   *NGC 869 and NGC 884, the Double Cluster. Photograph by George Viscome (14.5-inch f/6 Newtonian, hypered Tech Pan 2415 film, 45-minute exposure). South is up.*

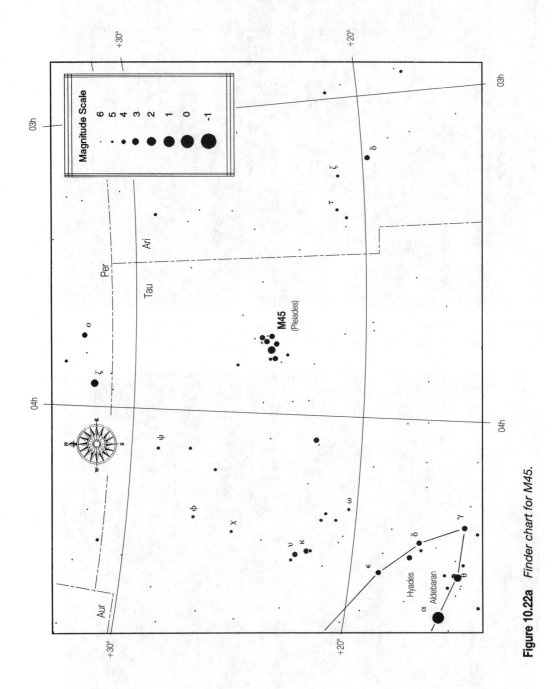

Figure 10.22a *Finder chart for M45.*

Description: I would like to nominate this as the grandest open star cluster of all for binoculars. Even the smallest glasses will cause a population explosion in the family of the Seven Sisters by revealing dozens of fainter stellar siblings (Figure 10.22b). The half-dozen stars visible to the naked eye increases by a factor of ten or more when M45 is seen through 7× binoculars. An outstanding vista awaits observers viewing with giant binoculars, as dozens of brilliant blue-white stellar sapphires fill the field. Current estimates place at least 100 stars within the Pleiades.

Under very dark, very clear conditions, it is also possible to glimpse some patches of reflection nebulosity that engulfs some of the cluster stars. The brightest patch is labeled NGC 1435 and surrounds Merope, the southernmost star in the Pleiades' bowl-shaped grouping. Look for it extending to the south of the star. (Here's a tip: Many first-time observers looking at the Pleiades immediately think they are seeing the nebulosity around all of the bright stars in the cluster. The Pleiades nebulosity is *not* that easy to see. To confirm your observation, turn your telescope toward the Hyades cluster, which forms the head of Taurus. If you see nebulosity there, too, then what you are witnessing is light-scatter in the telescope—the Hyades are nebula-free. If, on the other hand, the glow disappears, then you are probably seeing a portion of the Pleiades' nebulosity.)

M1 (NGC 1952) Supernova remnant in Taurus
R.A. 05h34m.5 Dec. +22°01′ Mag. = 8.2 Diameter = 6′ × 4′

Figure 10.22b *M45, the Pleiades or Seven Sisters. Photograph by George Viscome (8-inch f/5.6 Newtonian, hypered Tech Pan 2415 film, 80-minute exposure). North is up.*

Where to look: Begin at Zeta Tauri, which marks the tip of Taurus's southeastern horn. Looking through your finder, locate two 6th-magnitude stars just to its north that combine with Zeta to form a westward-pointing right triangle. Extend the leg of the triangle formed by the 6th-magnitude stars farther west for about half their distance. Through a low-power eyepiece, M1 should be visible in your telescope field just to the south of this point. (See Figure 10.23a.)

Description: M1 (Figures 10.23b and 10.23c) appears as an amorphous gray oval disk through 3- to 6-inch telescopes, but 8- to 12-inch instruments begin to show some of its irregularities. Larger apertures increase the mottled look of M1, with the biggest amateur telescopes revealing some of the crablike appendages that led to this object's nickname: the Crab Nebula.

M42   (NGC 1976)   Bright Nebula in Orion
R.A. 05^{h}35^m.4   Dec. −05°27′   Mag. = 2.9   Diameter = 66′ × 60′

Where to look: M42 may be seen with the naked eye as the middle star in Orion's sword, directly below the three stars in the Hunter's belt. (See Figure 10.24a)

Description: The Great Nebula in Orion, M42, is the finest deep-sky object visible from the Northern Hemisphere. All magnifications work well, with each offering a different perspective. Low powers (Figures 10.24b and 10.24c) are best for seeing the BIG picture; medium magnification reveals the nebula's complex structure and its varying colors and contrasts; high power works best for spying the intricate area in and around the nebula's center. On one especially transparent night a few years ago, my 13.1-inch f/4.5 reflector displayed a tremendous cloud with many stars embedded within. Tenuous curved fingers of glowing gas reached from the main body of the nebula to grasp many of the neighboring stars.

Even the smallest telescopes will show the Trapezium, four young, hot, blue-white stars buried within the center of M42. Their energy combines with that from other, fainter stars within the cloud to excite the Orion Nebula's hydrogen into luminescence. This results in the reddish color that is so vivid in photographs but tough to see otherwise. Hints of red can be seen along the misty fringes of M42 through large amateur instruments, but the overall coloring of the cloud appears blue-green.

Just north of M42 is a second, much smaller tuft of nebulosity cataloged separately as M43. Interestingly, although M42 was discovered telescopically in 1610, M43 was not recognized until 1731.

M35   (NGC 2168)   Open Cluster in Gemini
R.A. 06^{h}08^m.9 Dec. +24°20′   Mag. = 5.3   Diameter = 28′

Where to look: Figure 10.25a shows how M35 forms one corner of a triangle with Eta and Mu Geminorum, two stars in the foot of the twin brother Castor in Gemini. Recreate this triangle by aiming at the invisible third corner, where you should spot M35 through your finder. In fact, this cluster is so bright that it

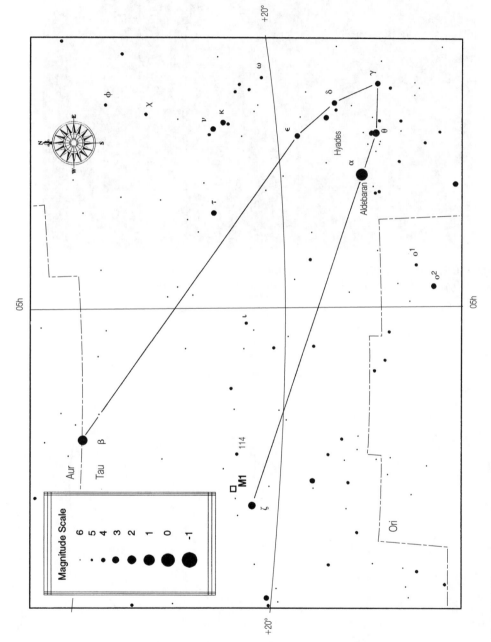

Figure 10.23a *Finder chart for M1.*

311

Figure 10.23b   *M1, the Crab Nebula. Photograph by George Viscome (8-inch f/5.6 New-tonian, hypered Kodak Tech Pan 2415, 70-minute exposure). South is up.*

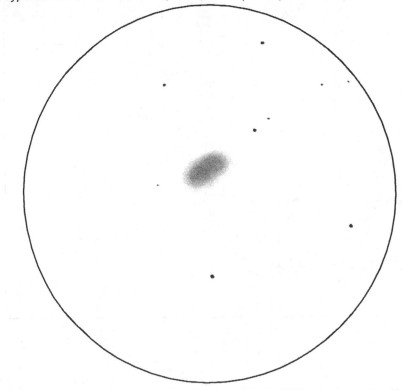

Figure 10.23c   *M1. Drawing by the author using an 8-inch f/7 Newtonian and a 12-mm Tele Vue Nagler eyepiece (119×).*

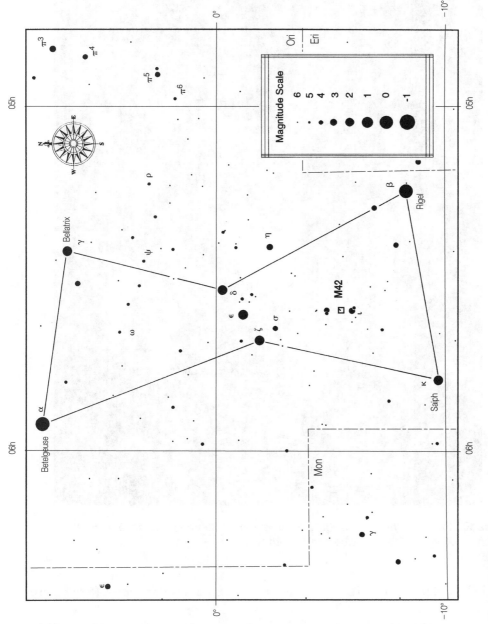

Figure 10.24a *Finder chart for M42.*

Figure 10.24b *M42, the Orion Nebula. Photograph by George Viscome (6-inch f/8 Newtonian, Tri-X film, 40-minute exposure). South is up.*

might be visible faintly to the naked eye under extremely dark skies. (See Figure 10.25a.)

Description: M35 (Figure 10.25b) is one of winter's finest open clusters. Binoculars begin to show some individual stars, but at least a 3-inch scope is needed for resolution of the cluster. Eight-inch and larger instruments resolve just about all of the 200 stars that call this group home. Most appear blue-

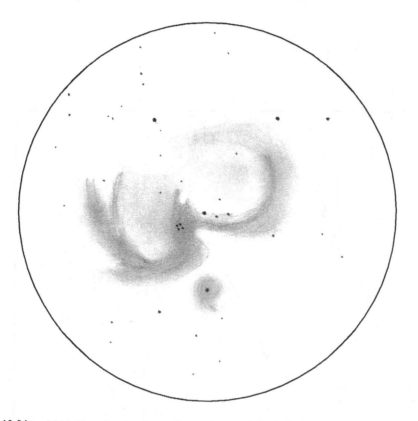

Figure 10.24c *M42. Drawing by the author using an 8-inch f/7 Newtonian, 24-mm Tele Vue Wide Field eyepiece, and a DayStar LPR filter (59×).*

white although some shine yellow and orange. Your lowest-power eyepiece will provide the best view.

Half a degree southwest of M35 is a second, much fainter, more distant open cluster: NGC 2158. Visible in a 3-inch, NGC 2158 requires at least a 6-inch telescope for partial resolution.

Astrophotography 101

One of the most popular pastimes for amateur astronomers is trying to capture the beauty of the universe on film. Just look at the superb photographs that highlight this book. Outstanding, aren't they? Although they rival the best photos from professional observatories, all were taken by amateur astronomers!

Yet in spite of its popularity, astrophotography can be one of the most time-consuming and frustrating aspects of our hobby. For those who have dabbled in it before, how often have you sent out a roll of film to be processed, confident in your success, only to get back pictures that are either out-of-focus, blurred, overexposed, underexposed, or just not right? An astrophotographer

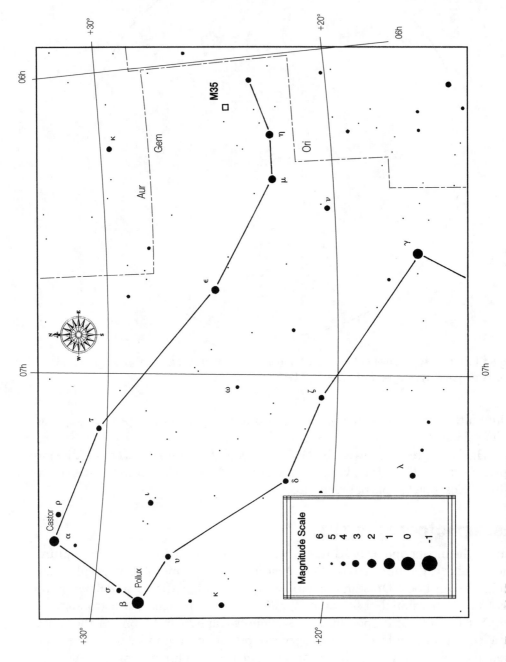

Figure 10.25a *Finder chart for M35.*

Figure 10.25b *M35. Photograph by George Viscome (8-inch f/5.6 Newtonian, hypered Tech Pan 2415 film, 40-minute exposure). South is up.*

friend once stated that even after years of perfecting his craft, he expects about five bad photographs for every one of high quality.

What makes astrophotography so difficult? Today, we live in a point-and-shoot world. Modern cameras are capable of determining exposure, setting the correct lens opening, focusing the lens, advancing the film to the next frame, and even telling the camera's light meter what kind of film is being used . . . all automatically! All of this is amazing, indeed, but applies only to terrestrial, not celestial, photography. By comparison, astrophotography is still in the Dark Ages, with good results coming only after much initial trial and error. Although it can be frustrating at times, perhaps this is also part of its appeal.

For the sake of this brief introduction, I have chosen to break the subject down into four broad categories, as shown below:

1. Fixed-camera
2. Short exposures through the telescope
3. Guided exposures
4. Long exposures through the telescope

This order was chosen as an approximate indicator of difficulty, ranging from low to high. It might be argued by some that items 2 and 3 could be reversed, but I will leave that judgment to you. For the moment, let's discuss the specifics.

Fixed-Camera

Fixed-camera photography requires four things: a manually adjustable camera and lens, a tripod on which to set it, a roll of fast (>ISO 200) film, and a locking cable release. That's all—no expensive telescope or elaborate clock-driven mounting is required. To take a photograph, aim the camera at the desired area of sky, set the camera's focus at infinity (∞), close the aperture one f-stop from the maximum (to lessen edge distortion), and set the shutter speed to B (bulb) or T (time). Open the shutter for anywhere from a few seconds to a half-hour or more. The net result is an accurate record of the night sky to at least the naked-eye limit.

Even though the camera is attached to a sturdy tripod, it is always moving because of the Earth's rotation. As a result, all of the stars in the photograph will be recorded as trails rather than points—the longer the exposure, the longer the trails. If star trails are the desired result, then exposure duration depends only on sky darkness. Under dark, rural conditions, exposures up to 30 minutes or more are possible, whereas suburban and urban photographs must probably be limited to no more than 5 or 10 minutes because of light pollution.

If you want to use a fixed camera to photograph the stars as points, then the exposure time must be limited even more. Just how limited depends on the focal length of the camera lens being used. Table 10.2 compares lens focal length to maximum exposure before trailing will become evident. This table also shows how wide a slice of sky is covered by a given focal-length lens, an important consideration for proper framing.

Figure 10.26 shows the effects of exposure duration on the constellation Ursa Major. Using a 50-mm lens, a 15-minute exposure was taken. Five minutes later, a 25-second exposure was made on the same frame. The stars were

Table 10.2   **Lens Focal Length versus Maximum Time Exposure**

(35-mm-format film)		
Focal Length of Lens mm	Maximum Exposure[1] sec	Sky Coverage[2] °
28	25	49 × 74
50	14	28 × 41
85	8	16 × 24
105	7	13 × 20
200	4	7 × 10
300	2	5 × 7

Notes:
1. The maximum duration of an exposure before stars begin to trail. These values are for stars at 0° declination; longer exposures are possible if the target area is above or below the celestial equator. Increase exposure by up to 50% at 45° declination, 100% above 60° declination.
2. Area covered by a frame of 35-mm film using a lens of the given focal length.

Figure 10.26 *A double exposure of Ursa Major (the Big Dipper) showing the difference between short- and long-exposure fixed-camera photography. Five minutes elapsed between the long 15-minute exposure and the shorter 25-second exposure. (Fujichrome 400 slide film, 50-mm f/1.8 lens set at f/2.8).*

recorded as points during the 25-second exposure but as trails in the 15-minute exposure.

Through the Telescope I: Short Exposures

The simplest method for photographing through a telescope, short-exposure telescopic work can be done with just about any type of telescope on just about any type of mounting (yes, even Dobsonian mounts, if balanced correctly). Subject matter appropriate for short exposures is restricted to the Sun, Moon, and the brighter planets, with exposure times ranging from $1/1000$th second to maybe four or five seconds. Other subjects and longer exposures are covered under a separate heading later on.

Just about any type and speed of film can be used to photograph the Sun, Moon, and planets, although slower films (for example, <ISO 200) are usually preferred. These offer finer grain structure than high-speed emulsions, an important consideration when trying to record subtle detail on a planet's surface or atmosphere.

As Figure 10.27 shows, there are four basic camera–telescope combinations available to the astrophotographer. Some are more appropriate than others for

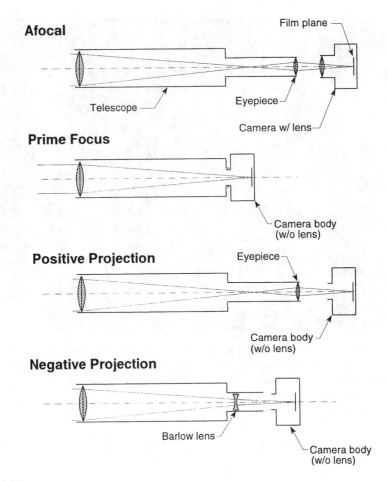

Figure 10.27 *Four possible camera/telescope configurations.*

capturing certain objects, but no single setup is best for everything. Table 10.3 compares all four options illustrated in Figure 10.27.

Afocal. The afocal system uses both a camera's lens and a telescope's eyepiece. This allows just about any single-lens reflex (SLR) camera to be used for celestial photography, regardless of whether its lens is removable or not. You'll recall from the discussion in chapter 6 that the viewfinder of an SLR camera looks through the camera lens and therefore shows exactly what the film will see. This ability makes life a lot easier when checking the camera's alignment with the telescope. While through-the-telescope astrophotography is possible with nonreflex cameras, it is not recommended.

Achieving a sharp focus is the biggest challenge facing the afocal photographer (as it is with the other methods as well). Place the camera on a separate tripod and align it to the telescope's eyepiece. With the camera's lens wide open

Table 10.3 **Camera-Telescope Configurations**

System Name	Camera Lens?	Telescope Eyepiece?	Best For
Afocal	Yes	Yes	Moon, Sun
Prime focus	No	No	Moon, Sun (whole-disk)
Positive projection	No	Yes	Moon, Sun (close-ups), planets
Negative projection	No	Yes (Barlow lens)	Moon, Sun (close-ups)

and focused at infinity, look through the viewfinder. Slowly turn the telescope's focusing knob in and out until the image is at its sharpest.

What exposure should be used? Over the years, many articles and books have been written on how to calculate proper exposure for the Moon, Sun, and planets. Unfortunately, most recommendations vary from one source to another, attesting to the fact that astrophotography is more art than science. However, as a guide, Table 10.4 offers suggested exposures for different subjects. Be sure to bracket the exposures *at least* one shutter speed either side of the suggestion.

Notice that these charts are based on knowing the *effective focal ratio* (EFR) of the camera-telescope combination. You DON'T know that value? Yes, you do! For the afocal method, the EFR can be calculated using the following formula:

$$\text{EFR} = \frac{(\text{camera lens focal length} \times \text{telescope magnification})}{\text{aperture of telescope}}$$

For example, consider taking a photograph of the first-quarter Moon using the afocal method, an 8-inch f/10 telescope, a 25-mm eyepiece, and a 50-mm f/1.8 camera lens. First, convert inches to millimeters or vice versa so that the units of measure are all the same. For the sake of this discussion, we will convert the telescope's 8-inch aperture to 203 millimeters (8 inches × 25.4 millimeters per inch). From the formula in chapter 1, the telescope's magnification is 80×. Plugging these values into the formula above yields

$$\text{EFR} = \frac{(50 \text{ mm} \times 80)}{203 \text{ mm}} = \text{f}/19.7$$

With the f-ratio known, an estimation for proper exposure may now be picked off the chart. Assuming ISO 100 film, the suggested exposure is $1/15$ of a second. But take more than one shot, varying the exposure of each.

Prime focus. Prime focus is the cleanest camera–telescope combination but it requires a single-lens reflex camera with a removable lens. This method couples the lensless camera body to the eyepieceless telescope, in effect making

Table 10.4 **Suggested Exposures for Selected Objects**

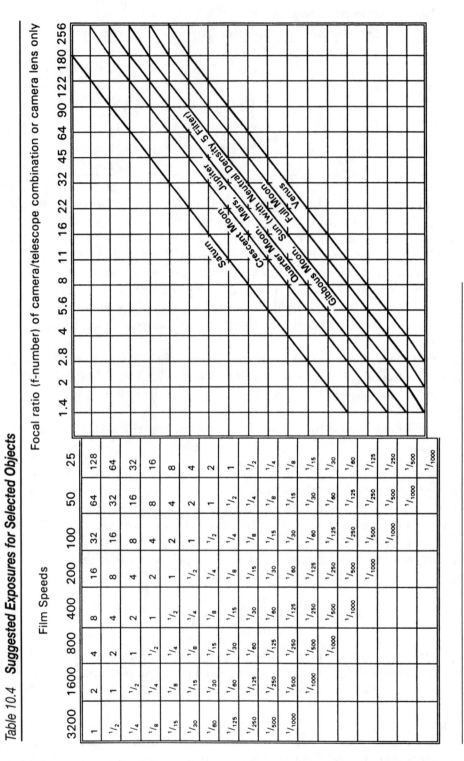

Recommended exposures for brighter sky targets. To read the table, find your camera/telescope focal ratio and your selected target. Follow the boxes to the left until you line up with your film's ISO column. The suggested exposure lies at their intersection.

the telescope itself a large telephoto lens. Determining the effective focal ratio requires no calculation; it is simply the telescope's own focal ratio. For example, the prime-focus EFR of an 8-inch f/10 telescope is f/10, while the EFR of a 14-inch f/5 telescope is f/5, and so on.

To take prime-focus photos through a telescope, the camera is attached directly to the telescope using a two-piece camera-to-telescope adapter. Focusing is done directly through the camera's viewfinder by turning the focusing knob(s) in and out until a sharp image is seen. Unfortunately, this is largely a hit-or-miss technique. To take the guesswork out of focusing, use one of the focusing aids reviewed in chapter 6. Though they may seem an unnecessary frill at first, there is no worse feeling than getting back a perfectly exposed, perfectly framed photograph that is out of focus!

Just about all modern refractors and catadioptrics are designed for prime-focus photography, but not so with some Newtonian reflectors. Many Newtonians are constructed in such a way that the prime-focus point lies too far down in the focusing mount to be accessible by a camera. In cases like this, the only alternative is to move the main mirror up the tube to reduce its distance to the eyepiece. Reducing this distance will push the prime focus out toward the end of the focusing mount.

If your Newtonian telescope will not permit prime-focus photography, measure the distance from the primary to the diagonal mirror. Add to this the distance from the diagonal's center to about the upper edge of the eyepiece holder's base. To access the prime focus, this total distance should be about an inch or so less than the telescope's focal length. If it is more, then subtract the instrument's focal length from your overall measurement plus 1 inch. Moving the primary up the tube by that difference (probably no more than a couple of inches) will push the prime focus farther up the focusing tube and make it accessible to the camera. For example, an 8-inch f/7 telescope has a focal length of 56 inches. Measuring from the primary to the diagonal and then from the diagonal to the eyepiece holder indicates a length of 58 inches. From the discussion above, the primary must be moved [(58 + 1) − 56], or 3 inches, up the tube to access the prime focus with a 35-mm single-lens reflex camera. (The extra inch is to allow for the distance from the eyepiece holder to the film plane in the back of the camera.) Do not move the mirror too far up, or the telescope may not focus when used visually. If this latter problem occurs, then an extension tube made from a 1.25-inch brass drainpipe (used in bathroom sinks) will be required. Already mentioned in Chapter 7, these drainpipes decrease from a 1.25-inch inside diameter to a 1.25-inch outside diameter. Note also that the diagonal mirror may also be too small to reflect all the light coming from the primary if its distance is reduced too much. In addition, when you move the mirror up, you will probably have to rebalance the telescope tube in its mount.

Positive projection. Sometimes called *eyepiece projection*, this arrangement uses an eyepiece to enlarge whatever is being photographed, in effect stretching a telescope's focal length. Like prime focus, a camera-to-telescope adapter is used, but this time with an extension tube between the telescope adapter and T-ring. An eyepiece is placed inside the extension tube (don't forget to lock the eyepiece in place using the side-mounted screw) to project the image from the telescope into the camera. Positive projection is ideal for solar and lunar close-ups as well as detailed photographs of the planets.

As with the afocal and prime-focus systems, the effective focal ratio of the camera-telescope team must be known before proper exposure can be estimated. The following equation may be used for this calculation:

$$EFR = \frac{f_t \times (L_e - f_e)}{f_e}$$

where f_t = telescope focal ratio
L_e = projection distance from eyepiece
f_e = eyepiece focal length

To illustrate this, consider taking a photograph of, say, Jupiter with an 8-inch (200-mm) f/10 telescope. Because of the planet's small apparent size, the positive-projection method will be used. A 12-mm eyepiece is selected for the task, with an eyepiece-to-film projection distance of 3 inches (75 mm). Plugging these values into the positive-projection equation above produces

$$EFR = \frac{f/10 \times (75 - 12)}{12} = f/10 \times 5.25 = f/52.5$$

Thus, the overall system has an effective focal ratio of f/52.5, yielding an effective focal length of 10,500 mm (2,000 mm × 5.25). An exposure estimate may now be gleaned from the accompanying tables.

Negative projection. Similar to positive projection, negative projection puts a negative, or concave, lens between the telescope and the lensless camera body. Negative projection does not extend a telescope's focal length to the great extent that positive projection does, making this method ideal for enlarged shots of the Moon and Sun but not as useful for photographing the planets.

The most commonly available negative lenses are the Barlow lens and the photographic teleconverter, with many photographers preferring the latter. To use a teleconverter on a telescope, simply connect it between the camera body and the T-ring/telescope adapter used for prime-focus work. A Barlow is used as a projection lens by inserting it into the eyepiece holder and affixing the camera to it using a telescope adapter/T-ring. The resulting increase in magnification will be equal to the power of the teleconverter or Barlow, usually either 2× or 3×. Therefore, using either a 2× teleconverter or a 2× Barlow with, for example, an f/10 telescope will double both the instrument's focal length and effective focal ratio (in this case, to f/20).

Telecompression. Although not shown separately in Figure 10.27, telecompression may be thought of as reverse projection. Rather than enlarging an image, placing a telecompressor or focal reducer between the telescope and lensless camera will lower the effective focal ratio. The net result is a negative magnification effect and a wider field of view. This is especially useful when trying to photograph the full disk of either the Moon or Sun through a long-focal-length telescope.

Most telecompressors/focal reducers state their *deflation factor* right on them. For instance, a Celestron or Meade Reducer-Corrector cuts the effective focal ratio of an f/10 telescope by 37% to f/6.3 (and, therefore, the effective focal length from 80 inches to 50 inches, in the case of an 8-inch f/10). That same telecompressor will reduce an f/6.3 instrument (such as some of Meade's Schmidt-Cassegrains) to f/4, and so on. Not only does lowering a telescope's focal ratio increase the field covered in a photograph, but it also decreases the exposure time needed to take the picture. To see how dramatic this effect is, look at any of the exposure charts in this chapter and compare the straight telescope's f-ratio with the same instrument telecompressed.

If you are considering using a telecompressor for photography, be sure to use a reducer/corrector. A telecompressor without the corrective optics causes much more distortion around the edges of the film frame.

Guided-Camera

Sometimes called *piggyback astrophotography,* this next step up requires the camera to be placed on an equatorial mounting and tracked with the stars. Frequently, the camera is mounted sidesaddle on an equatorially mounted telescope, giving rise to the piggyback nickname.

Both Meade and Celestron (as well as many aftermarket companies such as Orion) sell piggyback brackets that mount directly to their Schmidt-Cassegrain telescopes. Any camera with a standard tripod socket on its baseplate may then be attached to the bracket. Some equatorially mounted refractors and reflectors also come with provisions for attaching a camera for guided exposures.

An expensive clock-driven telescope mounting is not needed for guided-camera astrophotography. For about twenty dollars, you can build a Scotch mount, a hand-driven camera platform that will track the stars as accurately as a mounting costing 50 times as much. Plans for making a Scotch mount were included in this book's first edition and may be found on the Star Ware Home Page (see appendix C).

Preferred films for guided-camera astrophotography are any with an ISO rating of at least 200, with many photographers favoring films with ISO values in excess of 800. These allow the maximum amount of starlight to be recorded in the minimal amount of time.

Before taking the first picture, align the mounting with the celestial pole by following the directions given in the previous chapter.

With the camera firmly attached to the polar-aligned mounting and aimed at the desired area, begin the exposure. Follow the same advice given under the fixed-camera section earlier. Set the lens focus at infinity, the shutter to either B or T, and move the lens down one stop. Before opening the shutter, lock the camera's mirror up (if so equipped). Failing that, hold a black card in front of the lens with one hand while you trigger the shutter with the other. Count several seconds to let any mirror-slap-induced vibrations dampen out and then pull the card away. Finally, choose the exposure duration. Begin with 1 minute and double the exposure on each of the next five frames to 16 minutes. If the mount's polar axis is aimed properly at the celestial pole, the clock drive should track the stars accurately.

Through the Telescope II: Long Exposures

The most challenging of all astrophotos to take are those magnificent pictures of deep-sky objects that adorn the pages of every astronomy magazine and book (including those found in this one). How do these photographers do it?

To begin with, long-exposure telescopic photography employs either the prime-focus or telecompression methods highlighted in the short-exposure section. To take successful long exposures, however, requires more than a camera, telescope, and adapter. With exposure times that can extend for an hour or more, long-exposure astrophotography requires a rock-steady equatorial mounting precisely aligned to the celestial pole, sophisticated (in other words, *expensive*) accessories, and an overabundance of patience.

Unlike guided exposures, where the clock drive can run unattended during the exposure, through-the-telescope photography requires constant monitoring to make certain that the telescope is following the stars as it should. (Though modern-day clock drives are amazingly accurate—especially those featuring periodic-error-correction circuits—they still experience tracking errors.) To control the telescope's tracking rate, the photographer must place and keep a star in the center of an illuminated-reticle eyepiece.

But how can the photographer watch a guide star to make sure the telescope follows it correctly if the camera is looking through the eyepiece holder? There are two alternatives. Many astrophotographers opt for off-axis guiders, while others prefer side-mount, long-focus refractors. Both methods work the same way: The observer keeps his or her eye trained on that star to make sure it does not deviate from the eyepiece's crosshairs during the photograph. Any minor adjustments in the tracking rate of the clock drive are made using a handheld control box plugged into either a built-in or external dual-axis drive corrector. (Note that an increasing number of astrophotographers are using the autoguider capability of CCD cameras. This way, the telescope automatically compensates for any tracking errors while the photographer relaxes. Many observe side by side with binoculars or another telescope, but others

Astrophotography Log

Date	Frame #	Film Type	Object	Exposure	f/#	Lens/Telescope	Conditions	Comments

Figure 10.28 *Suggested log form for recording photographic details (exposures, film, etc).*

read or watch television!) Choice of film and exposure must be based on what is being photographed. In general, beginning astrophotographers will do best by selecting a fast (high-ISO) film. Take a look back at chapter 6 for some suitable choices. Also widely used today is hypersensitized Kodak Technical Pan 2415 black-and-white film. Normally operating at about ISO 25, TP 2415 jumps to ISO 200 when hypersensitized. Best of all, hypersensitizing the film significantly reduces reciprocity failure. This makes it ideal for the long exposures required for faint deep-sky objects.

Logging Off

Regardless of what kinds of astrophotos you take, it pays to record all the particulars in a permanent logbook. Be sure to include such items as date, subject, equipment (camera, lens, and/or telescope), film type and ISO value, frame number, length of exposure, and f-number used. One possible log format is included as Figure 10.28. By knowing all of this information, it will be easy to compare technique with the actual results once the film is returned. All the books in the world will not teach you as much about astrophotography as will learning from your mistakes.

Appendix A
Specs at a Glance

A. Binoculars

	Magnification × Aperture	Field of View (degrees)	Exit Pupil	Eye Relief	Weight[3] (oz.)	Tripod Socket?	BaK-4 Prisms?	Coatings
Price Range: Under $100								
Celestron Enduro	7 × 35	9.2	5	11	22	Yes	No	FC
Celestron Enduro	7 × 50	6.8	7.1	13	28	Yes	No	FC
Celestron Enduro	8 × 40	8.2	5	12	25	Yes	No	FC
Celestron Enduro	10 × 50	7	5	13	28	Yes	No	FC
Orion Scenix	7 × 50	7.1	7.1	17	28	Yes	No	FC
Orion Scenix	8 × 40	9	5	14	24	Yes	No	FC
Orion Scenix	10 × 50	7	5	14	28	Yes	No	FC
Price Range: $100 to $200								
Bausch & Lomb Legacy	7 × 35	11	5	9	22	Yes	Yes	MC
Bausch & Lomb Legacy	8 × 40	8.5	5	9	24	Yes	Yes	MC
Bausch & Lomb Legacy	10 × 50	7.5	5	9	28	Yes	Yes	MC
Bushnell Natureview	10 × 42	6	4.2	19	25	Yes	No	FC
Bushnell Natureview	8 × 42	8	6	19	26	Yes	No	FC
Celestron Pro	7 × 50	7.1	7.1	23	29	Yes	Yes	FMC
Celestron Pro	8 × 40	8.2	5	18	27	Yes	Yes	FMC
Celestron Pro	10 × 50	6.5	5	15	30	Yes	Yes	FMC
Celestron Pro	12 × 50	5.5	4.2	10	29	Yes	Yes	FMC
Edmund	7 × 50	7	7.1	17	29	No	No	FC
Meade Infinity	7 × 35	9.2	5	11	24	Yes	No	FC
Meade Infinity	8 × 40	8.2	5	11	26	Yes	No	MC
Meade Infinity	10 × 50	7	5	13	30	Yes	No	FC
Minolta Standard XL	7 × 35	9.3	5	18	27	Yes	Yes	MC
Minolta Standard XL	7 × 50	7.8	7.1	16	35	Yes	Yes	MC
Minolta Standard XL	10 × 50	6.5	5	18	35	Yes	Yes	MC
Nikon Sky & Earth	7 × 35	7.2	5	18	25	No	Yes	FMC
Orion Explorer	7 × 35	7.6	5	10	20	No	Yes	FC
Orion Explorer	7 × 50	6.8	7.1	17	26	No	Yes	FC
Orion Explorer	8 × 40	6.5	5	13	22	No	Yes	FC
Orion Explorer	10 × 50	5.3	5	14	26	No	Yes	FC
Orion UltraView	7 × 50	6.5	7.1	22	32	Yes	Yes	FMC

	Magnification × Aperture		Field of View (degrees)	Exit Pupil	Eye Relief	Weight[3] (oz.)	Tripod Socket?	BaK-4 Prisms?	Coatings
Orion UltraView	8 ×	42	8.2	5.25	22	27	Yes	Yes	FMC
Orion UltraView	10 ×	50	6.5	5	22	32	Yes	Yes	FMC
Pentax PCF III	7 ×	35	7.2	5	20	24	Yes	Yes	MC
Pentax PCF III	7 ×	50	6.2	7.1	20	34	Yes	Yes	MC
Pentax PCF III	8 ×	40	6.3	5	20	27	Yes	Yes	MC
Pentax PCF III	10 ×	50	5	5	20	34	Yes	Yes	MC
Pentax PCF III	12 ×	50	4.2	4.2	20	34	Yes	Yes	MC
Pentax PCF III	16 ×	50	3.1	4.2	20	35	Yes	Yes	MC
Pentax PCF III	20 ×	50	2.5	2.5	20	35	Yes	Yes	MC
Swift Vanguard	15 ×	60	4.2	4	10	40	Yes	No	MC

Price Range: $200 to $300

	Magnification × Aperture		Field of View (degrees)	Exit Pupil	Eye Relief	Weight[3] (oz.)	Tripod Socket?	BaK-4 Prisms?	Coatings
Celestron Ultima	7 ×	50	7	7.1	20	27	Yes	Yes	FMC
Celestron Ultima	8 ×	40	6.6	5	19	21	Yes	Yes	FMC
Celestron Ultima	8 ×	56	6.1	7	21	31	Yes	Yes	FMC
Celestron Ultima	10 ×	42	6.6	4.2	15	20	Yes	Yes	FMC
Celestron Ultima	10 ×	50	5	5	21	27	Yes	Yes	FMC
Meade Safari Pro	7 ×	36	9.3	5.1	19	24	Yes	Yes	MC
Meade Safari Pro	8 ×	42	8.2	5.3	19	26	Yes	Yes	MC
Minolta Standard XL	12 ×	50	5.5	4.2	14	35	Yes	Yes	MC
Nikon Sky & Earth	7 ×	50	6.2	7.1	16	35	No	Yes	FMC
Nikon Sky & Earth	8 ×	40	6.3	5	18	27	No	Yes	FMC
Orion Mini Giant	8 ×	56	5.8	7	18	32	Yes	Yes	FMC
Orion Mini Giant	9 ×	63	5	7	18	36	Yes	Yes	FMC
Orion Vista	7 ×	50	6	7.1	22	28	Yes	Yes	FMC
Orion Vista	8 ×	42	6.5	5.25	18	22	Yes	Yes	FMC
Orion Vista	10 ×	50	5.3	5	16	28	Yes	Yes	FMC
Swift Audubon	8.5 ×	44	8.2	5.2	9	29	Yes	Yes	MC
Swift Kestrel	10 ×	50	7	5	15	29	Yes	Yes	MC
Swift Ultra-Lite	7 ×	42	7	6	25	21	Yes	Yes	MC
Swift Ultra-Lite	8 ×	42	6.6	5.3	22	21	Yes	Yes	MC
Swift Ultra-Lite	10 ×	42	6.6	4.2	13	21	Yes	Yes	MC

Price Range: $300 to $500

	Magnification × Aperture		Field of View (degrees)	Exit Pupil	Eye Relief	Weight[3] (oz.)	Tripod Socket?	BaK-4 Prisms?	Coatings
Bausch & Lomb Custom	8 ×	36	6.5	4.5	19	22	Yes	Yes	MC
Bausch & Lomb Custom	10 ×	40	5.2	4	19	29	Yes	Yes	MC
Bausch & Lomb Discoverer	7 ×	42	8	6	20	29	No	Yes	MC
Bausch & Lomb Discoverer	10 ×	42	6.5	4.2	16	29	No	Yes	MC
Celestron ED	6.5 ×	44	7.3	6.8	20	26	Yes	Yes	FMC
Celestron ED	9.5 ×	44	6	4.6	16	26	Yes	Yes	FMC
Celestron Giant Standard	20 ×	80	3.5	4	14	74	Yes	Yes	MC
Celestron Giant Deluxe	20 ×	80	3.5	4	16	89	Yes	Yes	MC
Celestron Ultima	9 ×	63	5.4	7	21	35	Yes	Yes	FMC
Fujinon MT-SX	7 ×	50	7.5	7.1	12	45	Yes	Yes	FMC
Fujinon MT-SX	10 ×	70	5.3	7	12	76	Yes	Yes	FMC
Fujinon MT-SX	16 ×	70	4	4.4	12	76	Yes	Yes	FMC
Meade Safari Pro	10 ×	50	6.5	5	19	32	Yes	Yes	MC
Orion Giant	11 ×	80	4.5	7	16	77	Yes	Yes	FMC

	Magnification × Aperture	Field of View (degrees)	Exit Pupil	Eye Relief	Weight[3] (oz.)	Tripod Socket?	BaK-4 Prisms?	Coatings
Orion Giant	16 × 80	3.5	5	16	77	Yes	Yes	FMC
Orion Giant	20 × 80	3.5	4	15	77	Yes	Yes	FMC
Orion Little Giant II	11 × 70	5	7	18	47	Yes	Yes	FMC
Orion Little Giant II	15 × 70	4	5	18	47	Yes	Yes	FMC
Orion Little Giant II	20 × 70	3	3.5	18	47	Yes	Yes	FMC
Swift Audubon ED	8.5 × 44	8.2	5.2	14.5	29	Yes	Yes	FMC
Swift Observer	11 × 80	4.5	7.3	16	78	Yes	Yes	MC
Swift Satellite	20 × 80	3.5	4	15	80	Yes	Yes	MC

Price Range: $500 to $1,000

Fujinon FMT-SX	7 × 50	7.5	7.1	23	50	Yes	Yes	FMC
Fujinon FMT-SX	10 × 70	5.3	7	23	76	Yes	Yes	FMC
Fujinon FMT-SX	16 × 70	4	4.4	12	76	Yes	Yes	FMC
Orion/Vixen	36 × 80	1.1	2.2	20	11 lbs	Yes	Yes	MC
Nikon IF SP Prostars	7 × 50	7.3	7.1	15	48	No	Yes	FMC
Pentax PIF	7 × 50	7.3	7.1	20	56	Yes	Yes	FMC
Pentax PIF	10 × 50	6.5	5	15	56	Yes	Yes	FMC

Price Range: Over $1,000

Fujinon High Power	25 × 150	2.7	6	19	41 lbs	Yes	Yes	FMC
Fujinon High Power	40 × 150	1.7	3.8	19	41 lbs	Yes	Yes	FMC
Miyauchi BJ100iB	20 × 100	2.5	5	27	10 lbs	Yes	Yes	FMC
Miyauchi BS-77	20 × 77	2.5	3.9	20	6.6 lbs	Yes	Yes	FMC
Miyauchi BJ100iA	20 × 100	2.5	5	27	10 lb	Yes	Yes	FMC
Nikon IF SP Astroluxe	10 × 70	5.1	7	15	70	No	Yes	FMC
Orion SuperGiant	14 × 100	3.3	7	14	120	Yes	Yes	FMC
Orion SuperGiant	25 × 100	2.6	4	10	120	Yes	Yes	FMC
Orion/Vixen	20 × 125	3.0	6.2	20	24 lbs	Yes	Yes	MC
Orion/Vixen	30 × 125	1.6	4.2	20	24 lbs	Yes	Yes	MC
Zeiss B/GA T Dialyt	7 × 42	8.6	6	19	28	No	No	FMC
Zeiss B/GA T Dialyt	10 × 40	6.3	4	16	27	No	No	FMC
Zeiss B/GA Design Selection	8 × 56	7.5	7	17	51	No	No	FMC
Zeiss B/GA Design Selection	10 × 56	6.3	5.6	16	50	No	No	FMC

B. Achromatic Refractors

	Aperture	Focal Ratio	Tube Material[1]	Mounting Type[2]	Total Weight[3]	Heaviest Assembly
Price Range: Under $300						
Orion ShortTube 80	3.1	5	Aluminum	n/s	n/a	n/s
Price Range: $300 to $600						
Celestron Firstscope 80	3.1	11.4	Aluminum	Alt-az	18	n/s
Celestron Firstscope 80	3.1	11.4	Aluminum	GEM	24	n/s
Celestron Rich Field 80	3.1	5	Aluminum	n/s	n/a	2

	Aperture	Focal Ratio	Tube Material[1]	Mounting Type[2]	Total Weight[3]	Heaviest Assembly
Meade 390	3.5	11	Aluminum	Alt-az	17	17
Orion Observer 80	3.1	11.3	Aluminum	GEM	n/s	n/s
Orion/Vixen VX80	3.1	11.4	Aluminum	Alt-az	n/s	n/s
Price Range: $600 to $1,000						
Meade 395	3.5	11	Aluminum	GEM	27	27
Orion SkyView 90	3.5	11.1	Aluminum	GEM	n/s	n/s
Tele Vue Ranger (semi-apochromat)	2.8	6.8	Aluminum	n/s	n/a	3.5
Price Range: $1,000 to $2,000						
Celestron GP-C102	4	9.8	Aluminum	GEM	39	n/s
D&G Optical	5	12,15,or 20	Aluminum	n/s	n/a	19–20
D&G Optical	6	12,15,or 20	Aluminum	n/s	n/a	26–28
Orion/Vixen VX90	3.5	11.1	Aluminum	GEM	42	n/s
Orion/Vixen VX102	4	9.8	Aluminum	GEM	48	n/s
Tele Vue Pronto (semi-apochromat)	2.8	6.8	Aluminum	n/s	n/a	6
Price Range: Over $2,000						
D&G Optical	8	12,15,or 20	Aluminum	n/s	n/a	37–40
D&G Optical	10	12,15,or 20	Aluminum	n/s	n/a	75–80

C. Apochromatic Refractors

	Aperture	Focal Ratio	Tube Material[1]	Mounting Type[2]	Total Weight[3]	Heaviest Assembly
Price Range: $1,000 to $3,000						
Astro-Physics 105EDFS	4.1	6	Aluminum	n/s	n/a	9
Celestron GP-C102ED	4	9	Aluminum	GEM	39	n/s
Meade Apochromatic 102ED	4	9	Aluminum	GEM	69	23
Orion/Vixen VX90	3.5	9	Aluminum	GEM	41	n/s
Orion/Vixen VX102	4	9	Aluminum	GEM	48	n/s
Tele Vue Genesis-SDF	4	5.4	Aluminum	n/s	n/a	12
Tele Vue Renaissance-SDF	4	5.4	Aluminum	n/s	n/a	13
Tele Vue TV-101	4	5.4	Aluminum	n/s	n/a	12
Price Range: $3,000 to $5,000						
Astro-Physics 130EDFS	5.1	6	Aluminum	n/s	n/a	15
Astro-Physics 155EDFS	6.1	7	Aluminum	n/s	n/a	23
Meade Apochromatic 127ED	5	9	Aluminum	GEM	78	23
Meade Apochromatic 152ED	6	9	Aluminum	GEM	153	55
Takahashi FS-78	3	8	Aluminum	GEM	30	23
Takahashi FS-102	4	8	Aluminum	GEM	35	24
Price Range: Over $5,000						
Astro-Physics 155EDF	6.1	7	Aluminum	n/s	n/a	27 (OTA)
Meade Apochromatic 178ED	7	9	Aluminum	GEM	176	55
Takahashi FS-128	5	8.1	Aluminum	GEM	73	56
Tele Vue TV-140	5.5	5	Aluminum	n/s	n/a	16

	Aperture	Focal Ratio	Tube Material[1]	Mounting Type[2]	Total Weight[3]	Heaviest Assembly
Price Range: Way Over $5,000						
Takahashi FC-150	6	8	Aluminum	GEM	n/s	n/s
Takahashi FCT-150	6	7	Aluminum	GEM	n/s	n/s
Takahashi FCT-200	8	10	Aluminum	GEM	n/s	n/s

D. Newtonian Reflectors

	Aperture	Focal Ratio	Tube Material[1]	Mounting Type[2]	Total Weight[3]	Heaviest Assembly
Price Range: Under $300						
Celestron Star-Hopper	4.5	8	Aluminum	Alt-az	18	n/s
Coulter Odyssey 6	6	8	Cardboard	Dob	37	n/s
Edmund 3	3	6	Plastic	Fork	12	12
North Star 6	6	8	Cardboard	Dob	35	n/s
Orion ShortTube	4.5	8.8	Aluminum	GEM	n/s	n/s
Stargazer Steve Kit	4.25	10.5	Cardboard	Dob	16	16
Stargazer Steve SGR-3	3	10	Cardboard	Alt-az	7	7
Price Range: $300 to $500						
Celestron Firstscope 114	4.25	8	Aluminum	GEM	19	n/s
Celestron Star-Hopper 6	6	8	Cardboard	Dob	40	n/s
Coulter Odyssey 8	8	4.5 or 7	Cardboard	Dob	39–59	n/s
Edmund Astroscan	4.25	4	Plastic	Ball	13	11
Meade 4500	4.5	8	Aluminum	GEM	27	n/s
Meade Starfinder 6	6	8	Cardboard	Dob	34	20
Meade Starfinder 8	8	6	Cardboard	Dob	44	26
North Star 8	8	6	Cardboard	Dob	43	n/s
Orion Optics Europa 110 (UK)	4.5	8	Aluminum	GEM	26	n/s
Orion Deep-Space Explorer 6	6	8	Cardboard	Dob	33	18
Orion Deep-Space Explorer 8	8	6	Cardboard	Dob	41	21
Orion SkyView 4.5	4.5	8	Aluminum	GEM	n/s	n/s
Orion ShortTube Ultra	4.5	8.8	Aluminum	GEM	n/s	n/s
Starsplitter Gem 4.5	4.5	8	Aluminum	GEM	28	n/s
Starsplitter Gem 6	6	5	Aluminum	GEM	40	n/s
Price Range: $500 to $1,000						
Celestron C4.5	4.5	7.9	Aluminum	GEM	28	n/s
Celestron Star Hopper 8	8	6	Cardboard	Dob	56	n/s
Celestron Star Hopper 11	11	4.5	Cardboard	Dob	64	37
Coulter Odyssey 10	10.1	4.5	Cardboard	Dob	65	n/s
Coulter Odyssey 13	13.1	4.5	Cardboard	Dob	97	n/s
Dark Star Custom 6.25	6.25	7.7	PVC (plastic)	Dob	n/s	22
Dark Star Custom 8.75	8.75	5.8 or 7.3	PVC (plastic)	Dob	n/s	29
Meade Starfinder 10	10	4.5	Cardboard	Dob	62	32
Meade Newtonian 12.5	12.5	4.8	Cardboard	Dob	96	55
Meade Starfinder Equatorial 6	6	8	Cardboard	GEM	57	35
Meade Starfinder Equatorial 8	8	6	Cardboard	GEM	69	35
Meade Starfinder Equatorial 10	10	4.5	Cardboard	GEM	97	35
MorningStar MS-8	8	6 or 8	Cardboard	Dob	45	n/s

	Aperture	Focal Ratio	Tube Material[1]	Mounting Type[2]	Total Weight[3]	Heaviest Assembly
North Star 10	10	6	Cardboard	Dob	54	n/s
Orion Deep-Space Explorer 10	10	4.5	Cardboard	Dob	68	38
Orion Premium DS Explorer 10	10	5.6	Cardboard	Dob	77	44
Orion Optics Europa 150 (UK)	6	5	Aluminum	GEM	33	n/s
Orion Optics Europa 200 (UK)	8	6	Aluminum	GEM	47	n/s
Orion Optics SX150 (UK)	6	6	Aluminum	GEM	40	n/s
StarMaster Oak Classic	7	5.6	Composite	Dob	41	21
Starsplitter Compact 8	8	6	Two-pole	Dob	35	30
Starsplitter Gem 6 Premium	6	5	Aluminum	GEM	40	n/s

Price Range: $1,000 to $2,000

	Aperture	Focal Ratio	Tube Material[1]	Mounting Type[2]	Total Weight[3]	Heaviest Assembly
Celestron Star Hopper 14	14	4.5	Cardboard	Dob	105	60
Celestron Star Hopper 17.5	17.5	4.1	Cardboard	Dob	180	100
Dark Star Custom 10	10	5.1 or 6.4	PVC (plastic)	Dob	n/s	35
Dark Star Custom 12	12	4.3 or 5.3	PVC (plastic)	Dob	n/s	44
Dark Star Custom 14	14	5	PVC (plastic)	Dob	n/s	57
Mag One PortaBall 8	8	6	Alum. truss	Ball	30	24
Meade Starfinder 16	16	4.5	Cardboard	Dob	170	100
MorningStar MS-10	10	8	Alum. truss	Dob	75	30
Orion Premium DS Explorer 12.5	12.5	4.8	Cardboard	Dob	98	61
Orion Premium DS Explorer 16	16	4.5	Cardboard	Dob	175	95
Orion/Vixen R200SS	7.9	4	Aluminum	GEM	n/s	16
Orion Optics Europa 250 (UK)	10	4.8	Aluminum	GEM	57	n/s
Orion Optics SX200 (UK)	8	6	Aluminum	GEM	48	n/s
Orion Optics SX250 (UK)	10	4.5	Aluminum	GEM	55	n/s
Orion Optics GX150 (UK)	6	6	Aluminum	GEM	44	n/s
Orion Optics GX200 (UK)	8	6	Aluminum	GEM	53	n/s
StarMaster Large Aperture 10	10	6	Alum truss	Dob	74	44
Starsplitter Compact 10	10	6	Two-pole	Dob	40	35
Starsplitter Compact 12.5	12.5	4.8	Two-pole	Dob	55	50
Starsplitter Compact II 10	10	6	Alum truss	Dob	35	30
Takahashi MT-130	5.2	6	Aluminum	n/s	n/a	12

Price Range: $2,000 to $3,000

	Aperture	Focal Ratio	Tube Material[1]	Mounting Type[2]	Total Weight[3]	Heaviest Assembly
Dark Star CompuDob 10	10	5.1 or 6.4	PVC (plastic)	Dob	n/s	35
Dark Star CompuDob 12	12	4.3 or 5.3	PVC (plastic)	Dob	n/s	44
Dark Star Custom 16	16	5	PVC (plastic)	Dob	n/s	86
Mag One PortaBall 12.5	12.5	4.8	Alum. truss	Ball	67	55
Meade Starfinder Equatorial 16	16	4.5	Cardboard	GEM	247	62
MorningStar MS-10D	10	8	Alum truss	Dob w/drive	77	32
Optical Guidance Systems N-8	8	6	Aluminum	n/s	n/a	30
Orion Optics GX250 (UK)	10	4.5	Aluminum	GEM	59	n/s
Parallax Instruments PI200	8	7.5	Aluminum	n/s	n/a	42
Parallax Instruments PI250	10	6.5	Aluminum	n/s	n/a	52
Sky Designs 14.5	14.5	4.5	Alum. truss	Dob	88	n/s
Sky Valley Ultra Light 12.5	12.5	4.5 or 5.5	Alum. truss	Dob	60	40
StarMaster Large Aperture 12.5	12.5	5/5.6	Alum. truss	Dob	96	30
StarMaster Large Aperture 14.5	14.5	4.5 or 5.2	Alum. truss	Dob	128	41
Starsplitter Compact II 12.5	12.5	4.8	Alum. truss	Dob	40	35
Starsplitter Compact II 14.5	14.5	4.5	Alum. truss	Dob	55	50

	Aperture	Focal Ratio	Tube Material[1]	Mounting Type[2]	Total Weight[3]	Heaviest Assembly
Starsplitter Compact 14.5	14.5	4.5	Two-pole	Dob	67	60
Starsplitter II 12.5	12.5	4.8 or 6	Alum. truss	Dob	70	60
Takahashi MT-160	6.3	6	Aluminum	n/s	n/a	18
Tectron 15	15	5	Alum. truss	Dob	100	n/s

Price Range: $3,000 to $5,000

	Aperture	Focal Ratio	Tube Material[1]	Mounting Type[2]	Total Weight[3]	Heaviest Assembly
Dark Star CompuDob 14	14	5	PVC (plastic)	Dob	n/s	57
Dark Star CompuDob 16	16	5	PVC (plastic)	Dob	n/s	86
Dark Star CompuDob 500	20	4	Alum. truss	Dob	n/s	n/s
JMI NGT-12.5	12.5	4.5	Alum. truss	SREM	115	38
Obsession 15	15	4.5	Alum. truss	Dob	88	60
Obsession 18	18	4.5	Alum. truss	Dob	109	75
Obsession 20	20	5	Alum. truss	Dob	141	90
Optical Guidance Systems N-10	10	5	Aluminum	n/s	n/a	35
Orion Optics DX300 (UK)	12	5.3	Aluminum	GEM	26	n/s
Parallax Instruments PI320	12.5	6	Aluminum	n/s	n/a	70
Parallax Instruments PI370	14.5	6	Aluminum	n/s	n/a	85
Parallax Instruments PI400	16	6	Aluminum	n/s	n/a	100
Sky Designs 18	18	4.5	Alum. truss	Dob	138	n/s
Sky Designs 20	20	4	Alum. truss	Dob	166	n/s
Sky Valley Rotating Tube 12.5	12.5	4.5 or 5.5	Alum. truss	Dob	120	85
Sky Valley Rotating Tube 14	14	4.5 or 5.5	Alum. truss	Dob	130	75
Sky Valley Rotating Tube 16	16	4.5 or 5.5	Alum. truss	Dob	145	105
Sky Valley Rotating Tube 18	18	4.5 or 5.5	Alum. truss	Dob	160	120
Sky Valley Ultra Light 14	14	4.5 or 5.5	Alum. truss	Dob	75	53
Sky Valley Ultra Light 16	16	4.5 or 5.5	Alum. truss	Dob	100	60
Sky Valley Ultra Light 18	18	4.5 or 5.5	Alum. truss	Dob	115	75
StarMaster Large Aperture 16	16	4.5 or 5	Alum. truss	Dob	136	48
StarMaster Large Aperture 18	18	4.5	Alum. truss	Dob	158	58
StarMaster Large Aperture 20	20	4 or 4.5	Alum. truss	Dob	180	68
StarMaster Large Aperture 22	22	4.1 or 4.5	Alum. truss	Dob	205	82
Starsplitter II 14.5	14.5	5	Alum. truss	Dob	80	70
Starsplitter II 16	16	4.7	Alum. truss	Dob	90	80
Starsplitter II 18	18	4.5	Alum. truss	Dob	102	90
Starsplitter II 20	20	5	Alum. truss	Dob	137	125
Takahashi MT-200	7.9	6	Aluminum	n/s	n/a	32
Tectron 18	18	5	Alum. truss	Dob	105	n/s
Tectron 20	20	5	Alum. truss	Dob	150	n/s

Price Range: Over $5,000

	Aperture	Focal Ratio	Tube Material[1]	Mounting Type[2]	Total Weight[3]	Heaviest Assembly
JMI NGT-18	18	4.5	Alum. truss	SREM	230	75
JMI NTT-25	25	5	Alum. truss	Alt-az	350	275
JMI NTT-30	30	4.2	Alum. truss	Alt-az	425	350
Obsession 25	25	5	Alum. truss	Dob	240	110
Obsession 30	30	4.5	Alum. truss	Dob	395	175
Optical Guidance Systems N-12.5	12.5	4.8	Aluminum	n/s	n/a	55
Optical Guidance Systems N-14.5	14.5	5.5	Aluminum	n/s	n/a	90
Optical Guidance Systems N-16	16	5	Aluminum	n/s	n/a	105
Optical Guidance Systems N-20	20	4	Aluminum	n/s	n/a	160
Optical Guidance Systems N-24	24	4	Aluminum	n/s	n/a	230

	Aperture	Focal Ratio	Tube Material[1]	Mounting Type[2]	Total Weight[3]	Heaviest Assembly
Starsplitter II 22	22	4.5	Alum. truss	Dob	155	140
Starsplitter II 25	25	5	Alum. truss	Dob	238	220
Starsplitter II 30	30	4.5 or 5	Alum. truss	Dob	392	370
Takahashi CN-212	8.3	3.9 or 12.5	Aluminum	GEM	n/s	21 (OTA)
Takahashi Epsilon 160	6.3	3.3	Aluminum	GEM	n/s	17 (OTA)
Takahashi Epsilon 210	8.2	3	Aluminum	GEM	n/s	25 (OTA)
Takahashi Epsilon 250	9.8	3.4	Aluminum	GEM	n/s	42 (OTA)
Takahashi MT-250	9.8	6	Aluminum	n/s	n/a	n/s
Tectron 24	24	4.5	Alum. truss	Dob	160	n/s
Tectron 30	30	4.5	Alum. truss	Dob	270	n/s

E. Cassegrain Reflectors

	Aperture	Focal Ratio	Tube Material[1]	Mounting Type[2]	Total Weight[3]	Heaviest Assembly
Price Range: $5,000 to $10,000						
OGS Classic Cassegrain-10	10	15	Aluminum	GEM	n/a	35 (OTA)
OGS Classic Cassegrain-12.5	12.5	16	Aluminum	GEM	n/a	55 (OTA)
OGS Classic Cassegrain-14.5	14.5	16	Aluminum	GEM	n/a	85 (OTA)
OGS Ritchey-Chretien-10	10	8.5	Aluminum	GEM	n/a	30 (OTA)
OGS Ritchey-Chretien-12.5	12.5	9	Aluminum	GEM	n/a	50 (OTA)
Takahashi Mewlon M-210	8.3	11.5	Aluminum	GEM	n/a	18 (OTA)
Price Range: $10,000 to $20,000						
OGS Classic Cassegrain-16	16	14.25	Aluminum	GEM	n/a	95 (OTA)
OGS Ritchey-Chretien-14.5	14.5	7.9	Aluminum	GEM	n/a	70 (OTA)
OGS Ritchey-Chretien-16	16	8.4	Aluminum	GEM	n/a	95 (OTA)
Takahashi Mewlon M-250	9.8	12	Aluminum	GEM	n/a	28 (OTA)
Price Range: Over $20,000						
OGS Classic Cassegrain-20	20	16	Aluminum	GEM	n/a	160 (OTA)
OGS Classic Cassegrain-24	24	16	Aluminum	GEM	n/a	230 (OTA)
OGS Classic Cassegrain-32	32	10	Aluminum	GEM	n/a	350 (OTA)
OGS Ritchey-Chretien-20	20	8.1	Aluminum	GEM	n/a	150 (OTA)
OGS Ritchey-Chretien-24	24	8	Aluminum	GEM	n/a	210 (OTA)
OGS Ritchey-Chretien-32	32	7.6	Aluminum	GEM	n/a	340 (OTA)
Takahashi Mewlon M-300	11.8	11.9	Aluminum	GEM	n/a	55 (OTA)

F. Schmidt-Cassegrain Catadioptrics

	Aperture	Focal Ratio	Tube Material[1]	Mounting Type[2]	Total Weight[3]	Heaviest Assembly
Price Range: Under $1,000						
Celestron C5 Spotting	5	10	Aluminum	n/s	6	n/s
Price Range: $1,000 to $2,000						
Celestron C5+	5	10	Aluminum	Fork	23	23
Celestron Celestar 8	8	10	Aluminum	Fork	37	n/s

	Aperture	Focal Ratio	Tube Material[1]	Mounting Type[2]	Total Weight[3]	Heaviest Assembly
Celestron Celestar 8 Deluxe	8	10	Aluminum	Fork	40	n/s
Celestron Celestar 8 Computerized	8	10	Aluminum	Fork	37	n/s
Celestron GP-C8	8	10	Aluminum	GEM	45	n/s
Meade LX10	8	10	Aluminum	Fork	49	26
Meade LX50	8	10	Aluminum	Fork	71	38
Meade LX50	10	10	Aluminum	Fork	89	55
Orion/Vixen VC200L	7.9	9	Aluminum	GEM	n/s	15 (OTA)
Orion XSN200 (UK) Schmidt-Newt	8	4	Aluminum	GEM	n/s	48

Price Range: $2,000 to $3,000

	Aperture	Focal Ratio	Tube Material[1]	Mounting Type[2]	Total Weight[3]	Heaviest Assembly
Celestron Celestar 8 Comp Deluxe	8	10	Aluminum	Fork	40	n/s
Celestron CG-9¼	9.25	10	Aluminum	GEM	53	21
Celestron Fastar 8	8	10	Aluminum	Fork	42	n/s
Celestron Ultima 2000	8	10	Aluminum	Alt-az/Fork	39	28
Celestron Ultima 9¼	9.25	10	Aluminum	Fork	74	n/s
Meade LX200	8	6.3 or 10	Aluminum	Fork	69	41
Orion GPX200 (UK) Schmidt-Newt)	8	4	Aluminum	GEM	n/s	53

Price Range: Over $3,000

	Aperture	Focal Ratio	Tube Material[1]	Mounting Type[2]	Total Weight[3]	Heaviest Assembly
Celestron CG-11	11	10	Aluminum	GEM	100	31
Celestron Ultima 11	11	10	Aluminum	Fork	80	n/s
Celestron CG-14	14	10	Aluminum	GEM	135	31
Meade LX200	10	6.3 or 10	Aluminum	Fork	86	58
Meade LX200	12	10	Aluminum	Fork	120	70
Meade LX200	16	10	Aluminum	Fork	313	120

G. Maksutov-Cassegrain Catadioptrics

	Aperture	Focal Ratio	Tube Material[1]	Mounting Type[2]	Total Weight[3]	Heaviest Assembly
Price Range: Under $1,000						
Meade ETX Astro	3.5	13.8	Aluminum	Fork	9	9
Price Range: $1,000 to $2,000						
Ceravolo HD145 (Mak-Newt)	5.7	6	Aluminum	n/s	n/a	12
Intes MK 67	6	10	Aluminum	n/s	n/a	10
Intes MK 69	6	6	Aluminum	n/s	n/a	n/s
Intes Micro Alter 603	6	10	Aluminum	n/s	n/a	13
Meade LX50	7	15	Aluminum	Fork	82	49
Orion Argonaut 150	6	12	Aluminum	n/s	n/a	10
Price Range: $2,000 to $4,000						
Intes MK 91	9	13.5	Aluminum	n/a	n/a	27
Intes Micro Alter 809	8	10	Aluminum	n/s	n/a	22
Meade LX200	7	15	Aluminum	Fork	80	52
Questar	3.5	15	Aluminum	Fork	8	8

	Aperture	Focal Ratio	Tube Material[1]	Mounting Type[2]	Total Weight[3]	Heaviest Assembly
Price Range: Over $4,000						
Ceravolo HD216 Mak-Newt	8.5	6	Aluminum	n/s	n/a	35
Intes Micro Alter 1008	10	10	Aluminum	n/s	n/a	44

H. Hyperbolic Astrographs

	Aperture	Focal Ratio	Tube Material[1]	Mounting Type[2]	Total Weight[3]	Heaviest Assembly
Price Range: Over $3,000						
Takahashi Epsilon	6.3	3.3	Aluminum	GEM	n/s	17 (OTA)
	8.3	3	Steel	GEM	n/s	30 (OTA)
	9.8	3.4	Aluminum	GEM	n/s	42 (OTA)

Notes:

1. *Tube material:*
   Aluminum: Solid-wall aluminum tube.
   Alum. truss: Open truss built from 6 or 8 aluminum poles.
   Cardboard: Sonotube or similar concrete-form tubing commonly used in construction.
   Composite: Cardboard-like tube coated with a vinyl finish.
   Plastic: Solid-wall plastic tube.
   Steel: Solid-wall steel tube.
   Two-pole: Open design that uses two poles to support eyepiece mount and diagonal mirror assembly.

2. *Mounting type:*
   Alt-az: Altitude-azimuth mount.
   Dob: Dobsonian-style altitude-azimuth mount.
   Fork: Fork equatorial mount.
   GEM: German equatorial mount.
   SREM: Split-ring equatorial mount.
   n/s: Not supplied; mounting must be purchased separately.

3. *"Total Weight" includes both telescope and mounting, while "Heaviest Assembly" lists the weight of the heaviest single part that a user is likely to carry out into the field. If the weights are the same in both columns, then the telescope and mounting are usually transported together. In addition, the following abbreviations apply:*
   n/s: Not supplied by manufacturer.
   n/a: Not applicable, usually denoting a telescope that is sold without a mounting; instead, a mounting must be purchased separately, which, depending on the mount used, may cause the total instrument weight to vary. In these cases, the weight quoted in "Heaviest Assembly" is that of the optical tube assembly (OTA). Bear in mind that the mounting will probably be heavier.
   All weights are as supplied by the manufacturer and have not been independently verified. All are expressed in pounds (rounded to the nearest pound), except for binoculars, which are expressed in ounces. To convert, there are 16 ounces per pound and 2.2 pounds per kilogram.

Appendix B
Eyepiece Marketplace

Company	Series	FL	AFV	Ctg	ER	EC	Pf
0.965-inch Barrel: Under $50							
Celestron	Plössl	26	52	MC	22	No	n/s
Celestron	Plössl	17	52	MC	13	No	n/s
Celestron	Plössl	12.5	52	MC	8	No	n/s
Celestron	Plössl	10	52	MC	7	No	n/s
Celestron	Plössl	7.5	52	MC	5	No	n/s
Celestron	Plössl	6.3	52	MC	5	No	n/s
Celestron	SMA	25	52	C	n/s	No	n/s
Celestron	SMA	12	52	C	n/s	No	n/s
Celestron	SMA	10	52	C	n/s	No	n/s
Celestron	SMA	6	52	C	n/s	No	n/s
Meade	MA	40	36	C	18	No	No
Meade	MA	25	40	C	16	No	No
Meade	MA	12	40	C	8	No	No
Meade	MA	9	40	C	6	No	No
Orion	Explorer II	25	50	FC	15	No	Yes
Orion	Explorer II	17	50	FC	11	No	Yes
Orion	Explorer II	13	50	FC	7	No	Yes
Orion	Explorer II	10	50	FC	4	No	Yes
Orion	Explorer II	6	50	FC	3	No	Yes
0.965-inch Barrel: $50 and Above							
Orion	Plössl	26	50	FC	16	Yes	Yes
Orion	Plössl	20	50	FC	13	Yes	Yes
Orion	Plössl	17	50	FC	10	Yes	Yes
Orion	Plössl	12.5	50	FC	8	Yes	Yes
Orion	Plössl	10	50	FC	5	Yes	Yes
Orion	Plössl	7.5	50	FC	4	Yes	Yes
1.25-inch Barrels: Under $50							
Adorama	Plössl	25	50	FC	22	No	Yes
Adorama	Plössl	20	50	FC	n/s	No	Yes
Adorama	Plössl	17	50	FC	13	No	Yes
Adorama	Plössl	10	50	FC	6.5	No	Yes
Adorama	Plössl	7.5	50	FC	5	No	Yes
Celestron	Plössl	26	52	MC	22	No	n/s
Celestron	Plössl	20	50	MC	20	No	n/s
Celestron	Plössl	17	52	MC	13	No	n/s
Celestron	Plössl	12.5	52	MC	8	No	n/s
Celestron	Plössl	10	52	MC	7	No	n/s
Celestron	Plössl	7.5	52	MC	5	No	n/s
Celestron	Plössl	6.3	52	MC	5	No	n/s

Company	Series	FL	AFV	Ctg	ER	EC	Pf
Celestron	SMA	25	52	C	n/s	No	n/s
Celestron	SMA	12	52	C	n/s	No	n/s
Celestron	SMA	10	52	C	n/s	No	n/s
Celestron	SMA	6	52	C	n/s	No	n/s
Meade	MA	40	36	MC	18	No	No
Meade	MA	25	40	MC	16	No	No
Meade	MA	12	40	MC	8	No	No
Meade	MA	9	40	MC	6	No	No
Orion	Explorer II	25	50	FC	14.5	No	Yes
Orion	Explorer II	17	50	FC	11.5	No	Yes
Orion	Explorer II	13	50	FC	7	No	Yes
Orion	Explorer II	10	50	FC	5	No	Yes
Orion	Explorer II	6	50	FC	4	No	Yes

1.25-inch Barrels: $50 to $100

Company	Series	FL	AFV	Ctg	ER	EC	Pf
Adorama	Plössl	40	43	FC	31	No	Yes
Adorama	Plössl	32	50	FC	22	No	Yes
Adorama	Plössl	12.5	50	FC	8	No	Yes
Adorama	Plössl	6.3	50	FC	5	No	Yes
Celestron	Plössl	40	46	MC	31	No	n/s
Celestron	Plössl	32	50	MC	22	No	n/s
Celestron	Ultima	12.5	51	FMC	9	Yes	Yes
Celestron	Ultima	7.5	51	FMC	5	Yes	Yes
Celestron	Ultima	5	50	FMC	4	Yes	Yes
Edmund	RKE	28	45	C	25	Yes	Yes
Edmund	RKE	21	45	C	19	No	Yes
Edmund	RKE	15	45	C	13	No	Yes
Edmund	RKE	12	45	C	11	No	Yes
Edmund	RKE	8	45	C	8	No	Yes
Meade	Plössl	40	44	MC	29	No	No
Meade	Plössl	25	50	MC	16	No	Yes
Meade	Plössl	16	50	MC	10	No	Yes
Meade	Plössl	9.5	50	MC	6	No	Yes
Meade	Plössl	6.7	50	MC	4	No	Yes
Meade	Plössl	5	50	MC	3	No	Yes
Meade	Super Plössl	26	52	MC	18	Yes	Yes
Meade	Super Plössl	20	52	MC	13	Yes	Yes
Meade	Super Plössl	15	52	MC	9	Yes	Yes
Meade	Super Plössl	12.4	52	MC	7	Yes	Yes
Meade	Super Plössl	9.7	52	MC	5	Yes	Yes
Meade	Super Plössl	6.4	52	MC	3	Yes	Yes
Orion	Plössl	40	43	FC	22	Yes	No
Orion	Plössl	32	50	FC	25	Yes	Yes
Orion	Plössl	26	50	FC	18	Yes	Yes
Orion	Plössl	20	50	FC	14	Yes	Yes
Orion	Plössl	17	50	FC	13	Yes	Yes
Orion	Plössl	12.5	50	FC	7	Yes	Yes
Orion	Plössl	10	50	FC	7	Yes	Yes
Orion	Plössl	7.5	50	FC	5	Yes	Yes
Orion	Ultrascopic	25	52	FMC	17	Yes	Yes
Orion	Ultrascopic	20	52	FMC	13	Yes	Yes

Company	Series	FL	AFV	Ctg	ER	EC	Pf
Orion	Ultrascopic	15	52	FMC	10	Yes	Yes
Orion	Ultrascopic	10	52	FMC	6	Yes	Yes
Orion	Ultrascopic	7.5	52	FMC	5.3	Yes	Yes
Orion	Ultrascopic	5	52	FMC	6	Yes	Yes
Orion	Ultrascopic	3.8	52	FMC	5.3	Yes	Yes
Tele Vue	Plössl	15	50	FMC	10	Yes	Yes
Tele Vue	Plössl	11	50	FMC	8	Yes	Yes
Tele Vue	Plössl	8	50	FMC	6	Yes	Yes

1.25-inch Barrels: $100 to $200

Company	Series	FL	AFV	Ctg	ER	EC	Pf
Celestron	Ultima	42	36	FMC	32	Yes	No
Celestron	Ultima	35	49	FMC	25	Yes	No
Celestron	Ultima	30	50	FMC	21	Yes	Yes
Celestron	Ultima	24	51	FMC	18	Yes	Yes
Celestron	Ultima	18	51	FMC	13	Yes	Yes
Meade	Super Wide Angle	24.5	67	MC	19	Yes	Yes
Meade	Super Wide Angle	18	67	MC	14	Yes	Yes
Meade	Super Wide Angle	13.8	67	MC	10	Yes	Yes
Meade	Ultra Wide Angle	6.7	84	MC	11	Yes	Yes
Meade	Ultra Wide Angle	4.7	84	MC	7	Yes	Yes
Meade	Super Plössl	40	44	MC	30	Yes	No
Meade	Super Plössl	32	52	MC	20	Yes	Yes
Orion	Ultrascopic	35	49	FMC	25	Yes	No
Orion	Ultrascopic	30	52	FMC	20.5	Yes	Yes
Takahashi	LE	24	52	FMC	17	Yes	Yes
Takahashi	LE	18	52	FMC	13	Yes	Yes
Takahashi	LE	12.5	52	FMC	9	Yes	Yes
Takahashi	LE	7.5	52	FMC	10	Yes	Yes
Takahashi	LE	5	52	FMC	10	Yes	Yes
Tele Vue	Nagler	4.8	82	FMC	7	Yes	Yes
Tele Vue	Plössl	40	43	FMC	28	Yes	No
Tele Vue	Plössl	32	50	FMC	22	Yes	Yes
Tele Vue	Plössl	25	50	FMC	17	Yes	Yes
Tele Vue	Plössl	20	50	FMC	14	Yes	Yes
Vernonscope	Brandon	32	50	FC	26	Yes	Yes
Vernonscope	Brandon	24	50	FC	19	Yes	Yes
Vernonscope	Brandon	16	50	FC	13	Yes	Yes
Vernonscope	Brandon	12	50	FC	10	Yes	Yes
Vernonscope	Brandon	8	50	FC	6	Yes	Yes
Vixen	Lanthanum LV	15	50	FMC	20	Yes	Yes
Vixen	Lanthanum LV	12	50	FMC	20	Yes	Yes
Vixen	Lanthanum LV	10	50	FMC	20	Yes	Yes
Vixen	Lanthanum LV	9	50	FMC	20	Yes	Yes
Vixen	Lanthanum LV	6	45	FMC	20	Yes	Yes
Vixen	Lanthanum LV	5	45	FMC	20	Yes	Yes
Vixen	Lanthanum LV	4	45	FMC	20	Yes	Yes
Vixen	Lanthanum LV	2.5	45	FMC	20	Yes	Yes

1.25-inch Barrels: *More than $200*

Company	Series	FL	AFV	Ctg	ER	EC	Pf
Pentax	XL	28	65	FMC	20	Yes	Yes
Pentax	XL	21	65	FMC	20	Yes	Yes

Company	Series	FL	AFV	Ctg	ER	EC	Pf
Pentax	XL	14	65	FMC	20	Yes	Yes
Pentax	XL	10.5	65	FMC	20	Yes	Yes
Pentax	XL	7	65	FMC	20	Yes	Yes
Pentax	XL	5.2	65	FMC	20	Yes	Yes
Takahashi	LE	30	52	FMC	20	Yes	Yes
Tele Vue	Nagler	7	82	FMC	10	Yes	Yes
Tele Vue	Panoptic	19	68	FMC	13	Yes	Yes
Tele Vue	Panoptic	15	68	FMC	10	Yes	Yes
1.25-inch/2-inch Barrel: $200 to $300							
Meade	Ultra Wide Angle	14	84	MC	23	Yes	Yes
Meade	Ultra Wide Angle	8.8	84	MC	16	Yes	Yes
Tele Vue	Panoptic	22	68	FMC	15	Yes	Yes
Tele Vue	Nagler Type 2	12	82	FMC	11	Yes	Yes
Tele Vue	Nagler	9	82	FMC	12	Yes	Yes
1.25-inch/2-inch Barrel: More than $300							
Tele Vue	Nagler Type 2	16	82	FMC	10	Yes	No
Tele Vue	Nagler	13	82	FMC	18	Yes	No
2-inch Barrel: $100 to $200							
Meade	Super Plössl	56	52	MC	47	Yes	No
Orion	Plössl	50	50	MC	42	No	No
Vixen	Lanthanum LV	30	60	FMC	20	Yes	Yes
2-inch Barrel: $200 to $300							
Meade	Super Wide Angle 40		67	MC	27	Yes	No
Meade	Super Wide Angle 32		67	MC	20	Yes	
Tele Vue	Plössl	55	50	FMC	38	Yes	No
Vernonscope	Brandon	48	50	FC	38	Yes	Yes
2-inch Barrel: More than $300							
Pentax	XL	40	65	FMC	20	Yes	Yes
Takahashi	LE	50	50	FMC	n/s	Yes	No
Tele Vue	Nagler Type 2	20	82	FMC	12	Yes	No
Tele Vue	Panoptic	35	68	FMC	24	Yes	Yes
Tele Vue	Panoptic	27	68	FMC	19	Yes	Yes

Notes:

All technical specifications were supplied by the manufacturers and have not been independently verified.

Ser = Series: The optical design or trade name of the particular eyepiece.

FL = Focal Length: The eyepiece's focal length, expressed in millimeters.

AFV = Apparent FOV: The eyepiece's apparent field of view, expressed in degrees.

Ctg = Coatings: The optical coatings applied to the eyepiece's elements. C = coated (i.e., single-layer magnesium fluoride, likely only on the outer surfaces of the eye and field lenses); FC = fully coated (i.e., single coating on all optical surfaces); MC = multi-coated (i.e., multiple coatings on some optical surfaces); FMC = fully multicoated (i.e., multiple coatings on all optical surfaces).

ER = Eye Relief: The distance, in millimeters, that an observer's eye must be away from the eye lens in order to see the entire field.

EC = Eye Cups: Useful for preventing stray light from entering through the corner of the observer's eye.

Pf = Parfocal: The eyepiece focuses at nearly the same distance as other eyepieces in the manufacturer's series. Occasionally, some eyepieces in a particular line may not be parfocal with all others in the same line but instead with only one or two others.

Appendix C
The Astronomical Yellow Pages

Manufacturers

Here's a listing of all of the companies whose products are discussed throughout this book, along with a few other enterprises that, while unmentioned earlier, offer worthwhile services to the amateur astronomical community.

Adirondack Video Astronomy
  35 Stephanie Lane, Queensbury, NY 12804
  Phone: (888) 799-0107, (518) 793-9484
  Email: 72323.3043@compuserve.com
  Web page: http://ourworld.compuserve.com/
    homepages/avaastro
  Main product(s): Video cameras
Adorama
  42 West 18th Street, New York, NY 10011
  Phone: (800) 723-6726, (212) 647-9800
  Email: goadorama@aol.com
  Product line(s): Eyepieces
Aries Instruments Co.
  Kherson, Ukraine, Russia
  Phone: +38 (055) 227-9653
  Email: aries@public.kherson.ua
  Product line(s): Apochromatic refractors
Astronomy Book Club
  3000 Cindel Drive, P.O. Box 6020,
    Delran, NJ 08075
  Phone: (609) 786-9778
  Product line(s): Books
Astro-Physics, Inc.
  11250 Forest Hills Road, Rockford, IL 61115
  Phone: (815) 282-1513
  Web page: http://www.astro-physics.com
  Main product(s): Apochromatic refractors,
    mountings

AstroSystems, Inc.
  5348 Ocotillo Court, Johnstown, CO 80534-9322
  Phone: (970) 587-5838
  Email: astrosys@frii.com
  Web page: http://www.frii.com/~astrosys/index
    .htm
  Main product(s): Telescope kits, collimation
    tools, accessories
Bausch & Lomb/Bushnell Sports Optics Worldwide
  9200 Cody Street, Overland Park, KS 66214
  Phone: (800) 423-3537, (913) 752-3400
  Web page: http://www.bushnell.com/
  Main product(s): Binoculars
Bean, L.L.
  Freeport, ME 04033-0001
  Phone: (800) 341-4341
  Web page: http://www.llbean.com
  Main product(s): Clothing, outdoor accessories
Beattie Systems, Inc.
  P.O. Box 3142, Cleveland, TN 37311
  Phone: (800) 251-6333
  Web page: http://www.beattiesystems.com
  Main product(s): Intenscreen camera-focusing
    screens
Bike Nashbar
  4111 Simon Road, Youngstown, OH 44512
  Phone: (800) 627-4227
  Web page: http://www.nashbar.com

Main product(s): Cold-weather clothing and
accessories (bike stuff, too!)

Bogen Photo Corporation (Manfrotto)
565 East Crescent Avenue,
P.O. Box 506, Ramsey, NJ 07446
Phone: (201) 818-9500
Web page: http://www.manfrotto.it/bogen/
Main product(s): Tripods

Byers, Edward R., Company
29001 West Highway 58, Barstow, CA 92311
Phone: (619) 256-2377
Main product(s): Telescope mountings and drive
systems

Campmor
810 Route 17 North, P.O. Box 997A,
Paramus, NJ 07653
Phone: (800) 226-7667, (201) 825-8300
Email: customer-service@campmor.com
Web page: http://www.campmor.com
Main product(s): Clothing, outdoor accessories

Carina Software
12919 Alcosta Blvd., Suite #7,
San Ramon, CA 94583
Phone: (510) 355-1266
Main product(s): Voyager II software

Celestron International
2835 Columbia Street, Torrance, CA 90503
Phone: (310) 328-9560
Web page: http://www.celestron.com
Main product(s): Binoculars, refractors, reflectors,
catadioptric telescopes/eyepieces; accessories

Ceravolo Optical Systems
Box 1427, Ogdensburg, NY, 13669
Box 151, Oxford Mills, ON, K0G 1S0 Canada
Phone: (613) 258-4480
Web page: http://www.cyanogen.on.ca/ceravolo/
index.html
Email: ceravolo@fox.nstn.ca
Main product(s): Catadioptric telescopes

Coulter Optical (Division of Murnaghan Instruments)
1781 Primrose Lane, West Palm Beach, FL 33414
Phone: (561) 795-2201
Email: murni@bix.com
Web page: http://www.murni.com
Main product(s): Reflectors

D & G Optical
6490 Lemon Street, East Petersburg, PA 17520

Phone: (717)560-1519
Email: dgoptical@aol.com
Main product(s): Refractors, reflectors

Daisy Manufacturing Company, Inc.
2111 South 8th Street, Rogers, Arkansas 72757
Phone: (800) 713-2479
Main product(s): Daisy sight aiming device

Damart
3 Front Street, Rollinsford, NH 03805
Phone: (800) 258-7300
Web page: http://www.damartusa.com
Main product(s): Cold-weather clothing and
accessories

David Chandler Company
P.O. Box 309, La Verne, CA 91750
Phone: (800) 516-9756, (909) 988-5678
Email: dschandler@frumble.claremont.edu
Web page: http://www.csz.com/dschandler
Main product(s): Deep Space computer software

DayStar
P.O. Box 5110, Diamond Bar, CA 91765
Phone: (909) 591-4673
Main product(s): Nebula filters, hydrogen-alpha
solar filters

E.L.B. Software
8910 Willow Meadow Drive, Houston, TX 77031
Phone: (713) 541-9723
Email:elb@ix.netcom.com
Main product(s): Megastar computer software

Eastman Kodak
Kodak Park, Rochester, NY 14650
Phone: (800) 242-2424
Web page: http://www.kodak.com
Main product(s): Film

Edmund Scientific Co.
101 E. Gloucester Pike, Barrington, NJ 08007
Phone: (609) 547-8880
Email: scientifics@edsci.com
Main product(s): Binoculars, reflectors, eyepieces

Equatorial Platforms
11065 Peaceful Valley Road,
Nevada City, CA 95959
Phone: (916) 265-3183
Email: tomosy@nccn.net
Web page: http://www.astronomy-mall.com
Main product(s): Equatorial platforms for
alt-azimuth mounts

Functional Design & Engineering LTD
  Briar House, Foxley Green Farm, Ascot Road,
    Holyport, Berks, SL6 3LA, UK
  Phone: (0)1628-777126
  Email: tplatt@starlight.win-uknet
  Web page: http://www.ibmpcug.couk/~starlite/
  Main product(s): Starlight Xpress CCD camera
Fuji Photo Optical Co. Ltd. (Fujinon, Inc.)
  10 High Point Drive, Wayne, NJ 07470 USA
  1-324 Uetake, Omiya City, Saitama 330, Japan
  Phone: (201) 633-5600
  Web page: http://www.fujinon.com/jp/products/
    optical
  Main product(s): Binoculars, film
Grabber Mycoal
  (Division of John Wagner Associates, Inc.
  4600 Danvers Drive Southeast,
  Grand Rapids, MI 49512 USA
  Phone: (415) 680-0777, (800) 423-1233
  Email: warmer@GrabberMan.com
  Web Page: http://www.grabberman.com/
    warmers/index.htm
  Main product(s): Hand and pocket warmers
Intes Telescopes
  33 Bolshaya Ochakovskaya ul.,
  Moscow, Russia 119361
  Phone: (095) 430-56-20
  (Note: Unconfirmed address and phone number)
  Main product(s): Maksutov telescopes
Jim's Mobile, Inc./JMI
  810 Quail Street, Unit E, Lakewood, CO 80215
  Phone: (303) 233-5353, (800) 247-0304
  Main product(s): Reflectors, telescope
    accessories
K.C. Woodsmith
  P.O. Box 75033, Wichita, KS 67575
  Phone: (316) 721-1566
  Main product(s): Observing chair
Kalmbach Publishing Company
  21027 Crossroads Circle,
  P.O. Box 1612, Waukesha, WI 53187
  Phone: (800) 446-5489
  Web page: http://www.kalmbach.com/astro
  Main product(s): *Astronomy* magazine and related
    publications
Kendrick Studio
  2775 Dundas Street West,

Toronto, ON M6P 1Y4 Canada
  Phone: (800) 393-5456, (416) 762-7946
  Email: jkendrick@sympatico.com
  Web page: http://www.astronomy-mall.com
  Main product(s): Dew-prevention system
Kufeld, Steve
  P.O. Box 6780, Pine Mountain Club, CA 93222
  Phone: (805) 242-5421
  Main product(s): Telrad aiming device
Losmandy/Hollywood General Machining
  1033 North Sycamore Avenue,
  Los Angeles, CA 90038
  Phone: (213) 462-2855
  Main product(s): Telescope mounts
Lumicon
  2111 Research Dr., Suites 4–5,
  Livermore, CA 94550
  Phone: (510) 447-9570, 800-767-9576
  Email: Marling@Pacbell.net
  Web page: http://www.astronomy-mall.com/
  Main product(s): Nebula (LPR) filters, astrophotog-
    raphy accessories, finders, digital setting circles
Mag One Instruments
  16342 W. Coachlight Drive,
  New Berlin, WI 53151-1475
  Phone: (414) 785-0926
  Email: mag1inst@aol.com
  Web page: http://www.mag1instruments.com/
  Main product(s): Reflectors
Manfrotto Nord Srl
  Z.I. di Villapaiera, I-32032 Feltre BL Italy
  Phone: +39 +439 89945
  Email: nord@manfrotto.it
  Web page: http://www.manfrotto.it
  Main product(s): Tripods (see also Bogen)
Meade Instruments Corp.
  6001 Oak Canyon, Irvine, CA 92620
  Phone: (714) 756-2291
  Web page: http://www.meade.com/
  Main product(s): Binoculars, reflectors, refractors,
    catadioptric telescopes, eyepieces, filters,
    CCD cameras
Minolta Corporation
  101 Williams Drive, Ramsey, NJ 07446
  Phone: (201) 825-4000
  Web page: http://www.minolta.com
  Main product(s): Binoculars

Miyauchi
Phone: 81-494-62-3371
Fax: 81-494-62-4858
Main product(s): Binoculars

MorningStar Telescope Works
10200 SE Orient Drive, Boring, OR 97009
Phone: (503) 663-9630
Email: duke@starstuff.com
Web page: http://www.starstuff.com/scopes.html
Main product(s): Reflectors

Nikon
1300 Walt Whitman Road, Melville, NY 11747
Phone: (800) 645-6687, (516) 547-4200
Web page: http://www.nikonusa.com/
Main product(s): Binoculars, cameras

North Star Systems
P.O. Box 99, Ellenburg, NY 12935
Phone: (518) 594-7250
Main product(s): Reflectors

Nova Astronomics
P.O. Box 31013, Halifax, NS B3K 5T9 Canada
Phone: (902) 443-5989
Email: ecu@fox.nstn.ca
Web page: http://www.fox.nstn.ca/Necu/
ecu.html
Main product(s): *Earth-Centered Universe* computer software

Novak, Kenneth F., & Co.
Box 69, Ladysmith, WI 54848
Phone: (715) 532-5102
Main product(s): Telescope components

Obsession Telescopes
P.O. Box 804, Lake Mills, WI 53551
Phone: (414) 648-2328
Email: obsessiontscp@globaldialog.com
Web page: http://www.globaldialog.com/
~obsessiontscp/OBHP.html
Main product(s): Reflectors

Optical Guidance Systems
2450 Huntingdon Pike,
Huntingdon Valley, PA 19006
Phone: (215) 947-5571
Web page: http://www.nb.net/~ogs/
Main product(s): Reflectors

Orion Optics (UK)
Unit 21, Third Avenue, Crewe,
Cheshire CW1 6XU England

Phone: (0)1270-500089
Email: sales@orionoptics.co.uk
Web page: http://www.orionoptics.co.uk
Main product(s): Reflectors, catadioptric telescopes

Orion Telescopes (US)
P.O. Box 1158, Santa Cruz, CA 95062
Phone: (800) 447-1001
(800) 443-1001 (in California only)
Email: sales@oriontel.com
Web page: http://www.oriontel.com
Main product(s): Binoculars, reflectors, refractors, catadioptric telescopes, eyepieces, accessories

Parallax Instruments
8318 Pineville-Matthews Road, Suite 708–192,
Charlotte, NC 28226
Phone: 704-542-4817
Main product(s): Reflectors, mounts

Pentax Corporation
35 Inverness Drive East, P.O. Box 6509,
Englewood, CO 80155
Phone: (800) 877-0155, (303) 799-8000
Email: pentaxinfo@pentax.com
Web page: http://www.pentax.com
Main product(s): Binoculars, eyepieces

Performance Bicycle Shop
P.O. Box 2741, Chapel Hill, NC 27514
Phone: (800) 727-2453
Web page: http://www.performancebike.com
Main product(s): Cold-weather clothing and accessories (bike stuff, too!)

Project Pluto
Ridge Road, Box 1607,
Bowdoinham, Maine 04008
Email: pluto@projectpluto.com
Web page: http://www.projectpluto.com
Main product(s): *Guide* computer software

Questar Corporation
P.O. Box 59, New Hope, PA 18938
Phone: (215) 862-5277
Main product(s): Maksutov-Cassegrain catadioptric telescopes and accessories

Rigel Systems
26850 Basswood,
Rancho Palos Verdes, CA 90275
Phone: (310) 375-4149
Email: rigelsys@ix.netcom.com

Web page: http://pw2.netcom.com/~rigelsys/ Rigelsys.html

Main product(s): LED flashlights, collimation tools, aiming devices, accessories

Santa Barbara Instrument Group (SBIG)

1482 East Valley Road, #33,

Santa Barbara, CA 93108

Phone: (805) 969-1851

Email: sbig@sbig.com

Web page: http://www.sbig.com

Main product(s): CCD cameras and accessories

Schmidling Productions, Inc.

18016 Church Road, Marengo, IL 60152

Phone: (815) 923-0031

Email:arf@maxx.mc.net

Web page: http://www.astronomy-mall.com

Main product(s): Ronchi test eyepiece

Sienna Software, Inc.

105 Pears Avenue, Toronto, ON M5R 1S9 Canada

Email: contact@siennasoft.com

Web page: http://www.siennasoft.com/index. shtml

Main product(s): *Starry Night Deluxe* computer software

Sky Designs

4100 Felps, #C Colleyville, TX 76034

Phone: (817) 581-9878

Main product(s): Reflectors

Sky Publishing Corporation

P.O. Box 9111, Belmont, MA 02178

Phone: (800) 253-0245, (617) 864-7360

Web page: http://www.skypub.com

Main product(s): *Sky & Telescope* magazine, books

SkyMap Software

9 Severn Road, Culcheth, Cheshire WA3 5ED UK

Email: sales@skymap.com

Web page: http://www.skymap.com

Main product(s): *SkyMap* software

Sky Valley Scopes

9215 Mero Road, Snohomish, WA 98290

Phone: (360) 794-7757

Main product(s): Reflectors

Small Parts, Inc.

13980 N.W. 58th Court, P.O. Box 4650, Miami Lakes, FL 33014

Phone: (305) 557-7955

Main product(s): Nuts, bolts, fasteners, knobs, and other assorted hardware

Software Bisque

912 12th Street, Suite A, Golden, Colorado 80401

Phone: (800) 843-7599

Web page: http://www.bisque.com

Main product(s): *The Sky* computer software

Southern Stars Software

12525 Saratoga Creek Drive, Saratoga, CA 95070

Phone: (408) 973-1016

Email: info@southernstars.com

Web page: http://www.southernstars.com

Main product(s): *SkyChart 2000* computer software

Spectra Astro Systems

6631 Wilbur Avenue, Suite 30, Reseda, CA 91335

Phone: (800) 735-1352

Email: spectraast@aol.com

Web page: http://www.astronomy-mall.com

Main product(s): Focusing devices, astrophoto accessories

SpectraSource Instruments

31324 Via Colinas, Suite 114,

Westlake Village, CA 91362

Phone: (818) 707-2655

Main product(s): CCD cameras and accessories

Star Instruments

555 Blackbird Roost #5, Flagstaff, AZ 86001

Phone: (520) 774-9177

Main product(s): Mirrors and other telescope optics

Starbound

68 Klaum Avenue, North Tonawanda, NY 14120

Phone: (716) 692-3671

Main product(s): Observing chair

Stargazer Steve

1752 Rutherglen Crescent,

Sudbury, ON P3A 2K3 Canada

Phone: (705) 566-1314

Email: stargazr@isys.ca

Web page: http://ww2.isys.ca/stargazer/

Main product(s): Reflectors

StarMaster Telescopes

Rt. 1, Box 780, Arcadia, KS 66711

Phone: (316) 638-4743

Email: starmaster@ckt.net

Web page: http://www.icstars.com/starmaster/

Main product(s): Reflectors, observing chairs

The Starry Messenger
  PO Box 6552, Ithaca, NY 14851
  Email: starrymess@aol.com
  Web page: http://www.starrymessenger.com/
  Main product(s): Used-equipment newsletter

Starsplitter Telescopes
  3228 Rikkard Drive, Thousand Oaks, CA 91362
  Phone: (805) 492-0489
  Email: strspltr@aol.com
  Web page: http//www.ez2.net/starsplitter/
  Main product(s): Reflectors

Swift Instruments, Inc.
  952 Dorchester Avenue, Boston, MA 02125
  Phone: (617) 436-2960, (800) 446-1116
  Main product(s): Binoculars

Tech 2000
  3349 State Route 99,
  South Monroeville, OH 44847
  Phone: (419) 465-2997
  Main product(s): Dual-axis Dobsonian tracking
    system

Tectron Telescopes
  3544 Oak Grove Drive, Sarasota, FL 34243
  Phone: (941) 758-9890
  Web site: http://icstars.com/tectron
  Email: aatclark@aol.com
  Main product(s): Reflectors, collimation tools,
    Amateur Astronomy magazine

Tele Vue Optics
  100 Route 59, Suffern, NY 10901
  Phone: (914) 357-9522
  Web page: http://www.televue.com/
  Main product(s): Refractors, eyepieces

Thousand Oaks Optical
  Box 4813, Thousand Oaks, CA 91359
  Phone: (800) 996-9111, (805) 491-3642
  Main product(s): Glass solar filters, LPR filters

TL Systems
  2184 Primrose Avenue, Vista, CA 92083
  Phone: (619) 599-4219
  Email: tlsystem@ix.netcom.com
  Web page: http://www.sunspots.com/tlsystems/

Main product(s): Equatorial table kits, binocular
  viewer kits

Tuma, Steven
  1425 Greenwich Lane, Janesville, WI 53545
  Phone: (608) 752-8366
  Email: stuma@inwave.com
  Web page: http://www.deepsky2000.com
  Main product(s): *Deepsky 97* computer software

Tuthill, Roger W., Inc.
  11 Tanglewood La., Mountainside, NJ 07092
  Phone: (800) 223-1063, (908) 232-1786
  Main product(s): Mylar solar filters, dew caps,
    finders, eyepieces, other astronomical
    accessories

VERNONscope & Co.
  5 Ithaca Road, Candor, NY 13743
  Phone: (607) 659-7000
  Main product(s): Brandon eyepieces

Virgo Astronomics
  608 Falconbridge Drive, Suite 46,
  Joppa, MD 21085
  Phone: (410) 679-7055
  Web page: http://www.astronomy-mall.com
  Main product(s): Binocular mounts

Vixen Optical Industries Ltd.
  247 Hongo, Tokorozawa, Saitama 359 Japan
  Phone: 0429-44-4141
  Web page: http://www02.so-net.or.jp/~vixen/
  Main product(s): Refractors, eyepieces

Walker, John
  Web page: http://www.fourmilab.ch/index.html
  Main product(s): *Home Planet* computer software

Willmann-Bell, Inc.
  P.O. Box 35025, Richmond, VA 23235
  Phone: (800) 825-7827, (804) 320-7016
  Web page: http://www.willbell.com
  Main product(s): Books

Zeiss, Carl, Inc.
  1015 Commerce Street, Petersburg, VA 23803
  Phone: (800) 338-2984
  Web page: http://www.zeiss.de/products_e.html
  Main product(s): Binoculars

Dealers and Distributors

An important warning: If you are shopping by mail, ALWAYS ask about shipping charges BEFORE ordering. Some companies offer exceptionally low prices only to charge the consumer exorbitant (and unpublished) shipping and handling costs. Not only will these hidden costs offset any savings, but they may end up making your supposedly cheaper choice more expensive than others you passed up in search of the best deal!

Arizona
 The Astronomy Shoppe
  15826 North Cave Creek Road,
  Phoenix, AZ 85032
  Phone: (602) 971-3170
  Product line(s): Tele Vue, more
 Stellar Vision and Astronomy Shop
  1835 South Alvernon, #206,
  Tucson, Arizona 85711
  Phone: (520) 571-0877
  Web page: http://www.theriver.com/
   stellar_vision
  Product line(s): Tele Vue, Takahashi, Celestron, Meade, more
Arkansas
 Rex's Astro Stuff
  63 Observatory Lane, Dover, AR 72837
  Phone: (501) 331-3773
  Product line(s): Celestron, Intes
California
 Crazy Ed Optical
  P.O. Box 110566, Campbell, CA 95011
  Phone: (408) 364-0944
  Web page: http://www.crazyedoptical.com
  Product line(s): Astrosystems, HB2000 Publications, Rigel Systems, Telrad
 Earth and Sky Adventure Products
  2382 Leptis Circle, Morgan Hill, CA 95037
  Phone: (408) 778-1695
  Email: adastra@garlic.com
  Web page: http://www.astrosales.com
  Product line(s): Intes
 Lumicon
  2111 Research Drive, Suites 4–5,
  Livermore, CA 94550
  Phone: (800) 767-9576
  Web page: http://www.astronomy-mall.com
  Product line(s): Celestron, Lumicon, Meade, Tele Vue, more

Oceanside Photo and Telescope
  1024 Mission Avenue, Oceanside, CA 92054
  Phone: (800) 483-6287, (619) 722-3348
  Email: opt@optcorp.com
  Web page: http://www.optcorp.com
  Product line(s): Bausch & Lomb, Celestron, Meade, Orion Telescopes, Tele Vue, more
 Scope City
  679 Easy Street, P.O. Box 440,
  Simi Valley, CA 93065
  Phone: (805) 522-6670
  Product line(s): Celestron, Edmund, Meade, Questar, Tele Vue, JMI, more
 Spectra Astro-Systems
  6631 Wilbur Avenue, Suite 30,
  Reseda, CA 91335
  Phone: (800) 735-1352
  Email: spectraast@aol.com
  Web page: http://www.astronomy-mall.com
  Product line(s): Celestron, Fujinon, JMI, Losmandy, SBIG, Takahashi, Tele Vue, more
 Telescope and Binocular Center
  P.O. Box 1158, Santa Cruz, CA 95062
  Email: sales@oriontel.com
  Web page: http://www.oriontel.com
  Phone: (800) 447-1001
  (800) 443-1001 (in California only)
  Product line(s): Celestron, Fujinon, Orion Telescopes, Pentax, Tele Vue, Vixen
Colorado
 Jim's Mobile Industries (JMI)
  810 Quail Street, Unit E,
  Lakewood, CO 80215
  Phone: (800) 247-0304, (303) 233-5353
  Product line(s): Celestron, JMI, Meade, Tele Vue
 S & S Optika
  5174 South Broadway,
  Englewood, CO 80110
  Phone: (303) 789-1089

Product line(s): Bausch & Lomb, Celestron, DayStar, Pentax, Takahashi, Zeiss, more

Florida

Hirsch, Edwin

8740 Egret Isle Terrace,
Lake Worth, FL 33467
Phone: (561) 641-2851
Web page: http://www.astronomy-mall.com
Product line(s): Celestron, DayStar, Meade, Tele Vue

Sarasota Camera Exchange

1055 South Tamiami Trail,
Sarasota, FL 34237
Phone: (813) 366-7484
Product line(s): Bausch & Lomb, Celestron, JMI, Meade, Minolta, Nikon, Pentax, Tele Vue, more

Georgia

Camera Bug Ltd.

1799 Briarcliff Road, Atlanta, GA 30306
Phone: (800) 545-8509, (404) 873-4513
Product line(s): Celestron, Fujinon, Meade, Minolta, Nikon, Orion Telescopes, Pentax, Tele Vue, more

Illinois

Shutan Camera and Video

312 West Randolph Street,
Chicago, IL 60606
Phone: (800) 621-2248, (312) 332-2000
Web page: http://www.shutan.com
Product line(s): Celestron, JMI, Meade, Tele Vue, more

Maryland

Company Seven

Box 2587, Montpelier, Maryland 20708
Phone: (301) 953-2000
Email: info@company7.com
Web page: http://www.company7.com/
Product line(s): Astro-Physics, AstroSystems, Celestron, Fujinon, Lumicon, Questar, Tele Vue, more

Massachusetts

Hunt's Photo and Video

100 Main Street, Melrose, MA 02176
Phone: (800) 924-8682, (617) 662-6685
Email: info@wbhunt.com
Web page: http://www.wbhunt.com

Product line(s): Bausch & Lomb, Celestron, Minolta, Nikon, Swift

Meischner, F.C., Company, Inc.

182 Lincoln Street, Boston, MA 02111
Phone: (800) 321-8439
Product line(s): Celestron, Meade, Questar, Takahashi, Tele Vue, more

New Hampshire

International Optics, Inc.

P.O. Box 6475, Nashua, NH 03063
Email: 73554.3420@compuserve.com
Web page: http://www.astronomy-mall.com
Product line(s): Orion Optics (sold in the United States as "Europa")

Rivers Camera Shop

454 Central Avenue, Dover, NH 03820
Phone: (800) 245-7963, (603) 742-4888
Product line(s): Celestron, Meade, Questar, Tele Vue, more

New Jersey

Dover Photo Supply

25 East Blackwell Street, Dover, NJ 07801
Phone: (201) 366-0994
Product line(s): Celestron, Meade, Nikon, Swift, Tele Vue

Tuthill, Roger W., Inc.

11 Tanglewood Lane,
Mountainside, NJ 07092
Phone: (800) 223-1063, (908) 232-1786
Product line(s): Celestron, Meade, Tele Vue, more

New Mexico

New Mexico Astronomical

834 N. Gabaldon Road, Belen, NM 87002
Phone: (505) 864-2953
Product line(s): Celestron, Meade, Tele Vue, more

New York

Adorama

42 West 18th Street, New York, NY 10011
Phone: (800) 723-6726, (212) 647-9800
Product line(s): Celestron, Meade, Tele Vue, Thousand Oaks, Fujinon, Pentax, more

Berger Brothers Camera Exchange

209 Broadway, Amityville, NY 11701
Phone: (800) 262-4160, (516) 264-4160
Product line(s): Celestron, Meade, Tele Vue

Focus Camera, Inc.
  4419 13th Avenue, Brooklyn, NY 11219
  Phone: (800) 221-0828, (718) 437-8810
  Email: 75457.1403@compuserve.com
  Web page: http://www.focuscamera.com
  Product line(s): Celestron, Fujinon, JMI,
    Meade, Nikon, Tele Vue, Thousand Oaks,
    more
North Carolina
  Internet Telescope Exchange
    7151 Market Street, Wilmington, NC, 28405
    Phone: (910) 686-9617
    Email: cescoweb@aol.com
    Web page: http://www.burnettweb.com/ite/
      index.html
    Product line(s): Intes, more
Ohio
  Eastern Hills Camera and Video
    7875 Montgomery Road,
    Cincinnati, OH 45236
    Phone: (800) 863-6138, (513) 791-2140
    Product line(s): Celestron, Minolta, Pentax,
      Swift, Vixen
Oklahoma
  Astronomics
    2401 Tee Circle, Suites 105/106,
    Norman, OK 73069
    Phone: (800) 422-7876, (405) 364-0858
    Email: okastro@aol.com
    Product line(s): Celestron, Meade, Takahashi,
      Tele Vue, more
Pennsylvania
  Europtik Ltd.
    P.O. Box 319, Dunmore, PA 18512
    Phone: (717) 347-6049
    Email: Europtik@ptd.net
    Web page: http://www.europtik.com/
    Product line(s): Intes, Takahashi
  Pocono Mountain Optics
    104 NP 502 Plaza, Moscow, PA 18444
    Phone: (800) 569-4323, (717) 842-1500
    Email: pcomtnop@ptdprolog.net
    Web page: http://www.astronomy-mall.com
    Product line(s): AstroSystems, Celestron,
      Edmund, Fujinon, JMI, Meade, Miyauchi,
      Questar, Pentax, Starbound, Tele Vue,
      Zeiss, more

Texas
  Analytical Scientific
    11049 Bandera Road,
    San Antonio, TX 78250
    Phone: (210) 684-7373
    Email: 103100.752@compuserve.com
    Web page: http://www.astronomy-mall.com
    Product line(s): Celestron, JMI, Meade, Tele
      Vue
  Texas Nautical Repair (Land, Sea and Sky)
    3110 South Shepherd, Houston, TX 77098
    Phone: (713) 529-3551
    Web page: http://www.neosoft.com/~lsstnr
    Product line(s): Miyauchi, Takahashi
Wisconsin
  Eagle Optics
    716 South Whitney Way, Madison, WI 53711
    Phone: (608) 271-4751
    Web page: http://www.eagleoptics.com
    Product line(s): Celestron, Fujinon, Meade,
      Nikon, Questar, Tele Vue
Canada
  EfstonScience
    3350 Dufferin Street, Toronto,
    Ontario M6A 3A4
    Phone: (416) 787-4581
    Email: efston@idirect.com
    Web page: http://www.efstonscience.com
    Product line(s): Celestron, Meade, more
  Focus Scientific Ltd.
    1489 Merivale Road,
    Nepean (Ottawa), ON K2E 5P3
    Phone: (613) 723-1350
    Email: kentfsc@compuserve.com
    Product line(s): Bausch & Lomb, Celestron,
      Fujinon, Meade, Nikon, Orion Telescopes,
      Swift, Tele Vue, more
  Harrison Scientific Instruments, Ltd.
    2574 Granville Street,
    Vancouver, BC V6H 3C8
    Phone: (604) 737-4303
    Product line(s): Celestron, Meade, Pentax,
      Tele Vue, Zeiss
  Khan Scope Centre
    3243 Dufferin Street, Toronto, ON M6A 2T2
    Phone: (416) 783-4140

Product line(s): Bausch & Lomb, Celestron, JMI, Lumicon, Meade, SBIG, Tele Vue, Thousand Oaks, more

Lire La Nature, Inc.
100 Goyer Street, Store 110,
LaPrairie, QU J5R 5G5
Phone: (514) 463-5072, (514) 357-9626
Web page: http://www.stjeannet.ca/broquet/zindex.html
Product line(s): Celestron, Lumicon, Meade, Tele Vue, Thousand Oaks

Sky Optics
4031 Fairview Street, Burlington, ON L7L 2A4
Phone: (905) 631-9944
Email: sbarnes@worldchat.com
Web page: http://worldchat.com/commercial/skyoptics
Product line(s): Celestron, Lumicon, Meade, SBIG, Tele Vue, more

United Kingdom
Beacon Hill Telescopes
112 Mill Road, Cleethorpes,
South Humberside DN35 8JD
Phone: (0)1472-692959
Product line(s): Vixen, Clavé, and more

Broadhurst, Clarkson & Fuller, Ltd.
63 Farringdon Road, London EC1M 3JB
Phone: (0)1714-052156
Product line(s): Meade, Takahashi, Vixen, Tele Vue, Celestron, JMI, more

Orion Optics
Unit 21, Third Avenue, Crewe,
Cheshire CW1 6XU
Phone: (0)1270-500089
Email: sales@orionoptics.co.uk
Web page: http://www.orionoptics.co.uk
Product line(s): Celestron, Orion Telescopes, Vixen

SCS Astro
South View, Runnington, Wellington,
Somerset TA21 0QW
Phone: (0)1823-661544
Email: scsastro@mail.eclipse.co.uk
Web page: http://www.scsastro.co.uk
Product line(s): Celestron, Orion Telescopes

Venturescope
Wren Centre, Westbourne Road,
Emsworth, Hampshire PO10 7RN
Phone: (0)1243-379322
Email: orders@venturescope.co.uk
Web page: http://www.venturescope.co.uk
Product line(s): Apogee Instruments, Fujinon, Orion Optics, Tele Vue, Vixen

Australia
Advanced Telescope Supplies
P.O. Box 447, Engadine, NSW 2233
Phone: +61 2 548 2428
Product line(s): Losmandy

Astro Optical Supplies
9B Clarke Street, Crows Nest, NSW 2065
Phone: +61 2 436 4360
13 Lower Plaza, 131 Exhibition Street,
Melbourne, VIC 3000
Phone: +61 3 650 8072
Product line(s): Celestron, Australian-made AOS reflectors

Astronomy & Electronics Centre
P.O. Box 45, Cleve, SA 5640
Phone: +61 86 282 435
Product line(s): Astro-Physics, Intes, Lumicon, Meade, Takahashi, more

The Telescope & Binocular Shop
55 York Street, Sydney, NSW 2000
Phone: +61 2 9262 1344
Web page: http://www.bintel.com.au/
Product line(s): Celestron, Nikon, Orion Telescopes, Pentax, Tele Vue, Telrad, Thousand Oaks, more

Elken Ridge Pty. Ltd.
7–8 Williams Road, Olinda, VIC 3788
Phone: +61 3 751 2176
Product line(s): JMI, SBIG

York Optical and Scientific
316 St. Paul's Terrace,
Fortitude Valley, QLD 4006
Phone: +61 7 252 2061
7/270 Flinders Street, Melbourne, VIC 3000
Phone: +61 3 654 7212
939 Hay Street, Perth, WA 6000
Phone: +61 9 322 4410
Product line(s): Bausch & Lomb, Fujinon, Meade, Questar, Vixen

New Zealand

    Blaxhall Science Company

        P.O. Box 25094, Christchurch

        Phone: +64 3 366 2828

        Product line(s): Celestron

    Skylab

        172 St. Asaph Street, Christchurch

        Phone: +64 3 366 2827

        Product line(s): Celestron, Meade, and more

    Telescope & Optics

        Otaraoa Road, Waitara, Taranaki

        Phone +64 6 754 6434

        Product line(s): Locally made Newtonian
          reflectors

    Carl Zeiss Jena

        5 Wakefield Street, Lower Hutt

        Phone: +64 4 566 7601

        Product line(s): Zeiss binoculars

Appendix D
An Astronomer's Survival Guide

It's not always easy being an amateur astronomer. We are nocturnal creatures by nature, going outdoors when the rest of the world sleeps; braving cold, heat, bugs, and things that go bump in the night; always looking up when most everyone else is looking down (astronomers are the eternal optimists).

Here is a checklist of things that I like to bring along for a night under the stars. They make the experience much more pleasurable and the cold night a little warmer.

Astronomical

Telescope____   Binoculars____   Eyepieces____   LPR filter(s)____
Color filter(s)____   Star atlas____   List of things to look at____
Clipboard____   Pen/pencil____   Flashlight (red)____   Flashlight (white)___
Etc._____

_____

Photographic

Camera(s)_____   Auxiliary lenses_____   Film_____   Tripod_____
Camera-to-telescope adapters____   Scotch mount_____
Drive corrector_____   CCD Camera/Computer_____
Cable releases (always bring two, in case one breaks)_____
Etc._____

_____

Miscellaneous

Sweatshirt_____   Long underwear_____   Heavy socks_____   Jacket_____
Winter coat____   Gloves/mittens____   Boots____   Foot/hand warmers_____
Hat_____   Folding table_____   Chair____   Insect repellent_____
Something warm to drink (nonalcoholic)____   Food/snacks____   Radio____
Etc._____

_____

Appendix E
Astronomical Resources

Special Interest Societies

Amateur Telescope Makers Association
17606 28th Ave. SE, Bothell, WA 98012
Email: atmj1@aol.com.

American Association of Variable Star Observers
25 Birch St., Cambridge, MA 02138
Email: aavso@aavso.org
http://www.aavso.org/

American Lunar Society
P.O. Box 209, East Pittsburgh, PA 15112
Email: s002psc@paladin.wright.edu
http://www.media.wright.edu/studorgs/lunar/als.htm

American Meteor Society
Dept. of Physics-Astronomy, SUNY-Geneseo, Geneseo, NY 14454
Email: meisel@uno.cc.geneseo.edu
http://www.serve.com/meteors/

Antique Telescope Society
1275 Poplar Grove Lane, Cumming, GA 30131
http://www1.tecs.com/oldscope/

Association of Lunar and Planetary Observers
P.O. Box 16131, San Francisco, CA 94116

Astronomical League
2112 Kingfisher Lane E., Rolling Meadows, IL 60008

Email: 73357.1572@compuserve.com
http://www.mcs.net/~bstevens/al/

Astronomical Society of the Pacific
390 Ashton Ave., San Francisco, CA 94112
Email: asp@stars.sfsu.edu
http://www.aspsky.org/
International Dark-Sky Association
3545 N. Stewart, Tucson, AZ 85716
http://www.darksky.org/~ida/index.html

International Meteor Organization
Dept. of Physics, Univ. of Western Ontario,
London, ON N6A 3K7, Canada
http://www.imo.net/.

International Occultation Timing Association
2760 SW Jewell Ave., Topeka, KS 66611-1614
Email: david-dunham@jhuapl.edu
http://www.sky.net/~robinson/ and http://www.anomalies.com/

International Supernova Network
C.P. 7114, I-47100 Forli, 39-543-72456
http://www.queen.it/web4you/noprofit/isn/isn.htm

National Deep Sky Observer's Society
1607 Washington Blvd., Louisville, KY 40242
Email: deepskyspy@aol.com

World Wide Web Resources

The Internet holds a whole universe of resources for the amateur astronomer. Trying to come up with a brief list is impossible, but here are a few sites that might be of interest.

Astronomy-Related Newsgroups

alt.sci.planetary	Discussion of planetary research and space missions.
sci.astro	General discussion group on anything astronomical.
sci.astro.amateur	General discussion group with a slant toward amateur observations and equipment.
sci.astro.hubble	Discussions of the latest findings from the Hubble Space Telescope.
sci.astro.planetarium	Planetarium professionals compare notes on shows and equipment.
sci.astro.research	Researchers post questions and answers about various research-related topics.

Email Lists

Many free electronic mail lists exist on several topics of interest to the amateur astronomer. In the following table you'll find several active email lists as well as details on how to join them. (Digest versions are often available, as well; consult the instructions that you will receive via email after you subscribe initially.)

Subject	To join, send an email note to:	Include this following message (without the quotes)
Astrophotography	majordomo @nightsky.com	"subscribe astro-photo" in the *body* of the message (leave subject area empty).
Amateur Telescope Making	atm-request @shore.net	"subscribe" in the *body* of the message (leave subject area empty).
Astromart Used Equipment	astromart-request @lists.best.com	"subscribe" in the *body* of the message (leave subject area empty).
Celestron Owner's List	celestronuser request @lists.best.com	"subscribe" in the *body* of the message (leave subject area empty).
Meade Owner's List	mapug-request @shore.net	"subscribe" in the *body* of the message (leave subject area empty).
SBIG User's List	sbiguser-request @lists.best.com	"subscribe" in the *body* of the message (leave subject area empty).

Telescopes and Equipment

Bill Arnett's Planetarium Software Review Page
http://www.seds.org/billa/astrosoftware.html

Astronomy Mall
http://www.astronomy-mall.com

Astro-Mart
http://www.astromart.com

Mel Bartel's Home Page
http://www.efn.org/~mbartels/

Richard Berry's Cookbook Camera Home Page
http://wvi.com/~rberry/

Todd Gross's Weather and Astronomy Home Page
http://www.weatherman.com/

Matt Marulla's ATM Page
http://www.atmpage.com

Star Ware Home Page
http://www.ourworld.compuserve.com/
homepages/pharrington

United Kingdom Amateur Telescope Making Suppliers
http://www.aegis1.demon.co.uk/supplier.htm

Sky Information

Abrams Planetarium
   http://www.pa.msu.edu/abrams/
Astronomy magazine
   http://www.astronomy.com
Lunar Eclipse Observer Home Page
   http://www-clients.spirit.net.au/~minnah/LEO.
   html

Jet Propulsion Laboratory's Comet Observation
Home Page
   http://encke.jpl.nasa.gov/
Sky On-Line
   http://www.skypub.com
Solar Eclipse Home Page
   http://planets.gsfc.nasa.gov/eclipse/eclipse.html

Appendix F
Visibility of the Planets, 1998–2004

Which bright planets are visible tonight? This table will tell you where to look for Venus, Mars, Jupiter, and Saturn. The positions of the planets are listed for the middle of each month. To make the table a little more concise, each constellation name is given its standard three-level abbreviation; a translation table is given in appendix H.

		Venus	**Mars**	**Jupiter**	**Saturn**
1998	January	Sgr	Cap	Cap	Psc
	February	Sgr	Aqr	Aqr	Psc
	March	Cap	Psc	Aqr	Psc
	April	Aqr	Ari	Aqr	Psc
	May	Psc	Tau	Aqr	Psc
	June	Ari	Tau	Psc	Psc
	July	Tau	Gem	Psc	Psc
	August	Cnc	Gem	Psc	Ari
	September	Leo	Cnc	Aqr	Ari
	October	Vir	Leo	Aqr	Ari
	November	Lib	Leo	Aqr	Ari
	December	Sgr	Vir	Aqr	Ari
1999	January	Cap	Vir	Psc	Ari
	February	Psc	Lib	Psc	Ari
	March	Psc	Lib	Psc	Ari
	April	Tau	Lib	Psc	Ari
	May	Gem	Vir	Psc	Ari
	June	Cnc	Vir	Ari	Ari
	July	Leo	Lib	Ari	Ari
	August	Leo	Lib	Ari	Ari
	September	Leo	Oph	Ari	Ari
	October	Leo	Sgr	Ari	Ari
	November	Vir	Sgr	Ari	Ari
	December	Lib	Cap	Psc	Ari
2000	January	Oph	Aqr	Psc	Ari
	February	Sgr	Psc	Psc	Ari
	March	Aqr	Psc	Ari	Ari

		Venus	Mars	Jupiter	Saturn
	April	Psc	Ari	Ari	Ari
	May	Ari	Tau	Ari	Ari
	June	Tau	Tau	Tau	Tau
	July	Cnc	Gem	Tau	Tau
	August	Leo	Cnc	Tau	Tau
	September	Vir	Leo	Tau	Tau
	October	Lib	Leo	Tau	Tau
	November	Sgr	Vir	Tau	Tau
	December	Cap	Vir	Tau	Tau
2001	January	Aqr	Lib	Tau	Tau
	February	Psc	Lib	Tau	Tau
	March	Psc	Oph	Tau	Tau
	April	Psc	Oph	Tau	Tau
	May	Psc	Sgr	Tau	Tau
	June	Ari	Oph	Tau	Tau
	July	Tau	Oph	Gem	Tau
	August	Gem	Oph	Gem	Tau
	September	Leo	Sgr	Gem	Tau
	October	Vir	Sgr	Gem	Tau
	November	Lib	Cap	Gem	Tau
	December	Oph	Aqr	Gem	Tau
2002	January	Sgr	Psc	Gem	Tau
	February	Aqr	Psc	Gem	Tau
	March	Psc	Ari	Gem	Tau
	April	Ari	Tau	Gem	Tau
	May	Tau	Tau	Gem	Tau
	June	Cnc	Gem	Gem	Tau
	July	Leo	Cnc	Gem	Tau
	August	Vir	Leo	Cnc	Tau
	September	Vir	Leo	Cnc	Tau
	October	Lib	Leo	Cnc	Tau
	November	Vir	Vir	Cnc	Tau
	December	Lib	Lib	Leo	Tau
2003	January	Oph	Lib	Cnc	Tau
	February	Sgr	Oph	Cnc	Tau
	March	Cap	Sgr	Cnc	Tau
	April	Aqr	Sgr	Cnc	Tau
	May	Psc	Cap	Cnc	Tau
	June	Tau	Aqr	Cnc	Gem
	July	Gem	Aqr	Leo	Gem
	August	Leo	Aqr	Leo	Gem
	September	Vir	Aqr	Leo	Gem
	October	Lib	Aqr	Leo	Gem

		Venus	Mars	Jupiter	Saturn
	November	Oph	Aqr	Leo	Gem
	December	Sgr	Psc	Leo	Gem
2004	January	Aqr	Psc	Leo	Gem
	February	Psc	Ari	Leo	Gem
	March	Ari	Tau	Leo	Gem
	April	Tau	Tau	Leo	Gem
	May	Tau	Gem	Leo	Gem
	June	Tau	Gem	Leo	Gem
	July	Tau	Cnc	Leo	Gem
	August	Gem	Leo	Leo	Gem
	September	Cnc	Vir	Vir	Gem
	October	Leo	Vir	Vir	Gem
	November	Vir	Vir	Vir	Gem
	December	Lib	Lib	Vir	Gem

Appendix G
The Messier Catalog Plus

Here's a listing of 175 of the finest deep-sky objects visible through amateur telescopes.

M #	NGC #	Con[1]	Season	Type[2]	R.A. (2000) h	m	Dec. (2000) °	′	Mag.	Size[3]	Remarks
1	1952	Tau	Winter	SNR	05	34.5	+22	01	8.2	6′ × 4′	Crab Nebula
2	7089	Aqr	Autumn	GC	21	33.5	−00	49	6.5	13′	
3	5272	CVn	Spring	GC	13	42.2	+28	23	6.4	16′	
4	6121	Sco	Summer	GC	16	23.6	−26	32	6.0	26′	
5	5904	Ser	Summer	GC	15	18.6	+02	05	5.8	17′	
6	6405	Sco	Summer	OC	17	40.1	−32	13	4.2	15′	Butterfly Cluster
7	6475	Sco	Summer	OC	17	53.9	−34	49	3.3	80′	
8	6523	Sgr	Summer	EN	18	03.8	−24	23	5.8	90′ × 40′	Lagoon Nebula
9	6333	Oph	Summer	GC	17	19.2	−18	31	7.9	9′	
10	6254	Oph	Summer	GC	16	57.1	−04	06	6.6	15′	
11	6705	Sct	Summer	OC	18	51.1	−06	16	5.8	14′	Wild Duck Cluster
12	6218	Oph	Summer	GC	16	47.2	−01	57	6.6	15′	
13	6205	Her	Summer	GC	16	41.7	+36	28	5.9	16′	
14	6402	Oph	Summer	GC	17	37.6	−03	15	7.6	12′	
15	7078	Peg	Autumn	GC	21	30.0	+12	10	6.4	12′	
16	6611	Ser	Summer	EN + OC	18	18.8	−13	47	6.0	35′	Eagle Nebula
17	6618	Sgr	Summer	EN	18	20.8	−16	11	7.0	46′ × 37′	Omega Nebula
18	6613	Sgr	Summer	OC	18	19.9	−17	08	6.9	9′	
19	6273	Oph	Summer	GC	17	02.6	−26	16	7.1	14′	
20	6514	Sgr	Summer	EN + RN	18	02.6	−23	02	8.5	29′ × 27′	Trifid Nebula
21	6531	Sgr	Summer	OC	18	04.6	−22	30	5.9	13′	
22	6656	Sgr	Summer	GC	18	36.4	−23	54	5.1	24′	
23	6494	Sgr	Summer	OC	17	56.8	−19	01	5.5	27′	
24	—	Sgr	Summer	OC	18	16.9	−18	29	4.5	90′	Small Sgr Star cloud
25	IC 4725	Sgr	Summer	OC	18	31.6	−19	15	4.6	32′	
26	6694	Sct	Summer	OC	18	45.2	−09	24	8.0	15′	
27	6853	Vul	Summer	PN	19	59.6	+22	43	8.1	8′ × 4′	Dumbbell Nebula
28	6626	Sgr	Summer	GC	18	24.5	−24	52	6.9	11′	
29	6913	Cyg	Summer	OC	20	23.9	+38	22	6.6	7′	
30	7099	Cap	Autumn	GC	21	40.4	−23	11	7.5	11′	
31	224	And	Autumn	Gx	00	42.7	+41	16	3.5	160′ × 40′	Andromeda Galaxy
32	221	And	Autumn	Gx	00	42.7	+40	52	8.2	3′ × 2′	
33	598	Tri	Autumn	Gx	01	33.9	+30	39	6.3	60′ × 35′	
34	1039	Per	Autumn	OC	02	42.0	+42	47	5.5	35′	

M #	NGC #	Con[1]	Season	Type[2]	R.A. (2000) h	m	Dec. (2000) °	′	Mag.	Size[3]	Remarks
35	2168	Gem	Winter	OC	06	08.9	+24	20	5.3	28′	
36	1960	Aur	Winter	OC	05	36.1	+34	08	6.0	12′	
37	2099	Aur	Winter	OC	05	52.4	+32	33	5.6	24′	
38	1912	Aur	Winter	OC	05	28.7	+35	50	6.4	21′	
39	7092	Cyg	Summer	OC	21	32.2	+48	26	4.6	32′	
40	—	UMa	Spring	**	12	22.4	+58	05	9.0,9.3	50″	Also known as Winnecki 4
41	2287	CMa	Winter	OC	06	46.0	−20	44	4.6	38′	
42	1978	Orl	Winter	EN	05	35.4	−05	27	2.9	66′ × 60′	Orion Nebula
43	1982	Orl	Winter	EN	05	35.6	−05	16	6.9	20′ × 15′	
44	2632	Cnc	Spring	OC	08	40.1	+19	59	3.1	95′	Beehive or Praesepe
45	—	Tau	Winter	OC	03	47.0	+24	07	1.2	110′	Pleiades
46	2437	Pup	Winter	OC	07	41.8	−14	49	6.1	27′	
47	2422	Pup	Winter	OC	07	36.6	−14	30	4.5	30′	
48	2548	Hya	Spring	OC	08	13.8	−05	48	5.8	55′	
49	4472	Vir	Spring	Gx	12	29.8	+08	00	8.4	9′ × 7′	
50	2323	Mon	Winter	OC	07	03.2	−08	20	5.9	16′	
51	5194	CVn	Spring	Gx	13	29.9	+47	12	8.4	11′ × 8′	Whirlpool Galaxy
52	7654	Cas	Autumn	OC	23	24.2	+61	35	6.9	13′	
53	5024	Com	Spring	GC	13	12.9	+18	10	7.7	13′	
54	6715	Sgr	Summer	GC	18	55.1	−30	29	7.7	9′	
55	6809	Sgr	Summer	GC	19	40.0	−30	58	7.0	19′	
56	6779	Lyr	Summer	GC	19	16.6	+30	11	8.2	7′	
57	6720	Lyr	Summer	PN	18	53.6	+33	02	9.7	70″ × 150″	Ring Nebula
58	4579	Vir	Spring	Gx	12	37.7	+11	49	9.8	5′ × 4′	
59	4621	Vir	Spring	Gx	12	42.0	+11	39	9.8	5′ × 3′	
60	4649	Vir	Spring	Gx	12	43.7	+11	33	8.8	7′ × 6′	
61	4303	Vir	Spring	Gx	12	21.9	+04	28	9.7	6′ × 5′	
62	6266	Oph	Summer	GC	17	01.2	−30	07	6.6	14′	
63	5055	CVn	Winter	Gx	13	15.8	+42	02	8.6	12′ × 7′	
64	4826	Com	Spring	Gx	12	56.7	+21	41	8.5	9′ × 5′	Black-Eye Galaxy
65	3623	Leo	Spring	Gx	11	18.9	+13	05	9.3	10′ × 3′	
66	3627	Leo	Spring	Gx	11	20.2	+12	59	9.0	9′ × 4′	
68	4590	Hya	Spring	GC	12	39.5	−26	45	8.2	12′	
69	6637	Sgr	Summer	GC	18	31.4	−32	21	7.7	7′	
70	6681	Sgr	Summer	GC	18	43.2	−32	18	8.1	8′	
71	6838	Sge	Summer	GC	19	53.8	+18	47	8.3	7′	
72	6981	Aqr	Autumn	GC	20	53.5	−12	32	9.4	6′	
73	6994	Aqr	Autumn	OC	20	58.9	−12	38	9.0	3′	
75	6864	Sgr	Summer	GC	20	06.1	−21	55	8.6	6′	
76	650	Per	Autumn	PN	01	42.2	+51	34	11.4	3′ × 1′	Little Dumbbell Nebula
77	1068	Cet	Autumn	Gx	02	42.7	−00	01	8.9	6′ × 5′	
78	2068	Ori	Winter	RN	05	46.7	+00	03	8	8′ × 6′	
79	1904	Lep	Winter	GC	05	24.5	−24	33	8.4	3′	
80	6093	Sco	Summer	GC	16	17.0	−22	59	7.2	9′	

M #	NGC #	Con[1]	Season	Type[2]	R.A. (2000) h	m	Dec. (2000) °	′	Mag.	Size[3]	Remarks
81	3031	UMa	Spring	Gx	09	55.6	+69	04	7.0	26′ × 14′	
82	3034	UMa	Spring	Gx	09	55.8	+69	41	8.4	11′ × 5′	
83	5238	Hya	Spring	Gx	13	37.0	−29	52	7.6	11′ × 10′	
84	4374	Vir	Spring	Gx	12	25.1	+12	53	9.3	5′ × 4′	
85	4382	Com	Spring	Gx	12	25.4	+18	11	9.2	7′ × 5′	
86	4406	Vir	Spring	Gx	12	26.2	+12	57	9.2	7′ × 6′	
87	4486	Vir	Spring	Gx	12	30.8	+12	24	8.6	7′	
88	4501	Com	Spring	Gx	12	32.0	+14	25	9.5	7′ × 4′	
89	4552	Vir	Spring	Gx	12	35.7	+12	33	9.8	4′	
90	4569	Vir	Spring	Gx	12	36.8	+13	10	9.5	9′ × 5′	
91	4548	Com	Spring	Gx	12	35.4	+14	30	10.2	4′ × 3′	
92	6341	Her	Summer	GC	17	17.1	+43	08	6.5	11′	
93	2447	Pup	Winter	OC	07	44.6	−23	52	6.2	22′	
94	4736	CVn	Spring	Gx	12	50.9	+41	07	8.2	11′ × 9′	
95	3351	Leo	Spring	Gx	10	44.0	+11	42	9.7	7′ × 5′	
96	3368	Leo	Spring	Gx	10	46.8	+11	49	9.2	7′ × 5′	
97	3587	UMa	Spring	PN	11	14.8	+55	01	11.2	3′	
98	4192	Com	Spring	Gx	12	13.8	+14	54	10.1	10′ × 3′	
99	4254	Com	Spring	Gx	12	18.8	+14	25	9.8	5′ × 4′	
100	4321	Com	Spring	Gx	12	22.9	+15	49	9.4	7′ × 6′	
101	5457	UMa	Spring	Gx	14	03.2	+54	21	7.7	27′ × 26′	Pinwheel Galaxy
102	(mistaken duplicate observation of M101)										
103	581	Cas	Autumn	OC	01	33.2	+60	42	7.4	6′	
104	4594	Vir	Spring	Gx	12	40.0	−11	37	8.3	9′ × 4′	Sombrero Galaxy
105	3379	Leo	Spring	Gx	10	47.8	+12	35	9.3	5′ × 4′	
106	4258	CVn	Spring	Gx	12	19.0	+47	18	8.3	18′ × 8′	
107	6171	Oph	Summer	GC	16	32.5	−13	03	8.1	10′	
108	3558	UMa	Spring	Gx	11	11.5	+55	40	10.1	8′ × 3′	
109	3992	UMa	Spring	Gx	11	57.6	+53	23	9.8	8′ × 5′	
110	205	And	Autumn	Gx	00	40.4	+41	41	8.0	10′ × 5′	
	Iota	Cnc	Spring	**	08	46.7	+28	46	4.0,6.6	30″	Yellow/blue
	2516	Car	Spring	OC	07	58.3	−60	52	3.8	30′	
	2808	Car	Spring	GC	09	12.0	−64	52	6.3	14′	
	3114	Car	Spring	OC	10	02.7	−60	07	4.2	35′	
	3115	Sex	Spring	Gx	10	05.2	−07	43	9.1	8′ × 3′	Spindle Galaxy
	3132	Vel	Spring	PN	10	07.7	−40	26	8.2p	84″ × 53″	Eight-Burst Nebula
	3242	Hya	Spring	PN	10	24.8	−18	38	8.6p	16″	Ghost of Jupiter Nebula
	3293	Car	Spring	OC	10	35.8	−58	14	4.7	40′	
	3324	Car	Spring	OC + EN	10	37.3	−58	38	6.7	6′	
	Zeta	UMa	Spring	**	13	23.9	+54	56	2.3,4.0	14″	Alcor
	3372	Car	Spring	EN	10	43.8	−59	52	5.0	120′	Eta Carinae Nebula
	3532	Car	Spring	OC	11	06.4	−58	40	3.0	55′	
	3572	Car	Spring	OC	11	10.4	−60	14	6.6	20′	
	3766	Cen	Spring	OC	11	36.1	−61	37	5.3	12′	

M # NGC #	Con[1]	Season	Type[2]	R.A. (2000) h	m	Dec. (2000) °	'	Mag.	Size[3]	Remarks
Melotte 111	Com	Spring	OC	12	25	+26		1.8	275'	Coma star cluster
	Cru	Spring	Dk	12	53	−63		—	7° × 5°	Coalsack
4755	Cru	Spring	OC	12	53.6	−60	20	4.2	16'	Jewel Box Cluster
5128	Cen	Spring	Gx	13	25.5	−43	01	7	18' × 14'	Centaurus A
5139	Cen	Spring	GC	13	26.8	−47	29	3.7	36'	Omega Centauri
6210	Her	Summer	PN	16	44.5	+23	49	9.2	15"	
6231	Sco	Summer	OC	16	54.0	−41	48	2.6	15'	
6281	Sco	Summer	OC	17	04.8	−37	54	5.4	8'	
B59,65-7	Oph	Summer	Dk	17	21	−27			5° × 3°	
B78	Oph	Summer	Dk	17	33	−26			3° × 2°	Pipe Nebula
6369	Oph	Summer	PN	17	29.3	−23	46	10.4	30"	Little Ghost Nebula
6388	Sco	Summer	GC	17	36.3	−44	44	6.9	9'	
IC 4665	Oph	Summer	OC	17	46.3	+05	43	4.2	41'	
6572	Oph	Summer	PN	18	12.1	+06	51	9.0	7"	
B92	Sgr	Summer	Dk	18	15.5	−18	11		12' × 6'	See M24
6712	Sct	Summer	GC	18	53.1	−08	42	8.3	7'	
Collinder 399	Vul	Summer	OC	19	25.4	+20	11	3.6	60'	Coathanger Cluster
Beta	Cyg	Summer	**	19	30.7	+27	58	3.2,5.4	34"	Albireo, yellow/blue
6826	Cyg	Summer	PN	19	44.8	+50	31	9.8	30"	Blinking Planetary
6939	Lac	Summer	OC	20	31.4	+60	38	7.8	8'	
6960	Cyg	Summer	SNR	20	45.7	+30	43		70' × 6'	West half of Veil Nebula
6992	Cyg	Summer	SNR	20	56.4	+31	43		60' × 8'	East half of Veil Nebula
7000	Cyg	Summer	EN	20	58.8	+44	20		2°	North America Nebula
7009	Aqr	Autumn	PN	21	04.2	−11	22	8.4	26"	Saturn Nebula
7293	Aqr	Autumn	PN	22	29.6	−20	48	6.5	15' × 12'	Helix Nebula
7662	And	Autumn	PN	23	25.9	+42	33	8.9p	32" × 28"	
55	Sci	Autumn	Gx	00	14.9	−39	11	8.0	32' × 6'	
246	Cet	Autumn	PN	00	47.0	−11	53	8.5	240" × 210"	
253	Sci	Autumn	Gx	00	47.6	−25	17	7.1	22' × 6'	
288	Sci	Autumn	GC	00	52.8	−26	35	8.1	14'	
457	Cas	Autumn	OC	01	19.1	+58	20	6.4	13'	
Gamma	Arl	Autumn	**	01	53.5	+19	18	4.6,4.7	8"	Orange/green; outstanding
752	And	Autumn	OC	01	57.8	+37	41	5.7	50'	
Gamma	And	Autumn	**	02	03.9	+42	20	2.1,5.1	10"	Orange/yellow; beautiful
889	Per	Autumn	OC	02	19.0	+57	09	4.3	30'	Double Cluster (h Per)
884	Per	Autumn	OC	02	22.4	+57	07	4.4	30'	Double Cluster (Chi Per)
891	And	Autumn	Gx	02	22.6	+42	21	10.0	14' × 3'	
Melotte 20	Per	Autumn	OC	03	22	+49		1.2	185'	Alpha Per Cluster

M #	NGC #	Con[1]	Season	Type[2]	R.A. (2000) h	m	Dec. (2000) °	'	Mag.	Size[3]	Remarks
	1316	For	Autumn	Gx	03	22.7	−37	12	8.9	4′ × 3′	
	1360	For	Autumn	PN	03	33.3	−25	51		390″	
	1514	Tau	Winter	PN	04	09.2	+30	47	10.9	2′	
	1973	Ori	Winter	EN + RN	05	35.1	−04	44		40′ × 25′	Also NGC 1975, 1977
	2194	Ori	Winter	OC	06	13.8	+12	48	8.5	10′	
	Beta	Mon	Winter	**	06	28.8	−07	02	4.6,5.1	7′	
	2237	Mon	Winter	EN	06	32.3	+05	03		80′ × 60′	Rosette Nebula
	2244	Mon	Winter	OC	06	32.4	+04	52	4.8	24′	Rosette Nebula Cluster
	2261	Mon	Winter	EN + RN	06	39.2	+08	44	10.0	2′	Hubble's Variable Nebula
	2264	Mon	Winter	OC	06	41.1	+09	53	3.9	20′	Christmas Tree Cluster
	2359	CMa	Winter	EN	07	18.6	−13	12		8′ × 6′	
	2392	Gem	Winter	PN	07	29.2	+20	55	8.3	13″	Eskimo Nebula
	2403	Cam	Winter	Gx	07	36.9	+65	36	8.4	18′ × 11′	
	2539	Pup	Winter	OC	08	10.7	−12	50	6.5	22′	

Notes:

1. Constellation—see appendix F.
2. Object type:
    ** = Double star
    OC = Open cluster
    GC = Globular cluster
    EN = Emission nebula
    RN = Reflection nebula
    Dk = Dark nebula
    PN = Planetary nebula
    SNR = Supernova remnant
    Gx = Galaxy
3. Apparent size of object in either minutes of arc, seconds of arc, or degrees. Most measurements were made from photographs; visual appearance may be smaller. For double stars, this number is a measure of the stars' separation from one another.

Appendix H
The Constellations

Constellation	Abbr.	Genitive Form	Meaning
Andromeda	And	Andromedae	The Daughter of Queen Cassiopeia
Antlia	Ant	Antliae	The Air Pump
Apus	Aps	Apodis	The Bird of Paradise
Aquarius	Aqr	Aquaril	The Water-bearer
Aquila	Aql	Aquilae	The Eagle
Ara	Ara	Arae	The Altar
Aries	Ari	Arietis	The Ram
Auriga	Aur	Aurigae	The Charioteer
Boötes	Boo	Boötis	The Herdsman
Caelum	Cae	Caeli	The Chisel
Camelopardalis	Cam	Camelopardalis	The Giraffe
Cancer	Cnc	Cancri	The Crab
Canes Venatici	CVn	Canum Venaticorum	The Hunting Dogs
Canis Major	CMa	Canis Majoris	The Big Dog
Canis Minor	CMi	Canis Minoris	The Little Dog
Capricomus	Cap	Capricomi	The Sea-goat
Carina	Car	Carinae	The Keel (of the mythical ship *Argo*)
Cassiopeia	Cas	Cassiopeiae	The Queen
Centaurus	Cen	Centauri	The Centaur
Cepheus	Cep	Cephei	The King
Cetus	Cet	Ceti	The Whale
Chamaeleon	Cha	Chamaeleontis	The Chameleon
Circinus	Cir	Circini	The Compasses
Columba	Col	Columbae	The Dove
Coma Berenices	Com	Comae Berenices	The Queen Berenice's Hair
Corona Australis	CrA	Coronae Australis	The Southern Crown
Corona Borealis	CrB	Coronae Borealis	The Northern Crown
Corvus	Crv	Corvi	The Crow
Crater	Crt	Crateris	The Cup
Crux	Cru	Crucis	The Southern Cross
Cygnus	Cyg	Cygni	The Swan
Delphinus	De	Delphini	The Dolphin
Dorado	Dor	Doradus	The Swordfish
Draco	Dra	Draconis	The Dragon
Equuleus	Equ	Equulei	The Little Horse
Eridanus	Eri	Eridani	The River
Fomax	For	Fomacis	The Fumace
Gemini	Gem	Geminorum	The Twins

Constellation	Abbr.	Genitive Form	Meaning
Hercules	Her	Herculis	The Giant
Horologium	Hor	Horologil	The Clock
Hydra	Hya	Hydrae	The Water Snake (male)
Hydrus	Hyi	Hydri	The Water Snake (female)
Indus	Ind	Indi	The Indian
Lacerta	Lac	Lacertae	The Lizard
Leo	Leo	Leonis	The Lion
Leo Minor	LMi	Leonis Minoris	The Little Lion
Lepus	Lep	Leporis	The Hare
Libra	Lib	Librae	The Scales of Justice
Lupus	Lup	Lupi	The Wolf
Lynx	Lyn	Lyncis	The Lynx
Lyra	Lyr	Lyrae	The Lyre
Mensa	Men	Mensae	The Table
Microscopium	Mic	Microscopii	The Microscope
Monoceros	Mon	Monocerotis	The Unicorn
Musca	Mus	Muscae	The Fly
Norma	Nor	Normae	The Square
Octans	Oct	Octantis	The Octant
Ophiuchus	Oph	Ophiuchi	The Serpent-bearer
Orion	Ori	Orionis	The Hunter
Pavo	Pav	Pavonis	The Peacock
Pegasus	Peg	Pegasi	The Flying Horse
Perseus	Per	Persei	The Warrior
Phoenix	Phe	Phoenicis	The Phoenix
Pictor	Pic	Pictoris	The Painter
Pisces	Psc	Piscium	The Fishes
Piscis Austrinus	PsA	Piscis Austrini	The Southern Fish
Puppis	Pup	Puppis	The Stern (of the mythical ship *Argo*)
Pyxis	Pyx	Pyxidis	The Compass
Reticulum	Ret	Reticuli	The Reticle
Sagitta	Sge	Sagittae	The Arrow
Sagittarius	Sgr	Sagittarii	The Archer
Scorpius	Sco	Scorpii	The Scorpion
Sculptor	Scl	Sculptoris	The Sculptor
Scutum	Sct	Scuti	The Shield
Serpens	Ser	Serpentis	The Serpent
Sextans	Sex	Sextantis	The Sextant
Taurus	Tau	Tauri	The Bull
Telescopium	Tel	Telescopii	The Telescope
Triangulum	Tri	Trianguli	The Triangle
Triangulum Australe	TrA	Trianguli Australis	The Southern Triangle
Tucana	Tuc	Tucanae	The Toucan
Ursa Major	UMa	Ursae Majoris	The Great Bear
Ursa Minor	UMi	Ursae Minoris	The Little Bear
Vela	Vel	Velorum	The Sails (of the mythical ship *Argo*)
Virgo	Vir	Virginis	The Maiden
Volans	Vol	Volantis	The Flying Fish
Vulpecula	Vul	Vulpeculae	The Fox

Appendix I
English/Metric Conversion

Most amateur astronomers in the United States will speak of a telescope's aperture in terms of inches, while the rest of the world uses centimeters. The table below acts as a translation table to help convert telescope apertures from one system to another. Recall that there are 2.54 centimeters per inch.

English (in.)	Metric (cm)
2	5
3.1	8
4	10
6	15
8	20
10	25
12	30
14	35
16	40
18	45
20	50
24	60
30	75
32	80
36	90

Appendix J
Star Ware Reader Survey

To help prepare for the next edition of *Star Ware*, I'd like you to take a moment and complete the following survey. Photocopy it, fill it out, and send it to the address shown at the end of this appendix. An on-line version is also available at the Star Ware Home Page (http://ourworld.compuserve.com/homepages/pharrington) or via email by writing to starware@juno.com. Feel free to reproduce it in your club newsletter as well.

Please answer all (or as many) of the questions below as possible. If you own more than one telescope, I'd like to hear about each. *Use as much room as you want!* I'll save each response and reference them when it comes time to write the reviews. Please include your address in case I need to contact you for questions or clarifications. Rest assured that your address will not be given to anyone, and neither will any company ever see your survey.

Thanks in advance!

Your name: _____

Address: _____

City: _____ State/Country: _____ Zip: _____

Email:_____

How long have you been involved in astronomy? _____

Do you consider yourself a: Beginner   Intermediate   Advanced

TELESCOPE
How many telescopes do you own? _____

Telescope model: _____

How old is it? _____ Are you the original owner? _____

What do you like about it? _____

_____

_____

What don't you like about it?

_____

_____

Lived up to your expectations? _____ Buy it again? _____

Have you ever had to contact the company about a problem?

_____

Was it resolved? Explain.

_____

_____

EYEPIECES
What eyepieces do you own?

_____

How do they work? Any particular likes or dislikes?
(Please list your impressions separately for each.)

_____

_____

ACCESSORIES
What accessories do you own? (Anything ... binoculars, books, software, filters, finderscope, etc.)

_____

_____

_____

Any particular likes or dislikes?
(Please list your impressions separately for each.)

_____

_____

_____

_____

Your vote for the best telescope of yesteryear
(only models that are no longer made): _____

Anything you think should be included in the next edition?

_____

_____

Please mail this survey to Phil Harrington c/o John Wiley & Sons, 605 Third Avenue, New York, NY 10158

Index